Experiments in Biochemistry:
Projects and Procedures

Michael J. Minch

University of the Pacific

Prentice Hall, Englewood Cliffs, New Jersey   07632

Library of Congress Cataloging-in-Publication Data

Minch, Michael J.
   Experiments in biochemistry.

   Bibliography
   Includes index.
    1. Biochemistry—Laboratory manuals. I. Title.
QP519.M56 1989   574.1′92′078   88-32479
ISBN 0-13-295239-4

Cover design: *Ben Santora*
Manufacturing buyer: *Paula Massenaro*

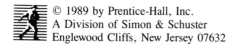 © 1989 by Prentice-Hall, Inc.
A Division of Simon & Schuster
Englewood Cliffs, New Jersey 07632

Printed in the United States of America

10 9 8 7 6 5 4 3 2 1

ISBN 0-13-295239-4

Prentice-Hall International (UK) Limited, *London*
Prentice-Hall of Australia Pty. Limited, *Sydney*
Prentice-Hall Canada Inc., *Toronto*
Prentice-Hall Hispanoamericana, S.A., *Mexico*
Prentice-Hall of India Private Limited, *New Delhi*
Prentice-Hall of Japan, Inc., *Tokyo*
Simon & Schuster Asia Pte. Ltd., *Singapore*
Editora Prentice-Hall do Brasil, Ltda., *Rio de Janeiro*
Whitehall Books Limited, *Wellington, New Zealand*

Dedicated in fond appreciation of the tireless efforts of
Gladys Murray whose constant cheerfulness in the face of
an unending work load is an inspiration to us all.

Helpful comments by
Richard Criddle, Carl Schmid, Paul Richmond, Mark Stolowitz,
Jaime Verela and Yong Ma are acknowledged.
Ingred Irelan and Sandy McNett-McGowen
deserve special praise as illustrators.

Most of all, thanks are due to the many outstanding students who worked
long into the night
developing and debugging the projects described
within

Contents

PART TWO: ADVANCED BIOCHEMICAL PURIFICATIONS AND THE CHARACTERIZATION OF BIOMOLECULES

Preface

To the Instructor

All textbooks are written with a philosophy in mind. This one was developed with the belief that the principal objectives of biochemistry laboratory are to introduce students to the highest practical level of sophistication possible in a teaching laboratory, to reinforce lecture material, to stress laboratory technique, and to develop further those communication, analytical, and manipulative skills learned in more basic laboratory courses. A new laboratory must expose students to most of the common laboratory methods of modern biochemistry, including those developed recently, and students must encounter a very wide range of biomolecules and biological sources in doing so. Of special importance are the new methods of DNA chemistry and molecular biology. In addition to new exclusively biochemical material, certain universal laboratory skills must be emphasized at every opportunity; each year as the student progresses toward graduation the standards for expected achievement must be raised. These skills include honest lab notebook keeping, accurate report writing, data analysis, the ability to make reproducible quantitative measurements, and perhaps most important, the ability to think through a procedure and adapt it towards new ends. In this author's opinion these skills define the usefulness of an undergraduate's scientific education and should never be lost sight of when designing a laboratory. A laboratory course that exposes students to sophisticated material while leaving them unable to take useful notes or to prepare a simple buffer solution on their own has failed.

In preparing this text the author has had to come to grips with five, sometimes contradictory, pedagogical requirements:

1. It is necessary that teaching experiments always "work." Anybody who has faced an angry class after the members have invested considerable time in a failed experiment will appreciate how important it is to debug an experiment and to free experimental protocols of vague, incomplete instructions. Hidden details necessary for success have no place in most undergraduate laboratory courses. Faculty are too hard pressed with other teaching duties to be expected to try everything out ahead of time. All the experiments described in this book have been tested in the teaching laboratory and were adopted only after years of trial and improvement. The first ten experiments represent a

corpus of standard biochemical methods that has been used in our laboratory for at least four years, and most for nearly a decade. The experimental procedures are written in full detail with extensive background sections. An instructor's guide giving exact amounts of chemicals and supplies required for each project is for the benefit of teaching personnel and, hopefully, students. The remaining twenty projects cover a wider range of more specialized techniques and problems encountered in biochemical research and for the most part are laboratory-tested both in the teaching and research laboratory. A number of them were first suggested by research associates of the author, then tested in the teaching laboratory. The experimental protocols for these more advanced projects are closer to those encountered in journal articles; trivial details are not included and descriptions make allowances for differences in equipment in different laboratories.

2. If the laboratory is seen as an extension of the lecture course then laboratory experiments should be selected to illustrate principles developed in lecture. This laboratory text is designed for a course emphasizing proteins the first semester and nucleic acids the second, and is compatible with current standard lecture textbooks by Stryer, Zubay, Lehninger, Rawn, or Devlin. The experiments were selected with this point in mind since they cover a very wide range of methods and illustrate most of the fundamental principles of biochemical separations and the properties of biomolecules. Kinetics, electrophoresis, chromatography and spectroscopy are prominently featured. Some of the first nine experiments are designed to enable the instructor to use them as short do-it-yourself lecture demonstrations emphasizing classical principles of protein chemistry, chromatography and electrophoresis. There are a number of classical experiments in the first section of indispensible historical as well as pedagogical value. In a number of instances, old and new techniques are included in the same chapter so that the instructor can pick and choose among the offerings to devise a laboratory project compatible with equipment and time limitations. The projects are written in as general a manner as possible so that the book will have lasting value. The description of each project includes a brief theoretical discussion of the methods used and a general account of the practical problems associated with the methods. The project descriptions provide enough background and theoretical principles to enable students to plan and execute the experiments within the project using the suggested methods and to understand the limitations and possible extensions of each method.

I have avoided tear-out pages and anything else that smacks of a workbook so that students are encouraged to read the background sections like a reference book. Tables of valuable information are spread throughout.

3. A laboratory should acquaint students with the methods and instrumentation they will be using in industrial, graduate or medical school laboratories. In the past two decades the methodogical basis of biochemistry has undergone a revolution. The bulk of experimentation is no longer focused on using the isolation and purification techniques of the organic chemist to work out the metabolic pathways interconnecting relatively abundant low-molecular-weight compounds. Biochemistry has moved towards elucidating the catalytic and regulatory mechanisms made possible by proteins and nucleic acids and discovering the detailed mechanisms by which genetic information is stored and utilized. The modern biochemist uses many experimental methods once only in the purview of the physical chemist. The use of X-ray crystallography, NMR spectroscopy, and rapid kinetics methods to study the structure and mechanisms of enzymes has transformed biochemistry from a descriptive to a predictive science. Modern biochemists work with nanogram quantities of substances isolated in high purity from mixtures of closely related compounds. New electrophoretic and

chromatographic techniques make possible isolations and purifications of macromolecules that were nearly impossible with "classical" methods. The recent technological innovations involved in recombinant DNA technology, especially sequencing methods and the use of restriction enzymes, have completely changed the approach biochemists take towards investigating the events within a cell. Tasks once requiring extraordinary effort, such as sequencing several thousand base pairs of DNA, have been rendered nearly routine. Progress will continue to be rapid; currently there are about 1000 papers published weekly on biochemistry and biochemistry-related fields. The changes in biochemistry have been accompanied by changes in its teaching, quickly in many graduate schools but more slowly in most undergraduate courses. The changes showed up first in lecture material, but some improvement has also been evident in biochemistry laboratory textbooks published in recent years. However, there has not been as radical a change in laboratory textbooks as one would expect given the enormous changes in what biochemists actually do in the laboratory. This laboratory text was written to bridge this gap. Novelty has not been embraced for its own sake but new techniques, employing sophisticated equipment, are introduced in a way that they can be used by students in their graduate or professional experiences in the twenty-first century.

4. Every new laboratory textbook must make allowances for the natural limitations of time and money available to a teaching laboratory and allow for a gradual adoption of new experiments requiring specialized equipment. Equipment budget limitations and time constraints are not the result of faculty sloth or indifference to the needs of students--they are facts of life with which all but the most fortunate are forced to live. These are limitations that can be accommodated, but never overcome. Many instructors are forced to select experiments that employ only marginally sophisticated equipment. Some are forced to use homemade equipment of questionable transferability to the real world. To cram exposure to as many methods, biological samples, and situations as possible in a short laboratory period, some shortcuts are inevitably made. This book contains sophisticated experiments that illustrate important principles with inexpensive and home-crafted equipment without budget crunching. There are also sophisticated experiments requiring sophisticated hardware (HPLC, fraction collectors, diafiltration equipment, liquid scintillation spectrometers) that can be phased in over a number of years. Information on the common suppliers of chemicals, supplies, and equipment is given in the instructor's guide. Only the most common equipment--centrifuges, power supplies, and chromatography columns are required for almost all experiments.

5. Finally, laboratories must emphasize problem solving abilities, self-direction, critical thinking, and data analysis. Such a goal is most often achieved by forcing students to contribute to the design of their own projects. The more advanced projects available for the second semester can be directed toward this goal. In a second twelve-week semester, three to five advanced projects can be completed; preferably, they should be linked to form one or two macroprojects.

A number of the projects require analysis of quantitative results and it is an assumption of the author that the instructor will require well written laboratory reports with conclusions based upon quantitative data analysis. A description of laboratory notebook and report writing is given in chapter 1.

DESCRIPTION OF THE PROJECTS IN THIS BOOK

This book employs a project-oriented approach to the undergraduate biochemistry laboratory that emphasizes the skills discussed above, while keeping in mind the practical limitations of time and budgets. The projects fall into three broad areas:

Basic Techniques: Chapters 2 through 10 illustrate basic techniques of experimental biochemistry, such as absorption spectroscopy, chromatography, liquid scintillation counting, and electrophoresis. The emphasis is on protein and nucleic acid purification and characterization. These projects represent a corpus of standard biochemical methods suitable for a one semester laboratory that meets twice a week (6 to 8 hours a week).

Biochemical Isolations: Chapters 11, 12, 14, 15, 16, and 19 through 26, and 28 illustrate the isolation of biochemically important compounds. This includes purifying an enzyme to homogeneity and quantitatively evaluating the purification at each step with specific activity measurements. Two different RNA, one carbohydrate, and three different DNA preparations are included.

Biochemical Characterizations: Chapters 13, 17, 18, 19, 23, 25, and 27-31 emphasize chromatographic, kinetic, spectroscopic, electrophoretic, and other physical-chemical techniques to characterize biological macromolecules.

HOW TO USE THIS BOOK

A book that only emphasizes principles must necessarily de-emphasize fine details or methological complexities, while a book that is exclusively technique-oriented will de-emphasize the variety of biological phenomena to which students are exposed. Neither extreme seems appropriate for a first-year laboratory course. For this reason I have written a book suitable for a two-semester sequence in which the first semester emphasizes basic techniques and universal principles of biochemical isolations while the second semester emphasizes a very wide range of biological phenomena and a number of advanced techniques usually reserved for research laboratories.

The first nine experiments are suggested for a one-semester course or for the first semester of a two-semester course. These chapters offer step-by-step instructions suitable for beginning students. Substitution of some advanced material for some of the basic projects also gives a good mix of advanced and basic project topics. For example, an enzyme kinetics project, either Chapter 17 or 18, employing commercial enzyme can also be included in a one semester course. Alternatively, the restriction enzyme experiment (Chapter 23) or the three hemoglobin experiments (Chapters 11-13), which also cover many of the same techniques introduced in the first section, can be used to replace one or more of the first nine projects to give a one-semester lab that introduces students to many biochemical techniques while introducing a project oriented flavor.

Instructors using this book for two semesters will find that during the second semester students are eventually weaned away from total reliance on recipes and are encouraged to attain a higher degree of self-sufficiency and inventiveness than in laboratories that use the "cookbook" approach both semesters. In order to stimulate student planning, the projects described in the more advanced chapters can be linked to form more extended macroprojects, emphasizing special methods or types of macromolecules. Project extensions are suggested at the end of most chapters but the author has found that, by the beginning of the second semester, students can see the more subtle connections between projects by themselves and that linking projects to emphasize one topic, e.g., DNA chemistry, often encourages students to integrate what they learn from each project into a larger scheme.

For instructors desiring an integrated approach, with each project contributing to a larger, semester-long effort, the following sets of individual chapter miniprojects have been joined as macroprojects to good advantage in our laboratory:

Enzymes and kinetics: 14, 15, 16 and 17 or 18
DNA Chemistry: 22, 23, and 25, then 24 or 26
RNA Chemistry: 20, 21, and 26.
Membrane Chemistry: 11, 24 (nucleus part), 29, and 31
Heme Chemistry: 11, 12, 13, and 31

The payoff to students for the obligation of planning their own experiments is a greater independence and the ability to carry over ideas and techniques from one project to another. This reinforces basic techniques and integrates material from superficially quite different projects.

A note of warning: Not all the advanced projects fit into 3-hour time frames! At UOP we have the luxury of running open-ended upper division laboratories so that students can work all night if they choose. This is not a reasonable option at larger, necessarily more inflexible institutions, consequently reasonable stopping points are indicated in each project. Parts of projects that are clearly unsuitable for large, necessarily highly structured laboratories are generally reserved as optional extensions, but the projects described in Chapters 29 and 31 are really suitable only for smaller, advanced laboratory sections.

Chapter 1

General Practice in the Biochemistry Laboratory

All students should read the following chapter before beginning work in the laboratory. The chapter begins with a general account of safe practices recommended for the biochemistry laboratory, followed by a description of the customs and courtesies that should prevail in an advanced laboratory that requires shared equipment and glassware. Suggestions for accurate notebook keeping and meaningful report writing are also included. Finally, because biological research often involves large samples of highly variable data, a brief introduction to statistics is given.

LABORATORY SAFETY

Safety in any laboratory depends primarily on awareness and responsibility. Questions about suspected hazards will never be treated with a lack of concern by the teaching and stockroom staff. If you can save your fellow students from harm, why not do so? The following rules are mandatory:

1. Never pipette by mouth; rubber bulbs are available.

2. Protective safety goggles must be worn by all students who do not use safety eyeglasses.

3. All poisonous and foul-smelling chemicals (pyridine, acrylamide, etc.) must be used under the fume hood. If the hood is occupied --wait--that is certainly better than breathing potentially hazardous compounds.

4. All spills on benchtops, around balances and within the centrifuge must be cleaned up immediately. Do not leave spilled water on the floor.

5. Label all samples, reagent bottles, and stock solutions with the chemical identity of their contents, your name, notebook page number, and date of preparation. All unlabeled samples will be discarded routinely by the teaching staff. Use radioactivity and biohazard warning labels for samples that warrent this caution, but do not misuse these important warnings.

1

6. Radioisotopes must only be used in designated laboratory areas under full supervision of the instructor. The disposal of radioactive samples can only be done by the laboratory instructor.

7. Report all accidents immediately to your teaching assistant. Appropriate forms are provided to protect you and the university from undue liability.

8. Be aware of the toxicity of all chemicals used in the laboratory and the appropriate steps that must be taken in case of accidental exposure. The teaching staff should have a toxic substance guide book which describes the safe handling of toxic substances and emergency first-aid information.[1]

USE OF SHARED EQUIPMENT AND GLASSWARE

Reserving Equipment

Equipment not in your laboratory drawer can be checked out from the biochemistry stockroom. An accurate record will be kept and students will be expected to pay for all unreturned equipment. Do not lend equipment checked out in your name to another student. Glassware broken or equipment damaged while checked out to you will be charged to you regardless of who actually did the damage.

Breakage

Report all breakage to the instructor, who will help you to obtain a replacement or make a repair. Many of the facilities in this laboratory such as the centrifuge and electrophoresis equipment are dangerous when in disrepair. Supplies can be obtained from the stockroom, but students are not allowed to enter it. The teaching staff will obtain all needed supplies that cannot be obtained from the stockroom clerk.

Scheduling Experiments

Your instructor may elect to allow you to work on advanced experiments at your own pace; if so, you will have to devise a detailed work plan and follow it. Students may reserve spectrometers, fraction collectors, and special chromatography equipment by using the scheduling board in the laboratory. Reservations may be made up to one week in advance. If an instrument is reserved but not used at the alloted time, the student loses the right to reserve the instrument a second time.

All students are expected to sign in and sign out in the laboratory logbook and to sign any instrument logbooks before using the instrument. All malfunctions must be recorded in these logs.

LABORATORY NOTEBOOKS

A systematic and accurate record of your observations and data is essential for all laboratory work. A major part of the success you will have in this course depends on your skill in writing a laboratory notebook. Observe the following rules:

1. Put your name on the cover and your name, address, and telephone number on the inside of the back cover.

2. Use a notebook with numbered pages. Reserve the first four or five pages for a Table of Contents.

3. Remember that your notebook belongs to your instructor and will be used as a reference source long after you have left the University. Write with this in mind. Pages should never be torn out. Changes and corrections should be made clearly with bad data or inappropriate notes crossed out but never obliterated. Sign and date every page. Use ink always.

There are two possible styles that can be used in a laboratory notebook:

1. Narrative or "diary" approach-every act in the lab is described in the chronological order in which it was done.
2. Integrative or "recipe book" approach--all related experiments appear together regardless of their temporal relationship.

Both approaches are reasonable; each has virtues and faults. The "diary" approach is the easiest style of notebook to keep, but day-by-day notations do not readily permit an overview or a comparison of results of similar experiments done at widely different times. The integrative approach is fine if the writer has the discipline to keep an accurate record of the details of a series of similar but not identical experiments. The integrative approach also presupposes that the relationships between experiments are clear cut or apparent beforehand - this is rarely the case.

The author recommends the "diary" approach, punctuated with summaries that integrate related experiments done on different days. There should also be a special section for often used formulations (gel recipes, etc.) so that the reader can be referred to these when needed. Departures from these formulations must be carefully spelled out in the notebook pages for the day in question; otherwise, a simple reference to the standard formulation is enough.

LABORATORY REPORTS

Laboratory work that has not been shared with the scientific community is of little value, regardless of how carefully the work has been executed or how cleverly it was conceived. Report writing is an important aspect of all scientific disciplines. Your instructor will probably obligate you to a certain number of formal reports describing the results of your laboratory efforts, and he or she will be the final arbitrator of what is expected in a laboratory report. Extensive treatments of scientific writing and good writing in general are available[2,3] but the following short list of suggestions should prove helpful

1. Remember that length alone does not constitute quality. Your instructor will specify how much experimental detail is required, but it is unlikely that a "beaker by beaker" account of all your efforts, successful and unsuccessful, will be required. A concisely written report that describes an experiment sufficiently accurately that one may repeat it, and supports all conclusions by referring to specific points in an organized body of data, is more useful than an exhaustive treatment of every detail.

2. Avoid wordiness; strip away all words that add nothing to the meaning of a sentence. Your instructor will recognize fuzziness of expression as a corollary of fuzziness of thinking.

3. Define specialized terms, but recognize that there is a difference between specialized terms and jargon. Specialized terms are more precise than ordinary words and more

economical. Jargon is used in lieu of ordinary words that would be more appropriate.

4. Spelling and punctuation are important.

5. All reports should be typed or printed. Computer printouts must be clearly readable with proper spacings between paragraphs and pages; a poor quality dot-matrix printout is generally not acceptable. The report should be on good quality 22 x 28 cm (8.5 x 11-inch) white paper, not erasable bond or onionskin. Script typefaces should be avoided.

6. The report should be assembled in the following order: title page, abstract, introduction, experimental procedure (materials and methods), results and discussion (data), and references.

7. The title must be brief and informative.

8. The abstract should briefly present the problem, the experimental approach, and the major findings and conclusions. It should never exceed 200 words.

9. The bibliography must contain references to all works used significantly in the preparation of the report but should not contain references to basic texts unless specific items are quoted or directly adapted. Bibliographies padded with unused references are not acceptable.

10. References should clearly identify the original contributers to the work being cited, the journal or book, the volume and page number and the publication year. The publisher of all books must also be indicated. The citation schemes, including punctuation, suggested by the editors of *Biochemistry* or the *Journal of the American Chemical Society* are recommended.

11. All copy should be double-spaced, including footnotes and bibliographies. The abstract, introduction, and bibliography sections should begin on new pages. Each page should be numbered consecutively from the title page.

12. Ample margins must be used, with the copy beginning 1.5 inches from the top and with at least 1 inch of margin on the left and 0.75 inch on the right and bottom. Copy should appear only on one side of the paper.

13. The student must make at least one photocopy of all reports submitted. One copy is retained and the original (and possibly one copy) are submitted to the instructor.

14. Complex, multicolumn data tables should be given short descriptive titles. They should be positioned in the text as close as possible to where the data are discussed. All column heads should be concise, with additional details discussed in footnotes to the table. Units must be specified.

15. All figures and graphs must have a short descriptive caption. Avoid caption material that more appropriately belongs in the text. Graphs must have clearly labeled axes, which specify units. Experimental points should be circled.

16. All mathematical expressions and chemical equations should be typed or clearly rendered in black permanent ink. Typed expressions should be centered, with spaces left around operation and relation signs but not between coefficients, exponents, and subscripts and their terms. If possible, the exponents and subscripts should be in reduced size. For example:

$$E = \sum m_1(C_i - V_o)^2 + m_2(C_i - V_o)^3 + \cdots \qquad (1\text{-}1)$$

17. Chemical equations and structures must be complete and include only standard symbols. Ring structures should be drawn with an adequate template.

18. The same unit system must be used consistently throughout the report. Your instructor may specify the units for all important variables.

19. Solution concentrations should be given in moles per liter, designated M, for concentrations above 0.01 M, and in units of millimoles per liter, designated mM, for more dilute solutions. In some instances concentrations can be given as percent by volume, % (v/v), or percent by weight, % (w/w), but never in mixed weight-volume percentages. Percent saturation will be given according to the accepted convention: volume percent dilution of saturated solution with pure solvent without considering volume changes (see Chapter 4). All temperatures will be given in degrees Celsius, centrifugation accelerations in multiples of the acceleration constant due to gravity, x g, with times in minutes or hours and sedimentation coefficients in Svedberg units, S. Electrical potentials will be in volts, v.

20. **Finally, all text must be the original composition of the student author.** Sections containing quotations must be clearly set aside with quotation marks and with clear credit to the original author and source. Adapted or para-phrased material must be referenced in the bibliography.

LIBRARY WORK IN BIOCHEMISTRY

The introductory experiments described in chapters 2 to 10 are accompanied with enough detail and background information to assure success but the project extensions and the advanced projects in chapters 11 to 31 may require additional information especially if the instructor expects a sophisticated interpretation of student data in the light of previous observations reported in the literature. Not infrequently, a trip to the library sheds more light on a subject than does the corresponding length of time in the laboratory. Repeating an experiment without reviewing what has been observed by others is not a good investment of time, and a good laboratory report must include references to observations reported in the literature. Each experiment contains literature references, but the instructor may require a more extensive literature search and prescribe in detail the level of references to be included in each laboratory report.

STATISTICS IN BIOCHEMISTRY

The following statistical formulae are useful when dealing with the large samples of highly variable data, inherent in biological investigations.

1. The *arithmetic mean* of a set of N numbers $X_1, X_2, X_3, ..., X_N$ is denoted as $\overline{X}$, defined as

$$\overline{X} = \frac{\Sigma X_i}{N} \qquad (1\text{-}2)$$

2. The *median* of a set of numbers arranged in order of magnitude is the middle value or the arithmetic mean of the two middle values.

3. The *mode* of a set of numbers is that value which occurs with the greatest frequency, i.e., the most common value. A mode may not exist.

4. The *mean deviation or average deviation* of a set of N numbers X_1, X_2, X_3, ..., X_N is denoted M.D. and is defined as

$$M.D. = \frac{\Sigma |X_i - \overline{X}|}{N} \tag{1-3}$$

5. The *standard deviation* of a set of N numbers X_1, X_2, X_3, ..., X_N is denoted by s and is defined by

$$s = \sqrt{\frac{\Sigma |X_i - \overline{X}|^2}{N}} \tag{1-4}$$

6. The *normal* or *Gaussian probability distribution* is defined by the equation

$$Y = \left(\frac{1}{\sigma \sqrt{2\pi}} \right) \exp\left[\frac{-0.5(X_i - \mu)^2}{\sigma^2} \right] \tag{1-5}$$

For a *normal or Gaussian* distribution, the standard deviation is designated σ and the mean deviation is 0.7979σ. A graph of this curve centered on μ is given in Figure 1-1.

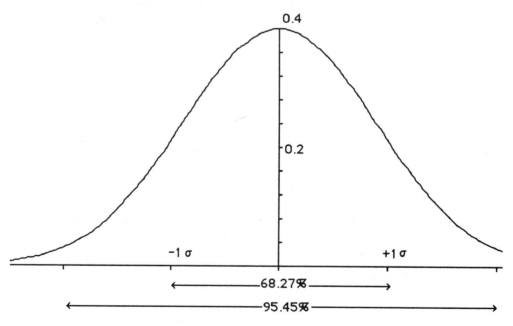

Figure 1-1 A normal or Gaussian distribution with ranges corresponding to first and second standard deviations indicated

This curve represents the distribution of measured values of X_i that range above and below the mean value $\overline{X} = \mu$, for a large number of measurements. Most X_i are close to $\overline{X}$, but some are far away. The percentages of X_i that lie within $\pm 1\sigma$, $\pm 2\sigma$ and $\pm 3\sigma$ of the mean are 68.27%, 95.45%, and 99.73%, respectively.

7. The *standard deviation of the mean* or **standard error** for a Gaussian distribution is defined by

$$\sigma_\mu = \frac{\sigma}{\sqrt{N}} \qquad\qquad (1\text{-}6)$$

As N, the number of samples points becomes larger, σ_μ becomes smaller and the precision is improved. Notice that the mean of a small number of measured values $\overline{X}$ is not equal to μ, but it approaches the "correct value" μ as N increases.

8. If we consider samples of size N drawn from a normal population, we can test the hypothesis that the normal population has mean μ by the *t score*, defined by

$$t = \frac{\overline{X} - \mu}{s} \sqrt{N-1} \qquad\qquad (1\text{-}7)$$

where $\overline{X}$ is the mean of a sample of size N and s is the sample standard deviation.

9. *Confidence limits* for the sample mean are related to the parameter t. One can define 95%, 99%, or other confidence intervals by using the table of student's t distribution cited in many books on statistics. For example for 10 measurements, t scores corresponding to 95% and 99% confidence are 1.83 and 3.25, respectively. For a table and example calculations see reference 4.

10. The *Chi-square distribution* is defined by

$$\chi^2 = \frac{\sum (X_i - \overline{X})}{\sigma^2} = \frac{Ns^2}{\sigma^2} \qquad\qquad (1\text{-}8)$$

Tables of χ^2 are used to test hypotheses, that is to determine how well theoretical distributions such as the normal Gaussian distribution fit those obtained from sample data. The χ^2 test can even be used to test whether the fit between observed and expected data is "too good," that is, better than one would expect given the random errors intrinsic to any measurements.

11. The statistical procedure for finding the best fitting line through a number of points amounts to finding a line through that set of data points that minimizes the deviations of the points from the prospective line. This can be done with a ruler and hopefully an objective eye, but a more formalized method minimizes subjectivity and affords an estimate of the goodness of fit. The *principle of least squares* amounts to choosing the line that minimizes the sum of squares of the deviations of the observed values of y from those predicted by the equation $Y = mX + b$, where m and b are estimated by a least squares algorithm: for example,

$$m = \frac{\sum (X_i - \overline{X})(Y_i - \overline{Y})}{\sum (X_i - \overline{X})^2} \qquad\qquad (1\text{-}9)$$

Fortunately most good pocket calculators are programmed to calculate all the above statistical parameters including m and b as well as a *correlation coefficient*, ρ. The term ρ measures the strength of the relationship between Y and X in a least squares fit. A value of $\rho = 0$ implies no linear correlation between Y and X; ρ is near 1.0 for a strong positive correlation, and r approaches -1.0 for a strong negative correlation.

REFERENCES

1. National Research Council, *Prudent Practices for Handling Hazardous Chemicals in Laboratories*, W. Spindel, Res. Director, Washington D.C: National Academy Press, 1981.

2. R. A. Day, *How to Write and Publish a Scientific Paper*, Philadelphia, PA.: ISI Press,1979.

3. W. Strunk and E. B. White, *The Elements of Style*, 3rd Ed., New York: Macmillan, 1979.

4. W. Mendenhall, *Introduction to Probability and Statistics*, 5th Ed., Duxbury   Press, 1979, p 280.

Chapter 2

Buffers, Titrations, and pH Measurement

In this experiment the general principles of quantitative solution preparation, buffers and pH measurements are demonstrated. A series of phosphate buffers are prepared and used to calibrate a pH meter. A comparison between the observed and calculated pH of a TRIS-HCl buffer will be made. Finally, a dilute solution of an unknown amino acid hydrochloride is titrated from pH 2 to 13 with carbonate-free NaOH solution. The titration curve can be used to limit the number of possible structures for the amino acid.

KEY TERMS

Buffers	*Gravimetric Methods*
Henderson-Hasselbalch equation	*pH standards*
Pipettors	*pK_a values*
Titration curves	*Volumetric Methods*

BACKGROUND

Most biologically important molecules contain acidic or basic groups that change charge as a function of pH. Acidic groups, such as carboxylic, phenolic, and phosphoric acid groups, are converted to anionic groups at pH levels above their pK_a values, while basic amino groups are cations at pH levels below their pK_a values. The charge on a molecule profoundly affects its chemical and biological properties; hence controlling solution pH is essential if one expects predictable and reproducible behavior from a biomolecule. Almost all solutions described in this book involve buffers, which are mixtures of weak acids and bases chosen to control the pH within limits governed by the pK_a values of the buffer components. The preparation of buffers is the logical first step in the long list of techniques that a biochemist must master. A biochemist must know the acid and base properties of common buffer components in order to select buffer formulations that will govern pH over the range desired and remain compatible with the chemical properties of all other solution components.

Buffers

Alkali metal hydroxides dissolve and dissociate completely to produce the same base OH^-. These metal hydroxides are all strong bases because OH^- ion will raise the pH of an unbuffered solution by combining with an exact equivalent of hydronium ions.

$$2H_2O \rightleftharpoons OH^- + H_3O^+ \qquad K_W = 10^{-14} \qquad (2\text{-}1)$$

Strong acids, such as HCl, HNO_3, and H_2SO_4, behave in an analogous way, dissociating completely in water to form hydronium ions, which combine with an equivalent of hydroxide ion by the same reaction. The concentrations of hydronium and hydroxide ions are related by the expressions

$$pH = -\log[H_3O+] \quad pOH = -\log[OH] \quad \text{and} \quad pH + pOH = 14 \qquad (2\text{-}2)$$

Weaker bases have less attraction for protons and combine with only a portion of the hydronium ions in a solution, the extent of the reaction being governed by its equilibrium constant. The strength of a weak base is, by convention expressed in terms of the acid dissociation constant of its conjugate acid (i.e., the acid formed from the base upon protonation). The ability of the conjugate acid to protonate the solvent water is described by the equilibrium constant K_a:

$$RNH_3^+ + H_2O = RNH_2 + H_3O^+$$

$$K_a = \frac{[RNH_2][H_3O^+]}{[RNH_3^+]} \qquad (2\text{-}3)$$

The stronger the base, the more tightly it will attract protons and the more stable will be its conjugate acid; hence the stronger the base, the smaller the K_a of its conjugate acid. The strength of a weak acid is defined directly in terms of its own K_a.

The stronger the acid, the less tightly will it hold onto protons and the larger will be its K_a. Because the strengths of acids and bases range over at least 10 orders of magnitude, the acid dissociation constants are customarily expressed as the logarithmic term, pK_a:

$$pK_a = -\log[K_a] \qquad (2\text{-}4)$$

A solution containing both a weak acid HA and its conjugate base A^- can regulate the pH of a solution. The addition of a small amount of hydroxide ion will not markedly change the pH because the acid will neutralize most of it; similarly hydronium ion will be consumed by the conjugate base. A weak base B and its conjugate acid BH^+ will also control pH the same way. This phenomenon is called buffering and the mixture is called a *buffer*. A solution of a strong acid or base alone has very little buffering capacity. The *Henderson-Hasselbalch equation* which follows, can be used to calculate the pH of a buffer solution given the concentration and pK_a values of the appropriate buffer components: a weak acid HA with its conjugate base A.

$$pH = pK_a + \log\left(\frac{[A]}{[HA]}\right) \qquad (2\text{-}5)$$

This equation gives the pH of a buffer solution if one knows the ratio of buffer components. The same equation can describe the pH changes accompanying a titration of HA with strong base. If one starts with the given amount of weak acid $[HA]_0$ and adds C_{OH} (moles/ liter) of strong base, which converts some HA into A, then the pH is given by the expression:

$$pH = pK_a + \log\left(\frac{C_{OH}}{[HA]_0 - C_{OH}}\right) \qquad (2\text{-}6)$$

Buffer solutions can be made by combining various proportions of the acid and base forms of the buffer in solution. If the acid and conjugate base forms are both readily available (e.g., acetic acid and sodium acetate), they can both be weighed out and added directly to water to give a buffer solution. An equivalent buffer can be prepared by adding less than an equivalent of the strong base sodium hydroxide to acetic acid. Similarly, a TRIS-HCl buffer can either be made by adding HCl to a solution of the buffer base TRIS (see Table 2-2) or by weighing out both the free base and its solid hydrochloride salt.

Buffer solutions must be made carefully if the pH is to be controlled accurately. The solvent must be free of extraneous acids or bases, including the gas CO_2, which forms carbonic acid. Solutions can be made by weighing out the appropriate amounts of all the components and dissolving them in distilled water. This method is convenient for large solution volumes, especially of concentrated solutions (100 mM or higher), but it is not convenient otherwise. The most common error the author has observed students make during this experiment is to use gravimetric approaches for dilute solutions when volumetric approaches are more convenient. If a series of very dilute solutions are required, it is faster to prepare them by diluting small amounts of a more concentrated stock solution rather than weighing out milligram quantities of material for each solution. On the other hand, for very accurate work, gravimetric methods may be preferable. All pipets should be calibrated by weighing delivered volumes of water.

Solution Preparation

The precision of quantitative bio-analytical work depends on the accuracy with which volumes and weights are measured and the reproducibility of these measurements. This means that solutions must be prepared with considerable care to ensure that concentration values are known and highly reproducible. Following is a brief discussion of quantitative solution preparation employing volumetric glassware and the analytical balance. For a more detailed treatment, the reader is referred to any of the standard textbooks on quantitative analysis.

Volumetric methods

Volumetric flasks are calibrated by the manufacturer to contain a given volume when filled at the calibration temperature (usually 20°C). The bottom of the meniscus should be tangent to the etched ring on the neck of the flask. Volumetric flasks should never be oven dried.

Pipets are of three types: volumetric, serological, and measuring. *Volumetric pipets* are calibrated to deliver a fixed volume. They can be identified by the fact that they have an enlarged midsection. The liquid is drawn into the pipet to a level above the mark on the stem with a rubber bulb. **Never use mouth suction!** The upper end is quickly closed with a dry index finger and the level is allowed to slowly drop to the mark by controlling the air entering the stem with the index finger. The pipet must be vertical. Any adhering liquid can be wiped from the tip of the pipet with a clean Kimwipe tissue (discard used tissues in an appropriate receptacle). The pipet will deliver an accurate volume into the receiving vessel if the tip is in contact with the vessel wall and the pipet is allowed to drain slowly. The liquid remaining in the tip is not blown out. *Serological pipets,* on the other hand, are made to deliver their total volume and the liquid remaining should be blown out. *Mohr Measuring pipets* are graduated and can deliver any known volume corresponding to the difference between two calibration marks. Measuring pipets can be less accurate for smaller volumes than either volumetric or serological pipets but they are more flexible in their use since a wide range of volumes can be delivered by any one pipet. Glass pipets must be washed with detergent solution, rinsed with distilled water, and dried with absolute methanol. Air drying in a dust-free environment is acceptable but oven drying for prolonged periods will ruin them.

Pipettors are designed to permit rapid reproducible deliveries of small volumes. Some pipettors are adjustable whereas others deliver only a single fixed volume. A pipettor is filled by pushing the plunger with gentle even thumb pressure while the plastic tip is below the liquid level. Releasing thumb pressure permits the tip to fill to the proper volume. Clean plastic tips must be used and each filled tip must be checked to ensure that air bubbles are not trapped within; this is a problem with "sudsy" protein solutions. Different types of automatic pipettors require different pipet tips. One must be sure to use the proper tip with each pipettor; they always should fit snugly but fill to below the stem of the pipettor. Consult your instructor to be sure. The pipettor delivers liquid accurately when the plunger is pressed to the first friction point. If contamination is a problem, the plastic should be discarded each time. A tip should never be reused after the solution within it has partially evaporated. Discard tips only in the proper receptacle. Common pipets and pipettors are illustrated in Figure 2-1.

Burets are used to deliver variable volumes in titrations. The titrant is delivered to the receiving vessel by repeatedly opening and closing the stopcock on the buret to release small portions at any one time.   To avoid pulling out the stopcock, it must be oriented to the right and operated with the left hand. This technique also leaves the right hand free to swirl the contents of the receiving flask after each addition. Burets are usually used for titrations (see Experimental Section) but they can also be used to prepare a series of similar solutions all at one time. For example, one buret can be used to prepare solutions containing: 1, 2, 5, 10, 15, and 20 mL of one component. When titrating to an indicator endpoint, the titrant can be added in reasonably large aliquots (>1 mL) until near the endpoint, then only very small aliquots (< 0.1 mL) are recommended. This may require a preliminary titration to get an estimate of the endpoint titrant volume. A 50-mL buret must never be emptied faster than 0.5 mL per second; otherwise, too much liquid adheres to the walls and the volume delivered will be less than the amount indicated.

When using a buret and a pH meter to obtain a titration curve, all aliquots should be as small as possible, given the time limitations, so that the maximum number of points defining the curve can be obtained. This is especially important near the equivalence point, where the pH is strongly dependent on titrant volume. Near the equivalence point the flow should be slowed to 1 drop or less at a time. A drop may be "split" by letting it form only partially before touching the buret tip to the wall of the receiving flask. Burets are read by keeping an eye at the meniscus level; sometimes a white card with a thick black horizontal line on it can be held behind the buret to improve the visualization of the meniscus. The buret stopcock must be lubricated with a water-insoluble stopcock grease but care must be taken not to over-grease the stopcock. Excess grease can become dislodged and plug the buret tip. Care must also be taken to avoid air bubbles in the buret tip.

Burets are cleaned with detergent solution, or for very dirty burets, H_2SO_4-$Na_2Cr_2O_7$ cleaning solution or alcoholic KOH can be used. Strong alkali or acid must be avoided if the buret will be used to deliver dilute solutions of acids or bases. The buret must be thoroughly rinsed with distilled water and the stopcock cleaned and dried. Lubricant must be applied only sparingly and only the proper kind should be used. Silicone or high-vacuum greases cannot easily be removed from the buret with aqueous cleaning solvents. Rapid cleaning can be accomplished by inverting the buret and drawing detergent solution into it; this avoids contact with the lubricated stopcock.

All well-cleaned volumetric glassware will drain without leaving visible drops on the glass. Glassware should be allowed to drain and dry on a drying rack in a dust-free environment then put away promptly.

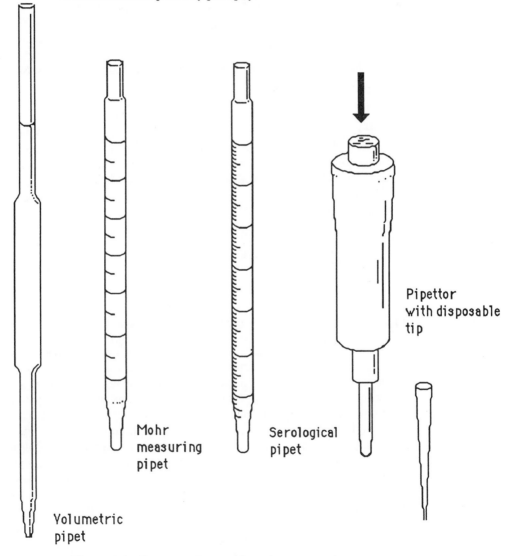

Figure 2-1 Common pipets and a pipettor

Gravimetric Methods

Solid and nonvolatile liquid buffer components should be weighed on an accurate balance. For large volumes of concentrated solutions, where weights to within ± 0.1 gram (g) are accurate enough, a triple-beam balance or a modern digital readout top-loading balance are faster and good enough. Dilute solutions or solutions with small volumes require the use of an *electronic analytical balance*. Electronic balances, when used properly, can be used to weigh samples to within 0.0001 g. The following precautions are obligatory to ensure such accuracy:

1. Never lean on the balance table or pile heavy objects on it when weighing.
2. Clean the balance pan and replace it properly.
3. Use clean weighing paper or a weighing boat.
4. Whenever possible, avoid fingerprints by using a pair of tweezers.
5. Place objects on the pan only after the balance is arrested.
6. Close the side windows when setting weights.
7. Set weights when the balance is semireleased and change weights slowly.
8. Arrest balance after setting weights; this stops the swinging of the pan.
9. Release balance fully and allow it to come to rest before determining the last two decimal places. Different balances have different scales, but most work on a Vernier principle. Consult your instructor on how to read the balance.
10. When finished, arrest the balance and remove the sample from the pan. Return all weight settings to zero and clean the pan. Close the side windows.
11. Never use analytical balances to weigh concentrated acid solutions, especially HCl or $HClO_4$.
12. Be sure to leave the Balance and the balance area clean.

Most laboratories will require a logbook entry whenever you use an expensive piece of shared equipment. If you have a log for the electronic balance, sign it each time you use it and note any potential problems in the log.

The pH Meter and pH Measurement

pH measurements are common practice in all chemical laboratories, but accurate pH measurements are especially important in biochemical experiments where the reactivity of enzymes and the stability of biomolecules may be optimum only over a very narrow pH range. Crude estimates of pH can be made with indicators or pH-indicator paper or strip; however, accurate estimates can only be conveniently obtained electrochemically with a glass electrode and a pH meter. Care should be taken in interpreting pH measurements since pH meter readings are not always directly convertible to hydrogen ion concentrations; the meter reading is sensitive to temperature and ionic strength. The common glass electrode assembly will be used in this laboratory to determine the hydrogen ion activity of dilute aqueous solutions of acids and bases under conditions where complications due to solvent, electrolyte, and temperature effects are not a problem. For an account of the general theory and practice of pH measurement, the reader is directed to the valuable book by Bates.[1]

The pH value is operationally defined in terms of the electromotiveforce E_x, measured with respect to a reference electrode.

$$pH = \frac{\mathcal{F}}{2.303RT}(\mathbf{E_x} - \mathbf{E_o}) \qquad (2\text{-}7)$$

where $\mathcal{F}$ is the charge of an Avogadro's number of electrons, 96,500 coulombs. All pH measurements require a reference electrode. Traditionally, a separate electrode containing saturated calomel·KCl has been used. The experimental electrode setup amounts to the galvanic cell:

reference electrode	reference solution	sat'd KCl	solution X	measuring electrode

The $\mathbf{E_o}$ value is defined by

$$\mathbf{E_o} = \mathbf{E_{st}} + \frac{2.303RT}{\mathcal{F}} + pH_{st} \qquad (2\text{-}8)$$

in which $\mathbf{E_{st}}$ is the electromotive force of the cell where solution X replaced by a standard pH solution. The controls on the pH meter are adjusted to read the appropriate value for the standard solution. This process is called *standardization*.

All pH measurements are made relative to pH values assigned for certain standard solutions. The student is referred to Bates[1] to learn the details of the assumptions behind such assignments. In general, the potential observed with a glass electrode is measured relative to a reference electrode. A glass electrode is placed in a solution of given acidity and the cell is completed with another reference electrode. The most common standard electrode is the silver-silver chloride electrode which consists of a silver wire immersed in a potassium chloride solution saturated with silver chloride. The electrochemical process, corresponding to the reference half-cell, is the reduction of silver ion to silver metal. The glass pH electrode consists of a silver-silver chloride electrode immersed in a pH 7 buffer saturated with silver chloride (Figure 2-2). The thin, ion-selective glass membrane develops an electrochemical potential when placed in a solution containing hydrogen ions. The hydrated outer layer of the glass membrane acts as a cation-exchanger that preferentially binds protons by displacing lithium ions from the negative binding sites on the hydrated silica membrane surface. This electrode corresponds to the half-cell:

Ag(s) | AgCl [sat'd], Cl$^-$ (inside), H$^+$ (inside) | glass membrane | H$^+$ (outside)

The difference in binding of protons to the inside and outside of the membrane is the origin of the electromotive force (EMF) measured with this electrode. The EMF of the cell depends on the difference in hydronium ion activity on the two sides of the glass membrane. Ions do not penetrate from one side to the other, but the difference in concentration of ions occupying silica lattice sites on the membrane surfaces results in a potential. The effect depends on the concentration of Na^+ and K^+ ions. It is also temperature dependent. The drift in strong acid or alkaline soutions is the result of changes in the composition of the surface layer of the glass electrode; consequently electrodes should never be stored in highly acidic or basic solutions.

In practice the pH meter is adjusted to read an "assigned pH value" for a given standard solution and once calibrated, the meter gives operationally useful pH measurements for solutions with hydrogen ion activities within ± 2 pH units of the standard solution. The pH of the standard solution must be close to the center of the proposed working pH range. Some meters can be calibrated to read pH values between two set values established by two standard solutions. The meter is first calibrated in one standard buffer with a calibration knob, then adjusted to a second pH value in a second solution with a drift knob. Modern pH meters with microchip memories warn the user if the sample is outside the pH range or temperature limits established by the calibration. For most practical applications pH can be equated to $-\log [H^+]$, but this assumption does not always hold up, for both theoretical and practical reasons. The pH of standard solutions is defined for a fixed temperature (usually 25^oC) and the pH meter reading will not correspond to this standard value at temperatures above and below this temperature. Corrections for temperature effects can be made either with the temperature control of the pH meter or by correcting the observed pH readings with the temperature coefficient (found in Bates[1] or the *CRC Handbook for Chemistry and Physics*). For example, the temperature coefficient of 0.025 M Na_2HPO_4 + 0.025 M KH_2PO_4 is -0.0028 pH unit per ^oC.

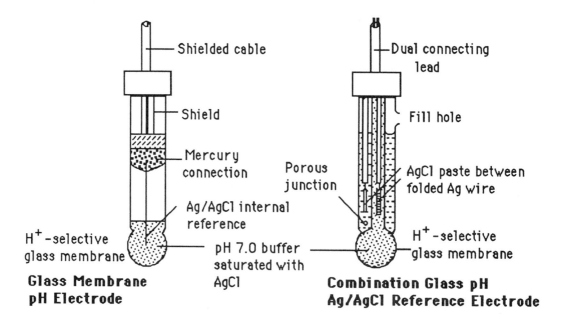

Figure 2-2 Electrodes used for pH measurement

Choice of a Buffer

The choice of a buffer must take into account the desired pH working range, but also any special characteristics of the buffer that may interfere with the planned experiment. Poisonous buffer components cannot be used for biological media; buffer components that inhibit, participate in, or otherwise or interfere with an enzyme reaction cannot be used in kinetics experiments. Buffer components that precipitate in the presence of other solution components are not good choices either. Buffers should also be chemically stable and easy to prepare.

The practical pH standards and the working ranges described in Table 2-1 should be remembered.

Table 2-1
pH STANDARDS AND WORKING RANGES

Calibration pH (25°C)	Standard Solution Composition	Useful Working Ranges
1.00	0.1 M HCl	0.7 - 2
4.008	0.05 M K acid phthalate	2
6.865	0.025 M Na_2HPO_4	5 -8
9.180	0.01 M borax	8 -10

The pH working range is determined by the pK_a of the buffer components. Inspection of the Henderson-Hasselbalch equation will indicate that the pH = pK_a when the conjugate pairs A^- and HA are equal in concentration (i.e., the midpoint in the titration of HA with base). This pH is the center of the working range of this buffer and buffering will be significant ±1 pH unit from this point. The practical working range depends on the concentration of the buffer, more concentrated buffers extend buffering over a wider pH range.

When choosing a buffer, one must pay attention to the pK_a value of the acid form. In Table 2-2 are given the pK_a values of many buffers used in biochemistry. Many of these acids have more than one acid dissociation, hence the Roman numerals indicating the deprotonation step defining the indicated pK_a. A number of the buffers are known as "Good" buffers because they contain relatively non-nucleophilic amine buffer components developed by the chemist N. E. Good.[2,3] Many of these buffers are zwitterionic over their useful pH range.

Tradition also plays a role in choice; buffers that are widely used include phosphate, which is excellent for most uses near pH 7, and TRIS-HCl, which buffers over a higher pH range. *Note*: Buffers are frequently referred to in terms of their conjugate acids, even though it it is understood that both conjugate species must be present for adequate buffering. For example, [0.1 M TRIS-HCl, pH 8.0] buffer contains the equivalent of 0.1 M TRIS with enough HCl to bring the pH to the specified range; some of the TRIS becomes TRIS-HCl, but the stoichiometric concentration of TRIS remains constant: 0.1 M. Phosphate buffers can buffer over a very wide range because of the second and third ionization of phosphoric acid. The molarity of a phosphate buffer refers to the stoichiometric phosphate concentration, regardless of its form, not the concentration of the metal ions. Table 2-3 is a recipe for making phosphate buffers over the pH range 6.0 to 8.0.

Table 2-3 can be used to prepare two stable buffer solutions for pH meter calibration. For pH values between 5.8 and 8.0, one uses 50 mL of 0.1 M KH_2PO_4 (13.60 g/L) and X mL 0.1 M NaOH. For pH values above 8.0, a mixture 0.1 M in both KCl and H_3BO_3 (7.455 and 6.184 g/L, respectively) with X mL 0.1 M NaOH is called for. All solutions should be diluted to 100 mL with freshly boiled, distilled deionized water. Carbonate-free 0.1 M NaOH is prepared by diluting a concentrated stock solution (>10%) with freshly boiled deionized water. Concentrated NaOH solutions should be stored in tightly stoppered Nalgene bottles. Do not use glass stoppered bottles for strong base, the glass stoppers will freeze tight.

Table 2-2

pK$_a$ VALUES OF COMMON BUFFERS[a]

pK$_a$	Name	pK$_a$	Name
1.23	oxalic acid (I)	7.1	imidazole
1.96	phosphoric acid (I)	7.2	phosphoric acid (II)
2.0	versene (I)	7.2	MOPS* [morpholino-
2.1	arsenic acid (I)		propane sulfonic acid
2.35	glycine	7.4	ethylene diamine (I)
2.7	alanine	7.5	TES N-tris (hydroxymethyl)*
2.7	versene (II)		methyl-2-aminoethane sulfonic
2.96	tartaric acid (I)		acid]
3.1	citric acid (I)	7.55	HEPES * [N-2-
3.39	malic acid (I)		hydroxyethylpiperazine N'-2-
3.4	dimethylglutaric acid (II)		aminoethane sulfonic acid
3.67	thioglycolic acid (I)	8.0	veronal
3.76	formic acid	8.08	TRIS * [tris-(hydroxymethyl
3.83	glycolic acid		amino methane]
4.16	tartaric acid (II)	8.35	BICENE * [N,N-bis(2-
4.19	oxalic acid (II)		hydroxyethyl glycine]
4.2	succinic acid (I)	8.4	TAPS *[tris(hydroxymethyl)
4.4	citric acid (II)		methylaminopropane sulfonic
4.8	acetic acid		acid]
4.86	propionic acid	8.7	2-amino-2-methyl-1,3-
5.05	malic acid (II)		propanediol
5.3	pyridine	8.7	diethanolamine
5.5	citric acid (III)	9.22	dimethylaminoethanol
5.6	succinic acid (II)	9.2	arsenic acid (III)
5.86	malonic acid (II)	9.2	boric acid
5.92	piperazine (I)	9.2	histidine
6.1	histidine	9.4	ammonium chloride
6.15	MES * [2-(N-morpholino)	9.5	ethanolamine
	ethanesulfonic acid]	9.78	glycine
6.2	versene (III)	9.81	piperazine (II)
6.5	pyrophosphoric acid (II)	10.2	ethylene diamine (II)
6.5	3,3-dimethylglutaric acid (II)	10.4	CAPS * [cyclohexylamino
6.8	arsenic acid (II)		propane sulfonic acid]
6.8	PIPES [piperazine-N,N'-	10.6	methylamine
	bis(2-ethanesulfonic acid)]	10.68	thioglycolic acid(II)
		10.72	dimethylamine
		12.44	phosphoric acid (III)

[a]Good's Buffers indicated with *

Table 2-3
CONVENIENT BUFFER SOLUTIONS[a]

pH 25°C	mL 0.1 M NaOH added
Phosphate Buffers	
5.80	3.6
6.00	5.6
6.20	8.1
6.40	11.6
6.60	16.4
6.80	22.4
7.00	29.1
7.20	34.7
7.40	39.1
7.60	42.4
7.80	44.5
8.00	46.1
Borate Buffers	
8.50	10.1
9.00	20.8
9.50	34.6
10.00	43.7

[a] From references 4,5

EXPERIMENTAL SECTION

Laboratory Assignments

1. Students work in pairs. Each pair is asked to prepare two <u>phosphate</u> buffer solutions corresponding to pH 7.0 and 8.5, respectively and to use these to calibrate their pH meter. *Note*:: One buffer can be prepared according to Table 2-3 but the second cannot. The instructor will check the calibration with some NBS (National Bureau Standards) standard buffer. Acceptable buffers must be within a total range of 0.05 pH unit.

2. The instructor will assign each pair a TRIS-HCl buffer of specified pH and ionic strength. Students are expected to calculate the concentrations of TRIS and HCl required by using the Henderson-Hasselbalch equation. The buffer is then prepared according to the calculation and the pH is measured. Students are expected to compare the observed pH with the desired one.

3. Each pair of students will then be given an unknown amino acid to characterize by pH titration which affords the pK_a values of the ionizable groups. If one knows the weight of the amino acid (as hydrochloride salt), the molecular weight can be calculated from the equivalent volume of base used to titrate it. Since the amino acid is available as the hydrochloride salt and is likely to be hydroscopic, it must be carefully weighed to the nearest tenth of a milligram in a sealed container and dissolved quickly in 1 mL of distilled water in a titration vessel or Erlenmeyer flask. The titration requires

carbonate free 0.001 M NaOH, and the pH and titrant volume must be recorded after each addition of base. For an amino acid of molecular weight 200, 2 mg corresponds to <u>at least</u> 10^{-5} mol of titratable material which is the neutralization equivalent of one mL of 0.01 M NaOH. Consequently, a 10 mg sample of amino acid will require about 5 mL of 0.01 M NaOH titrant to titrate the carboxyl group and one or more equivalent volumes to titrate the amino and side chain groups.

If a significant but unknown excess of HCl were used, the equivalent volume V_E would also include the volume of base required to titrate this excess acid. Therefore if very accurate pK_a values are to be determined, it is recommended that the excess HCl be removed by lyophilization. The pK_a values of the amino acid can be determined from the titration curve illustrated in Figure 2-3. The endpoint of the titration V_E corresponds to the inflection point in the curve. With very good data, the endpoint can be obtained by plotting the first derivative ($\partial pH/\partial V$) of the curve vs. volume titrant. The endpoint corresponds to a maxima of the first derivative curve. The pK_a value corresponds to the pH in which the acid and base forms of the titrated species are in equal concentration. (The student should prove this using the Henderson-Hasselbalch equation.) This corresponds to the pH observed at one-half equivalence volume $V_E/2$. For amino acids with acidic or basic sidechains, overlapping or nonsymmetrical titration curves are a frequent occurrence. In such cases pK_a values determined from titration curves must be interpreted with caution.

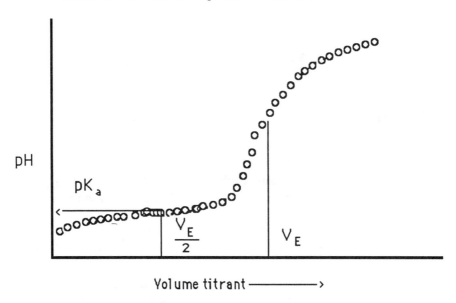

Figure 2-3 Peptide titration curve

DISCUSSION QUESTIONS

1. Calculate the pH of the following solutions:
   a. 0.01 M HCl
   b. 0.01 M HCl, 0.01 M TRIS
   c. 0.01 M HCl, 0.02 M TRIS

      d. 0.01 M HCl, 0.02 M Na_2HPO_4
      e. 0.01 M NaOH, 0.02 M Na_2HPO_4

2. What is the concentration of hydronium ions at pH 5.6? What is the concentration of hydroxide ions?

3. The concentration of hydroxide ions at pH 6.0 is 10^{-8} M; therefore, explain why the pH of 10^{-8} M NaOH in pure water is not 6.

4. Derive the Henderson-Hasselbalch equation starting from the definitions of pH and pK_a.

5. Why is it better to use 0.001 M NaOH to titrate your peptide rather than 0.01 M NaOH or 0.0001 M NaOH?

6. Titrations employing a pH indicator cannot be used to determine accurate equivalence points if the indicator concentration is more than a few percent of the peptide concentration. Explain.

7. Why would storage of a glass electrode in concentrated acid or base be an unwise practice?

REFERENCES

1. R. G. Bates, *Determination of pH: Theory and Practice*, (2nd ed.), New York: Wiley, 1964.

2. N. E. Good et al., *Biochemistry*, **1966**, *5*, 467.

3. W. J. Ferguson and N. E. Good, *Anal. Biochem.*, **1980**, *104*, 300.

4. R. Bower and R. G. Bates, *J. Res. Natn. Bur. Stand.* **1955**, *55*, 197.

5. R. M. C. Dawson et al., *Data for Biochemical Research*, 2nd ed. Oxford: Clarendon Press, 1969, p. 490.

Chapter 3

Peptide Characterization: Paper and Thin Layer Chromatography

In this experiment an unknown tripeptide is hydrolyzed by boiling in 20% hydrochloric acid and the total amino acid composition is determined by either paper chromatography or thin-layer chromatography (TLC) on cellulose sheets. Ninhydrin spray is used to detect the amino acid "spots." The amino acid components are identified by their R_f values for the paper chromatography approach or by comparison to the mobility of known amino acids on the cellulose sheet. The N-terminus of the tripeptide is identified by Sanger's method of DNP derivatization with FDNB. The derivatized amino acids are identified by TLC on polyamide sheets and comparison with authentic DNP amino acids. From these results students may postulate two possible tripeptide sequences for each unknown peptide.

KEY TERMS

Azeotropic drying
Lyophilization
N-terminal analysis
Peptide hydrolysis
Thin-layer chromatography

DNP derivatives
Ninhydrin visualization
Paper chromatography
R_f values

BACKGROUND

The sequence of amino acids in a polypeptide is called its primary structure. The sequence analysis of a polypeptide completely chemically defines the compound unless one of the amino acids has been chemically modified. A common strategy is employed for determining most amino acid sequences, with the usual approach involving the following steps:

1. Reductive cleavage of all Cys-S-S-Cys bonds.

2. Determination of the amino acid composition (after hydrolysis).

3. Identification of the N- and C-terminal amino acids.

4. Either direct sequencing by the Edman method, *or*

5. Cleavage of the peptide at specific amino acid residues, by chemical or enzymatic means, and characterization of these fragments by steps 2, 3 and possibly 4.

Considerable progress has been made in automating the characterization of polypeptides. Microgram quantities of polypeptides containing 50 or more residues can now be sequenced within a day. However, there is still educational utility in repeating classical sequence analysis on an unknown polypeptide, if for no other reason than to appreciate more fully the herculean efforts of Frederic Sanger and the first generation of modern protein chemists.[1]

Peptide Hydrolysis

Moderately concentrated hydrochloric acid is generally used and reaction times of 12 to 24 hours are required.

This hydrolysis is often accompanied by complex side reactions. The nature of these side reactions depends on the composition of the polypeptide or protein and the presence of impurities, especially metal ions, in the acid. Even if highest-quality hydrochloric acid is used, serine and threonine are usually destroyed. Tryptophan is largely destroyed, particularly if the hydrolysis is carried on in the presence of oxygen. Long reaction times (more than 24 hours) may also partially destroy cysteine and tyrosine. Incomplete hydrolysis is not a problem for peptides with very bulky amino acid side chains (valine, isoleucine).

The procedure adopted for this project is a compromise between practical considerations of experimental convenience (sample size, available equipment, etc.) and the requirements for good recovery of the more acid labile amino acids. The reader is urged to consult the recent literature before adopting this procedure when quantitative recovery of all amino acids is required.

Paper and Cellulose Chromatography of Amino Acids

Ordinary chromatography paper contains about 20% adsorbed water, which is equivalent to two molecules of water hydrogen bonded to each hexose subunit of cellulose. This water is immobile. In paper chromatography, the adsorbed water is the stationary phase and the organic solvent is the mobile phase. A solute is partitioned between these two phases and the rate at which the solute travels depends on the extent to which it is taken into the mobile phase. The solute partition coefficient depends upon the balance between the hydrophilic and hydrophobic characteristics of the solute.

The rate of flow or R_f *value* of a particular solute is defined by the ratio of D_1 to D_2 illustrated in Figure 3-1. This ratio is a constant, characteristic of the solute, solvent, temperature, and type of paper.[2] Both distances are measured from the origin (the spot where the mixture is applied to the paper or cellulose sheet). The distance traveled by the solvent D_2 is taken as the distance from the origin to the solvent front (the leading edge of wetness). The solvent front must be marked with a pencil as soon as the chromatogram is removed from the tank because the paper soon dries. The distance traveled by the solute D_1 is taken as the distance from the origin to the solute spot. If the mixture contains several components, each will have its own R_f value. When spots are too close to measure independently, their combined R_f values may be recorded. The spots must be kept small for accurate measurement. For broad spots the measurement of D_1 is to the center of color density or, if that is ambiguous, to the leading edge. The setup for paper chromatography is shown in Figure 3-2.

$$R_f = \frac{D_1}{D_2} = \frac{\text{distance traveled by the solute}}{\text{distance traveled by the solvent}} \qquad (3\text{-}1)$$

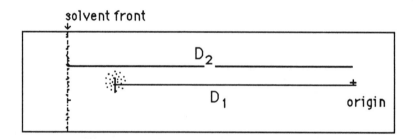

Figure 3-1 Illustration of R_f values

Note that even though paper chromatography is done on a solid matrix, it is really a type of liquid-liquid partition chromatography. It can be performed by ascending or descending solvent flow. The latter method is faster because gravity aids the solvent flow. In either type the chromatography chamber must be equilibrated to the solvent vapor. Hence the process must be run in some sort of chamber. A simple wide-mouth, screw-top 1-gallon jar can be used for small paper sheets. A cylinder or tank with a ground-glass edge and a glass plate top can be used for larger runs.

Amino acids are colorless and can be visualized only by spraying the dried chromatogram with ninhydrin or $KMnO_4$-Na_2CO_3 solutions. The chromatogram must be completely dry before spraying, and the spray must be evenly applied with no running or the spots will spread and accurate R_f values cannot be determined. Ninhydrin produces a purple spot with all amino acids but proline, which affords a yellow spot. (Permanganate gives a yellow spot on a pink background for methionine, cysteine, and tyrosine.) Both types of spots must be circled with pencil before they fade.

In Table 3-1 are given the R_f values for amino acids <u>on paper</u> in two solvent systems. These R_f values cannot be used for cellulose TLC. If the faster running cellulose strip method is used, students are expected to prepare their own table of R_f values from the observed mobilities of known amino acids.

Table   3-1
R_f VALUES FOR AMINO ACIDS ON PAPER

Amino Acid	R_f (Phenol-H_2O)[a]	R_f (Butanol:Acetic acid)[b]
Alanine	0.58	0.11
Arginine HCl	0.56	0.03
Asparagine	0.44	0.03
Aspartic acid	0.18	0.03
Cysteine	0.16	
Glutamic acid	0.31	0.05
Glutamine	0.31	0.04
Glycine	0.39	0.06
Histidine	0.64	0.03
Isoleucine	0.84	0.36
Leucine	0.84	0.34
Lysine-HCl	0.48	0.02
Methionine	0.77	
Phenylalanine	0.84	0.29
Proline	0.86	0.14
Serine	0.36	0.05
Threonine	0.49	0.09
Tryptophan	0.80	0.22
Tyrosine	0.67	0.16
Valine	0.78	0.24

[a]In phenol saturated with water. [b] butanol-acetic acid (9:1) saturated with water; see Ref. 3.

Sanger's Method of N-terminal Residue Analysis

Labeling the N terminus of a protein or polypeptide with 1-fluoro-2,4-dinitrobenzene (FDNB) is a method developed primarily by Frederick Sanger.[1] The α-amino group of the N-terminal amino acid displaces the fluorine from the reagent to yield a 2,4-dinitrophenyl or **DNP derivative** of the intact peptide. This derivative is called a DNP-peptide or DNP-protein. The DNP derivative is then cleaved by hydrolysis under conditions that leave the bond to the DNP group intact, so that the liberated N-terminal amino acid remains flagged with the DNP group, which imparts a yellow color to any chromatographic spot due to this residue. This amino acid can then be identified by comparing its R_f value with that of authentic DNP-amino acids. The procedure involves three steps:

1. Preparation of the DNP-protein or DNP-peptide is carried out under mildly alkaline conditions (pH 8 to 9), where the N-terminal amino group can displace the fluorine of FDNB. Because the α-amino group is a weaker base than the side-chain amino group of lysine, the latter remains largely protonated below pH 9 and is not a competitive nucleophile.

$$O_2N-\underset{\underset{NO_2}{|}}{\bigcirc}-F + H_2N-\underset{\underset{R_1}{|}}{\overset{\overset{H}{|}}{C}}-\underset{\overset{O}{\|}}{C}-N-\underset{\underset{R_2}{|}}{\overset{\overset{H}{|}}{C}}-\underset{\overset{O}{\|}}{C}-N-\underset{\underset{R_3}{|}}{\overset{\overset{H}{|}}{C}}-\underset{\overset{O}{\|}}{C}-etc. \longrightarrow$$

$$O_2N-\underset{\underset{NO_2}{|}}{\bigcirc}-\underset{\underset{H}{|}}{N}-\underset{\underset{R_1}{|}}{\overset{\overset{H}{|}}{C}}-\underset{\overset{O}{\|}}{C}-\underset{\underset{H}{|}}{N}-\underset{\underset{R_2}{|}}{\overset{\overset{H}{|}}{C}}-\underset{\overset{O}{\|}}{C}-\underset{\underset{H}{|}}{N}-\underset{\underset{R_3}{|}}{\overset{\overset{H}{|}}{C}}-\underset{\overset{O}{\|}}{C}-etc.$$

Unreacted FDNB can be discarded by extracting the alkaline solution of the DNP-peptide with ether. The anionic peptide remains in the aqueous phase while the neutral FDNB goes into the organic solvent (upper layer). A common side product will be 2,4-dinitrophenol which cannot be entirely removed by extraction. The dinitrophenol remains anionic at pH levels above 5.

2. The DNP-peptide is hydrolized in acid and the liberated α-DNP-amino acid is recovered by ether extraction from the acidic solution.

$$O_2N-\underset{\underset{NO_2}{|}}{\bigcirc}-\underset{\underset{H}{|}}{N}-\underset{\underset{R_1}{|}}{\overset{\overset{H}{|}}{C}}-\underset{\overset{O}{\|}}{C}-\underset{\underset{H}{|}}{N}-\underset{\underset{R_2}{|}}{\overset{\overset{H}{|}}{C}}-\underset{\overset{O}{\|}}{C}-\underset{\underset{H}{|}}{N}-\underset{\underset{R_3}{|}}{\overset{\overset{H}{|}}{C}}-\underset{\overset{O}{\|}}{C}-etc. \xrightarrow{HCl}$$

$$O_2N-\underset{\underset{NO_2}{|}}{\bigcirc}-\underset{\underset{H}{|}}{N}-\underset{\underset{R_1}{|}}{\overset{\overset{H}{|}}{C}}-\underset{\overset{O}{\|}}{C}-OH + H\overset{+}{}\underset{\underset{H}{|}}{N}-\underset{\underset{R_2}{|}}{\overset{\overset{H}{|}}{C}}-\underset{\overset{O}{\|}}{C}-OH + etc.$$

Following the hydrolysis of the DNP-peptide, the only α-DNP-amino acid will be derived from the original amino terminus. The other DNP-amino acids (DNP-ε-lysine, O-DNP tyrosine, DNP-histidine) will have free α-amino groups, which will be protonated under the extraction conditions, and they will not be extracted into ether. The Sanger method cannot be used for amino acids with DNP-amino acids destroyed under acid hydrolysis; DNP-glycine, DNP-cysteine and DNP-proline all require special techniques for optimum recovery of the derivative. DNP-arginine and diDNP-histidine are only partially soluble in ether. These complications can be minimized (in a teaching laboratory) by a judicious choice of unknowns.

3. Chromatographic identification of the DNP-amino acids is made by comparison with authentic DNP-amino acids. The DNP-amino acids, in acetone, can be spotted on Schleicher and Schull polyamide thin-layer chromatographic (TLC) plates and the spots developed with water : formic acid (97%) : acetic acid (9:6:5). The commercial plates are plastic sheets coated on both sides with a thin layer of polyamide, which permits spotting the knowns on one side and the unknown peptide derivative on the other. If all the unknown samples are placed exactly opposite the known samples then one of the unknown spots will end up exactly opposite one of the standards on the reverse side. This permits a direct comparison and avoids the time-consuming equilibration of the plates to solvent.

Polyamide plates have the advantages of speed, sensitivity, and reusability. None of these useful features are shared by paper, the original matrix used by Sanger to sequence insulin protein. The polyamide plates are easily dented and students should not make any mark on the plates either before or after chromatography. Spots are applied by use of a paper card template with holes punched in it. The thin-layer plates can be cleaned and reused by soaking them overnight in a 25:1 mixture of methanol and concentrated ammonia followed by rinsing twice in methanol and air drying. No significant changes in their properties are observed.

Because of the problems mentioned above and the large sample sizes required for the Sanger method, this method has largely been replaced by the dansyl method for N-terminal analysis which is more sensitive because the dansyl-amino acid derivative is fluorescent.[3] A combination of the Dansyl method and the Edman Degradation can be used to sequence manually up to 36 residues in a single day.[4] Automated sequence analysis of 100 to 200 nmol of peptide (1 to 2 mg of a 100-residue peptide) can be run routinely and sequences have been obtained on as little as 20 pmol of protein. The chromatographic analysis employs high performance liquid chromatography (HPLC) with fluorescence detection rather than the slower and less sensitive TLC method. The dansyl group is attached to the α-amino residue by a nucleophilic displacement on the sulfur of the sulfonyl group.

Dansyl chloride   Dansyl amino acid

Although the Sanger method has been largely superseded by the Edman degradation method, the Sanger method is quick and easy to do and has considerable educational utility because the derivatives are colored and can be followed through extraction and chromatography. Because the polyamide TLC method can be used with very small spots, it is nearly as sensitive as the upper end of the dansyl method.

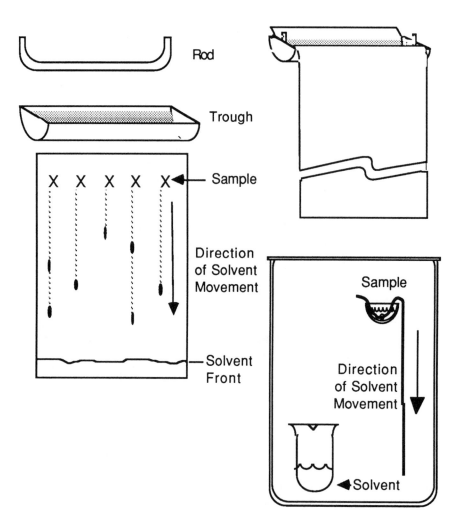

Rod

Trough

Sample

Direction of Solvent Movement

Solvent Front

Sample

Direction of Solvent Movement

Solvent

Figure 3-2 Descending paper chromatography

EXPERIMENTAL SECTION

Each student will be given a few milligrams of peptide to characterize and he or she must work independently during the hydrolysis, spotting and staining portions of this experiment. However, several chromatograms may be run together in the same tank, and students may divide up the chores associated with this portion in any mutually acceptable way. Some instructors may choose to avoid the lengthy hydrolysis step and give students amino acid hydrochloride samples corresponding to those in the peptide.

The following procedure for the hydrolysis is followed by procedures for descending paper chromatography and ascending cellulose TLC. The instructor will select between these two characterization options.

Hydrolysis of the Peptide

The tripeptide (1 to 2 mg) is dissolved in 100 μL of 20% HCl and transferred to a small-bore glass tube or ampule. The tube is sealed by the instructor and heated at 100°C for 12 hours.

CAUTION: WEAR SAFETY GLASSES
WHEN HANDLING SEALED TUBES OR AMPULES.

Tubes with cracks or bubbles in the glass must be avoided, and one should wrap samples in a wire screen before transferring them to the oven. The <u>cooled</u> tube must be carefully opened and the contents transferred to a small glass test tube and frozen in dry ice acetone or liquid nitrogen. Several tubes, each capped with parafilm that has been perforated with a needle, can be used in the same lyophilizer flask, as illustrated in Figure 3-3. The lyophilizer flask must be attached to a vacuum pump <u>with a KOH trap</u>. The volume is reduced to a thin wet film. Sometimes, if the HCl solution is frozen at dry ice temperatures, it just becomes very viscous.

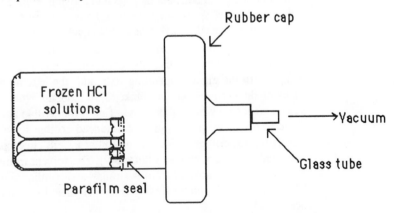

Figure 3-3 Lyophilization flask with sample tubes

If lyophilization apparatus is not available, the sample can be poured into a small round-bottomed flask, sealed with a one-hole rubber stopper, and attached to an aspirator system, but this is much slower. The sample must be gently shaken to ensure that it is spread to a thin film on the bottom of the flask. It will take well over an hour to remove most of the HCl solution at aspirator pressures even at room temperature. Care must be taken to avoid contact of the solution with the rubber stopper. The remaining liquid can be reduced to dryness by *azeotropic drying* by adding 5 mL of acetone : cyclohexane (1:1) and evacuating the flask again. Gently warming the flask while evacuating at full aspirator force will quickly remove the solvent.

If a lyophilizer is used, the flask should remain at room temperature and the organic liquids must not come in contact with the rubber cap. The solutions within the tubes will remain frozen, even at room temperature, as long as the pump is drawing off considerable water vapor. **The lyophilizer vacuum system pump oil must be changed and the trap must be cleaned out immediately after this experiment. The lyophilizer flask and rubber cap must be put away clean and dry. Discard the KOH carefully.** The resulting dry film of amino acid hydrochlorides is dissolved in 25 µL of water and characterized by either paper or cellulose chromatography.

Paper Chromatography of Amino Acids

This method of characterization takes less than 1 hour to set up, about 6 hours to run the two chromatograms simultaneously, 15 minutes to remove them from the tank and hang them up to dry overnight, and 30 minutes for staining and analysis. This method is more practical for laboratories that start in the early morning and can be opened briefly in the late afternoon to remove the paper from the tanks. Otherwise, the lab must be opened for 30 minutes late at night.

For each chromatogram, a large 46 x 57 cm sheet of Whatman No. 1 medium chromatographic paper sheet should be marked with a series of penciled X-marks about 15 cm from one of the short edges (see Figure 3-2). The marks should be at least 4 cm apart. Fingerprints and dust must be avoided when handling the sheet, and the tabletop must be covered with a clean sheet of paper or plastic. A tiny spot of hydrolyzed peptide amino acid solution should be applied to every other X-spot. A glass capillary tube or NANOCAP applicator must be used to ensure that the wetted area is only 1 to 2 mm in diameter. The spots may be more visible from the underside of the paper. On the intervening X-spots, small spots of known amino acid residues selected from Table 3-1 should be placed to give reasonable checks on whether the R_f values reported in the table are reliable for the solvent and paper used in this experiment.

After the spots are dry, the chromatography paper is secured onto the solvent trough with the glass suspending rods and the assembly is mounted into one of the racks of the chromatography tank. Two students working with forceps are required since one person working alone will find it impossible not to touch the paper and leave fingerprints. The paper must be balanced so that it will not slip out of the trough during the experiment. For this reason at least 8 cm of paper must protrude from the back end of the trough (see Figure 3-2). The solvent trough must be filled quickly and evenly so that the solvent front is straight and parallel to the leading edge of the paper. The solvent can be either freshly prepared phenol (saturated with water) or 1-butanol: acetic acid (9:1).

CAUTION: PHENOL CAUSES SEVERE SKIN BURNS.

In either case the chromatography tank should be reserved exclusively for that solvent system and the tank. For very accurate R_f measurements, the paper and atmosphere within the chromatography tank should be equilibrated to the solvent vapors for several hours before the solvent is added to the trough. It takes several hours for the solvent front to reach the bottom quarter of the paper. As soon as it is within 5 to 10 cm of the bottom, the paper must be removed. Care must be taken to avoid spilling solvent over the chromatogram when removing the upper end from the trough. **Also, the paper is very soft when wet.** The solvent front must be carefully marked with a soft pencil before the paper has dried appreciably. Once the paper is dry, the solvent front often cannot be seen except under ultraviolet illumination, .

As soon as the paper has partially dried, it can be hung on the drying lines in the hood. Clothespins are necessary and a weight must be suspended from the bottom of the paper to prevent the paper from flapping in the draft of the hood. Excess phenol is removed by allowing several hours for evaporation in the hood. A stock solution containing 400 mg of ninhydrin and 1.5 ml of s-collidine in 100 mL of 95% ethanol is prepared by one student but used by all. The dried chromatogram is sprayed with ninhydrin-collidine and allowed to dry. Commercially available chromatographic spray units give the best results, but any fine-mist spray unit is satisfactory. Large drops and splattering must be avoided.

NINHYDRIN MIST MUST NOT BE INHALED.

The spots are developed with a heat gun (hair dryer) held 12 inches from the paper. The spots that form do fade and must be circled with a soft pencil or covered with a clear lacquer to prevent fading. A staining procedure employing a spray of 1% potassium permanganate and 2% sodium carbonate in water is sensitive to methionine, cysteine, and tyrosine and can be used as a confirmatory test if a second chromatogram is run. Each student is expected to determine the R_f values for each principal spot. The average value of three or four determinations as well as the deviation from the mean are to be recorded.

The paper chromatography method often leaves students trying to decide between four or five amino acids as possible components of the tripeptide, therefore an estimate of experimental uncertainty in R_f is very helpful. When spots are too close to measure independently, their combined R_f values may be recorded. Tentative assignments for each spot can be made from the data given in Table 3-1.

Cellulose TLC of Amino Acids

This procedure takes about as much startup time as paper chromatography, but the chromatogram is developed within an hour, so it is much faster than paper chromatography and generally gives better resolved spots.

A commercial 20 x 20 cm cellulose precoated plastic sheet is cut into two 10 x 20 cm sheets and the amino acids from the peptide hydrolysis are placed in a row 2 cm from the 10-cm edge. When all the samples have dried, the TLC sheet is lowered into a TLC tank containing solvent-soaked towels but no solvent on the bottom. The sheet is leaned against one of the walls of the tank and the opposite wall is lined with solvent-soaked white paper towels. The solvent is butanol: acetone: water: acetic acid (10:10:5:2).

After 30 minutes of equilibration, the chromatography is initiated by pipetting in enough solvent to give a depth of 8 to 10 mm. The solvent must be added slowly and evenly so that an even solvent front is obtained. The solvent front should be allowed to reach about 70% of the way to the top of the sheet. The spots on the dried sheet are visualized by ninhydrin spraying, as with paper, but care must be taken to avoid overheating the plastic backing.

Preparation of the DNP-peptide

This portion of the experiment can be done while the first chromatograms are being developed. A small portion of peptide unknown (2 mg) is transferred to a centrifuge tube and combined with 250 μL of 2% $NaHCO_3$ and 400 μL of 2% FDNB in 95% ethanol. The pH must be adjusted to 8.0 to 9.0 with small drops of acid or base and the use of a pH-indicator strip. Any undissolved solid must be removed at this point by centrifugation at low speed (tabletop centrifuge). After 1 hour reaction time, the remaining solution is diluted with a twofold excess of 2% $NaHCO_3$ and the resulting yellow solution is extracted with peroxide-free ether. (The presence of peroxides is indicated by the formation of a red color when a test portion of the ether is shaken with an aqueous solution of ferrous ammonium sulfate and potassium thiocyanate. Peroxide-free water saturated ether is prepared by adding 0.25 g of ferrous sulfate and 2 mL of water to 100 mL of ether and stirring thoroughly. After the residue settles, the ether can be removed and used.) Unreacted FDNB will be taken into the upper ether layer, so the extraction must be repeated until the upper ether layer remains colorless after the extraction. This ether extract can be discarded. The remaining aqueous solution should be acidified to pH 1 with a few small drops of 6 N HCl and extracted two or three times with equal volume portions of peroxide-free ether. The combined ether layers from this extraction are transferred to an ampule and evaporated to dryness in a gentle stream of dry, oil-free air.

The dried extract contains DNP-peptide and dinitrophenol. One cautiously adds 500 μL of 6 N HCl to the ampule, which is then sealed, wrapped in a wire screen, placed upright in a beaker, and heated for 12 hours at $100^{\circ}C$.

CAUTION: WEAR SAFETY GLASSES

It is wise to check the ampule 15 minutes after placing it in the oven; an improperly sealed ampule will result in evaporation of the solvent and the destruction of the DNP derivative.

After 12 hours, the cooled ampule is opened and the contents diluted with an equal volume of water. The aqueous solution is extracted with several 1-mL portions of ether and the combined ether extracts evaporated to an oily film on a warm-water bath.

Polyamide TLC of DNP-Amino Acids

The DNP-peptide can be redissolved in 500 μL of dry acetone and applied as a series of spots on one side of a polyamide TLC plate. The spots are aligned with a punched paper card template provided by the TA.[5] Polyamide sheets are very easily overloaded and each spot should be very small (1 mm) and just barely visible under ordinary room light. It may be necessary to dilute the DNP-amino acids before spotting. Repeated applications with a NANOCAP applicator should not be necessary.

Spots of known DNP-amino acids are applied to the reverse side of the plate so that their Rf values can be determined under identical running conditions. The spots can be developed by setting the sheets upright in a TLC tank containing enough solvent [water: formic acid (97%): acetic acid (9:6:5)] to cover the lower edge of the plate up to about 1.5 cm below the original position of the spots. One must not mark the TLC plates with pencils; enough residual spot will remain (due to polar contaminates) to identify the origin. Scratches and dents on the polyamide surface will distort the chromatogram. The TLC-tank must be tightly stoppered throughout the run. The spots for the DNP-amino acids can be seen by their yellow color and they can be distinguished from dinitrophenol because the latter appears black under UV illumination. The dinitrophenol is colorless after the chromatogram is exposed to the vapors of concentrated HCl.

POSSIBLE RESULTS AND A PROJECT EXTENSION

The net amino acid composition can be deduced from chromatograms of the HCl hydrolysis products. If paper chromatography is used, then the R_f values, observed for two solvent systems, are usually enough to narrow the amino acid composition to three or four possibilities. If cellulose chromatography is used with known amino acids as internal markers, the composition is usually narrowed to three. The DNP derivatization followed by polyamide TLC will identify the N-terminal amino acid of the tripeptide. Hence the tripeptide sequence will be narrowed to two possibilities. Amino acids can be separated by two-dimensional chromatography on silica gel or cellulose and identified unambiguously if two widely different solvent systems are used for the two dimensions.[5,6]

DISCUSSION QUESTIONS

1. Why are concentrated amino acid HCl solutions much harder to freeze at dry ice temperatures than is pure water?

2. What is the purpose of the KOH trap between the lyophilizer flask and the vacuum pump?

3. Why will a solution remain frozen as long as it is losing water vapor to the vacuum pump?

4. Explain why R_f values and the time required for a paper chromatogram to be developed depends on how dry the paper is.

5. What is the chemical reaction that produces the characteristic purple spot for amino acids with ninhydrin?

6. Why would you expect large differences in the R_f values of amino acids observed with phenol:water and with butanol:acetic acid? Why are amino acids more mobile in the first system?

7. Why are the side-chain DNP derivatives of amino acids not a serious source of false N-termini in this procedure?

8. Would the Sanger method work with an N-acetyl terminus?

9. What is the color of 2,4-dinitrophenol in base, in acid? How is this difference useful in ruling out false N-termini spots?

REFERENCES

1. F. Sanger, *Ann. Rev. Biochem.*, **1988**, *57*, 1.

2. E. Lederer and M. Lederer, *Chromatography*, (2nd ed.), New York: Elsevier 1957; pp 312, 328.

3. B. S. Hartley, *Biochem. J* , **1970**, *119,* 805-822

4. J. M. Walker and W. Gaastra, *Techniques in Molecular Biology*, New York: Macmillan, 1983; pp 89-99.

5. R. Criddle, *Biochemistry Laboratory Manual*, courtesy of the Department of Biochemistry and Biophysics, 1982, University of California at Davis.

6. A. R. Fahmy, A. Niederwiser, G. Pataki and M. Brenner, *Helv. Chim. Acta,* **1961**, *44,* 2022.

7. E. von Arx and R. Neher, *J. Chromatog.* , **1963**, *12,* 329.

Chapter 4

Fractionation, Centrifugation, and Colorimetric Determination of Proteins

In this experiment a solution containing a mixture of proteins extracted from mung bean sprouts is fractionated by a series of ammonium sulfate precipitations. The protein precipitated in each fraction is collected by centrifugation and redissolved in the appropriate buffer. The protein concentrations in both the supernatant and redissolved precipitate from each fraction are quantitatively determined by the Bradford protein assay.

KEY TERMS

Albumins and globulins
Centrifugation
Percent saturation
Protein solubility

Beer-Lambert law
Hofmeister series
Protein assays
Salting in and salting out

BACKGROUND

The solubility of most proteins is a sensitive function of the nature and concentration of added salts. The solubility can be expressed as a function of ionic strength, which is defined by the equation

$$\mu = \frac{1}{2} \sum C_i Z_i^2 \qquad (4\text{-}1)$$

where C_i is the molarity of the ith ionic component and Z_i is its charge.[1] The solubility of proteins is usually increased by the addition of some salt and the protein is said to be *salted in* at low ionic strengths. At higher ionic strengths the protein is less soluble and is *salted out*.

In the high-salt region, protein solubility is logarithmically related to μ by the Setschenow equation

$$\log S = \beta - K_s \cdot \mu \qquad (4\text{-}2)$$

where S is the solubility (grams/liter), β is the zero ionic strength intercept, and K_S is termed the *Setschenow* or salting out constant. At salt concentrations in the 5 mM range or higher, this equation is a reasonable approximation, but β is probably higher than the true solubility at zero ionic strength.

Salting out is really a dehydration process because added ions require solvation by water molecules, leaving less water available to solvate the protein. The protein-protein interaction, which leads to precipitation, becomes more important as the protein-water interaction becomes less likely. In some cases protein denaturation may accompany the precipitation; in other cases, the protein may be precipitated as crystals of the native protein.

The constant K_S depends both on the nature of the protein and upon the salt used. Traditionally, water-soluble proteins are classified into two categories, depending on how salts influence their solubility. The classification scheme is limited to the effect of neutral salts, which do not markedly alter the pH. *Albumins* are readily soluble in aqueous solutions of low ionic strength and are "salted out" only by very high ionic strengths of neutral salts. *Globulins* are insoluble in deionized water but are salted in by low ionic strengths ($0 < \mu < 0.2$ M). Globulins are salted out at around 50% saturation of $(NH_4)_2SO_4$. Many proteins have solubility properties between these two extremes and cannot be clearly classified in either class. Salts differ markedly in their power to salt in or out. Some salts of ions like SO_4^{2-} are effective in precipitating native proteins whereas other ions, such as ClO_4^- or guanidinium cation, tend to dissolve proteins. Guanidine hydrochloride is an effective denaturing agent only at higher concentrations.[2] In 1888, Hofmeister first showed that certain salts enhance the solubility of proteins in water, whereas others precipitate proteins; he was able to establish a ranking of salts now known as the *Hofmeister* or *lyotropic* series:

(*precipitate*) SO_4^{2-}, HPO_4^{2-}, F^-, Cl^-, Br^-, NO_3^-, I^-, CNS^-, ClO_4^- (*solubilize*)

Our understanding of the origin of the Hofmeister series has not expanded much in the 100 years since Hofmeister's original observation, but we do know that the effect involves more than electrostatics. The series probably reflects differences in the specific interactions between these ions and water; some ions are referred to as "water structure breakers" and others as "water structure formers" because ions exhibit their effect on proteins by altering its interaction with water.[3] Some cations, such as tetra-alkylammonium and quanidinium ions, solubilize proteins by a different mechanism that probably involves hydrophobic interactions with nonpolar amino acid side chains.

Ammonium sulfate is widely used for protein precipitation because it is so soluble and effective. At $25^{\circ}C$ saturated $(NH_4)_2SO_4$ is 4.1 M ($\mu = 12.3$ M) and contains 767 g of salt per liter of water. There are two ways to use ammonium sulfate to precipitate proteins:

1. In large batch preparations, the salt is added in dry form, but this addition must be gradual to avoid premature precipitation. Also care must be taken to control the pH; a weak base like ammonium hydroxide or TRIS (see Chapter 3) may have to be added along with the salt.

2. When working with small volumes and with higher protein concentrations, a more generally applicable method is the dropwise addition of a saturated ammonium sulfate solution with a previously adjusted pH.

Concentrations of $(NH_4)_2SO_4$ solutions are commonly expressed in terms of percentage saturation. The percentage saturation is calculated <u>without</u> allowing for the volume change upon mixing. One volume of water plus one volume of 100% saturated $(NH_4)_2SO_4$ is said to be 50% saturated, regardless of the fact that the volume of the final solution is less than the combined volumes of the starting solutions.

One can calculate the volume V_x of ammonium sulfate solution of saturation S_x that must be added to V_1 milliliters of solution of saturation S_1 to yield a solution with saturation S_2 by equation 4-3.

$$V_x = \frac{V_1(S_2 - S_1)}{S_x - S_2} \qquad (4\text{-}3)$$

The saturation levels S_1 and S_2 are given as decimal equivalents. The solubility of ammonium sulfate does not change appreciably over a narrow temperature range of 5 to 10°C. The properties of saturated solutions of ammonium sulfate in water at several temperatures are given in Table 4-1.

Table 4-1
SATURATED SOLUTIONS OF AMMONIUM SULFATE
PROPERTIES AT SEVERAL TEMPERATURES[a]

	Temperature °C				
	0	10	20	25	30
Weight percentage	41.42	42.22	43.09	43.47	43.85
Moles of $(NH_4)_2SO_4$ in 1000 mL of solution	3.90	3.97	4.06	4.10	4.13
Moles of $(NH_4)_2SO_4$ in 1000 grams of water	5.35	5.53	5.73	5.82	5.91
Grams of $(NH_4)_2SO_4$ in 1000 mL of water	514.72	525.00	536.34	541.24	545.88
Grams of $(NH_4)_2SO_4$ into 1000 mL of water	706.86	730.53	755.82	766.80	777.55
Density	1.2128	1.2436	1.2447	1.2450	1.2449
Apparent specific volume	0.5262	0.5357	0.5414	0.5435	0.5458

[a] Calculated from values in *International Critical Tables*.

The solubility of proteins is also a function of pH and the concentration of any organic cosolvents. The solubility of most proteins shows a pronounced minimum at pH levels near the isoelectric point (pI). This is a much used principle, either to precipitate desired proteins or to remove unwanted ones. Representative plots of solubility as a function of pH are given in Figure 4-1. The U-shaped curves are all for the same hypothetical protein but for different ionic strengths. Even though all the curves are centered over pI, the shape of the curves and the minimum solubility are strongly dependent on ionic strength. It can be seen from the figure that the pH must be carefully controlled if salts are used to fractionate a protein mixture and that any systematic study of salts on protein solubility must be done at constant pH.

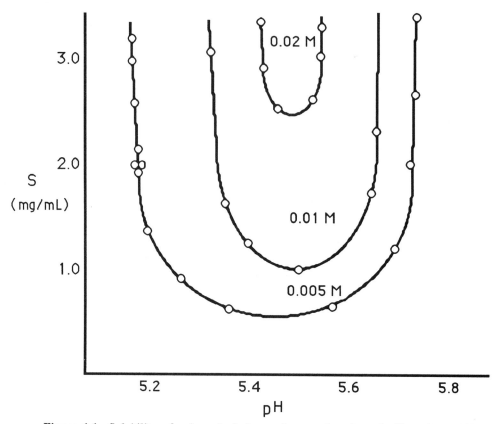

Figure 4-1 Solubility of a hypothetical protein as a function of pH at three different salt concentrations

Most organic solvents are protein precipitating agents, so that protein mixtures can be fractionated by increasing concentrations of the water-soluble solvents acetone or ethanol. Such solvents decrease the capacity of aqueous solvents to solvate charged amino acid side chains and also reduce the hydrophobic forces that stabilize protein tertiary structure. Urea is also a powerful protein denaturating agent because it decreases the hydrophobic interaction. By carefully controlled addition of salts and organic cosolvents, proteins can be precipitated in the native, active form and subsequently redissolved in cold aqueous buffer. This method is used to purify proteins and is demonstrated in this experiment. With rapid addition of organic solvents, especially at room temperature, proteins are often precipitated as denatured, inactive proteins.

Centrifugation   Theory

The centrifugal force at maximum speed depends on the rotor and the distance from the axis of rotation. The relative centrifugal force (RCF) developed within a centrifuge tube is related to the angular velocity ω (in radians/second) by the following formula, which expresses the force as multiples of **g** , the acceleration constant due to gravity:

$$F = \frac{\omega^2 r}{9800} \qquad (4\text{-}4)$$

The value of ω is directly related to the rpm value (0.00472 rpm), and the value of r is the distance (cm) from the axis of rotation (i.e., how far down the centrifuge tube one is interested); r can range from the top of the tube r_{min}, through the average value r_{av}, to the maximum value at the bottom of the tube, r_{max}. The RCF can be calculated as a function of the revolutions per minute (rpm) by the equation.

$$RCF = 1.118 \times 10^{-5} \times r \ (rpm)^2 \qquad (4\text{-}5)$$

When describing a centrifugation step, specifying rpm by itself is not a complete description. One should define the rotor type or, better still, the average gvalue. In this book g values are calculated from r_{av}. The experimental parameter that best correlates with sedimentation efficiencies is g values x time; two runs at different g values and different running times are equivalent if they have equal numbers of "g minutes." Although sedimentation is a complicated phenomenon,[2] g x minutes is the best handle on reproducible results, especially between runs with different rotors at different rpm values.

Sedimentation can be understood in terms of the centrifugal force due to rotor rotation working in opposition to the frictional and buoyancy effect of the sedimentation medium.

The frictional force acting on a rigid sphere of radius r in a medium of viscosity η is given by the Stokes equation.

$$F = 6\pi\eta r \frac{ds}{dt} \qquad (4\text{-}6)$$

where ds/dt is the velocity of the particle. There is also a buoyancy effect proportional to the difference in the density of the medium ρ_o and the particle ρ_p. During sedimentation in a rotor of radius r, the frictional force and buoyancy effect are balanced by the centrifugal force so that the velocity of the particle is given by the expression.

$$\frac{ds}{dt} = \frac{2r^2(\rho_o - \rho_p)\omega^2 r}{9\eta} \qquad (4\text{-}7)$$

The expression above holds only for spherical particles, but a correction can be applied for nonspherical particles, which can have frictional coefficient (f) up to 10 times larger than the frictional coefficient (f_o) of a sphere of equal volume.

$$\frac{ds}{dt} = \frac{2r^2(\rho_o - \rho_p)\omega^2 r}{9\eta} \left(\frac{f}{f_o} \right) \qquad (4\text{-}8)$$

The sedimentation property of a macromolecule is usually expressed in terms of its sedimentation coefficient (S) with units of s^{-1} or the Svedberg unit of $10^{-13} \ s^{-1}$.

$$S = \frac{\left(\dfrac{ds}{dt} \right)}{\omega^2 r} \qquad (4\text{-}9)$$

The sedimentation coefficient of a particle is independent of the speed and rotor geometry and can be used as an estimate of molecular weight (see chapter 31 and Ref 2, page 132).

Practical Centrifugation

There are several types of centrifuges in any well-equipped biochemistry laboratory. Small low-speed tabletop centrifuges are used for small volume samples, including freshly prepared protein solutions. Large low-speed floor-mounted centrifuges are used for large quantities of easily precipitated material, such as bacteria, blood, or yeast cell suspensions or stock solutions containing suspended solids, and similar materials. Both types of centrifuges operate at a maximum rotor speed of a few thousand rpm. **Even though these are "slow" centrifuges, a serious accident can result from their improper use.**

CONTACT YOUR INSTRUCTOR BEFORE USING ANY CENTRIFUGE.

High-speed microvolume tabletop centrifuges (e.g., Eppendorf Model 5412 Microfuge) are often used for volumes of a few hundred microliters at moderately high speeds (15,000 rpm). The refrigerated preparative high-speed centrifuge is used for larger volumes when high speeds (up to 25,000 rpm and 60,000 x g RCF) and temperature control are necessary. The performance specifications for a number of representative types are given in Table 4-2. Ultracentrifuges operate at much higher speeds (40,000 to 60,000 rpm) and afford very high relative centrifugal forces (up to 600,000 x g).

The choice of rotor (Figure 4-2) can be a serious limitation to the maximum run speed. **One must never attempt to exceed the manufacturer's recommended upper limit.** The speed control sets the actual speed within 200 rpm when the speed is less than 15,000 and within 300 rpm for faster speeds. The tachometer indicates actual speed within 300 rpm. After calibration the temperature control sets the rotor temperature $\pm 1°C$ of set temperature. In some models the rotor chamber is partially evacuated to minimize air-rotor friction at high speeds. Even at reduced pressures this friction leads to heating which is compensated for by the temperature compensation control. The rotor will not start to spin if its temperature is more than a few degrees from the set temperature.

Centrifuge tubes are specially constructed to withstand the high g forces generated in a high-speed centrifuge. Usually, tubes or bottles are designed for a specific rotor type and are not interchangeable. Tubes should fit snugly but not so tight that they are unable to move up or down or to rotate within the tube cavity in a rotor. Some of the lip should protrude above the cavity so that the tube can be removed with a gentle pull. Tools should usually not be necessary. Conical and pointed centrifuge tubes are designed to fit into shock-absorbent cushions in a low-speed centrifuge and should never be used in a high-speed rotor designed for rounded-bottom tubes or flat-bottom bottles. Conventional test tubes, bottles, or other containers- even if made of plastic- should not be used in place of the appropriate centrifuge tube or bottle. At the very least this can lead to the loss of a valuable sample. Broken tubes within a centrifuge can lead to rotor imbalance, putting dangerous stresses on a centrifuge. Plastic containers can collapse, expelling their contents into the centrifuge rotor chamber. Large centrifuge bottles can be protected with support rings to prevent deformations that can lead to bottle collapse. Centrifuge tubes for high-speed preparative centrifuges are almost always made of plastic. Translucent polypropylene is the most common material and is resistant to all aqueous solvents, but it swells and softens in purely hydrocarbon solvents. Polycarbonate, which is a transparent plastic, is unstable in acids, bases, and certain organic solvents, especially phenol solutions. This material will soften quickly in these solvents and collapse within the rotor even at relatively low speeds. Even momentary exposure to the incorrect solvent will dangerously weaken these tubes.

Cross-sectional diagrams of a fixed angle and a swinging bucket rotor are given in Figures 4-2 and 4-3 respectively.

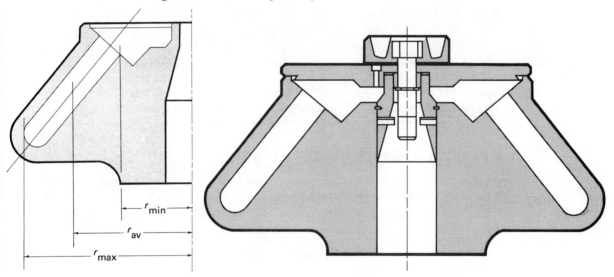

Figure 4-2 Cross-sectional view of a fixed angle centrifuge rotor showing r_{min}, r_{av} and r_{max} (Courtesy Beckman Instruments)

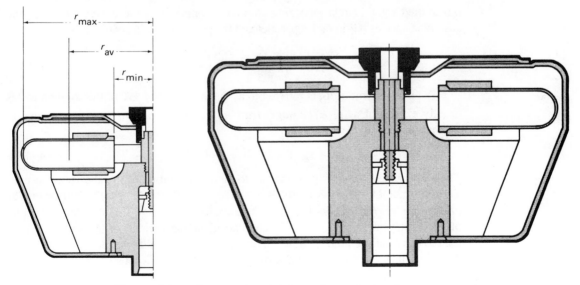

Figure 4-3   Cross-sectional view of a swinging bucket rotor (Courtesy Beckman Instruments)

Table 4-2

PERFORMANCE SPECIFICATIONS OF REPRESENTATIVE CENTRIFUGES AND ROTORS

Centrifuge	Common Rotors	
Beckman J-21C	JA-21 Rotor *Fixed angle*	JA-14 Rotor *Fixed angle*
Maximum run speed	21,000 rpm	14,000 rpm
Approximate acceleration. time	1.5 min	4 min
Number of tubes	18 x 10 mL	6 x 250 mL
Maximum capacity	180 mL	1,500 mL
Beckman J2-21	JA-20.1 Rotor *Fixed angle*	JS-13 Rotor *Swinging bucket*
Maximum run speed	20,000 rpm	13,000 rpm
Approximate. acceleration. time	2.5 min	3.8 min
Number of tubes	32 x 15 mL	4 x 50 mL
Maximum capacity	480 mL	200 mL

Absorption Spectroscopy Theory

All spectroscopic transitions are characterized by at least two parameters: (1) the *wavelength of maximum absorbance* λ_{max} and (2) the amount of light absorbed for a given concentration and pathlength. The first parameter is related to the energy difference between the absorbing and excited states:

$$\lambda = \frac{c}{\nu_{max}} \qquad \text{where} \qquad \Delta E = \hbar\nu_{max} \qquad (4\text{-}10)$$

The second parameter is defined by the *Beer-Lambert Law* which relates the intensity of the incident light I_0 (light entering the sample) to the intensity of the transmitted light I (light emerging from the sample) and the concentration of the absorbing species [C] in moles per liter and the path length l in centimeters.

$$\log \frac{I_0}{I} = \varepsilon\, l\, [C] \qquad (4\text{-}11)$$

The intensity ratio I/I_0 is referred to as the transparency %T of the solution and is usually expressed as a *percentage transmittance*. The *absorbance* is related to the transparency by the expression.

$$A = -\log\,(\%T) = \varepsilon\; l\; [C] \qquad (4\text{-}12)$$

The absorbance is directly proportional to the concentration of the absorbing species. The constant of proportionality ε is called the *extinction coefficient* or *molar absorption coefficient*, with units of liters x $mole^{-1}$x cm^{-1}; it is the slope of a plot of A vs [C] for a path length of 1 centimeter. Measurements of absorbance can be used to determine [C] if e is known. For applications in which the molecular weight of the absorbing species is unknown, a weight absorption coefficient $E^{1\%}$ can be used.

Practical Absorption   Spectroscopy

One generally uses a standard curve, obtained by plotting the observed absorbance values for samples of known concentration of colored species, as a function of these concentrations. Observed absorbances for samples of unknown concentration can then be used to determine such concentrations by simple interpolation (see Figure 4-4). One should establish the standard curve with enough points to ensure accurate interpolation, and the curve should cover the same absorbance range as the unknown since nonlinear behavior is often observed at higher concentrations. Note that the line will go through the origin if the reference blank solution is comparable to the solvent used for the absorbing sample. Every care should be taken to ensure that the points on the standard curve are obtained with exactly the same protocol as the absorbance measurements of unknown samples.

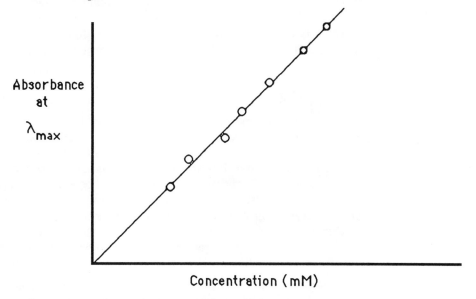

Figure 4-4 Standard curve obeying the Beer-Lambert Law.

Protein   Assays

A rapid and sensitive assay for protein is an essential part of any protein purification procedure. The following five assay methods can be used to determine the protein concentration of a sample or monitor the changes in protein concentration during a purification procedure, by measuring the absorption of light A at a specific wavelength in a 1-cm cell. The absorption is correlated to protein concentration by a standard curve made with known concentrations of a protein. The slope of the standard curve can be taken as the extinction coefficient used to calculated the protein concentration from the Beer-Lambert law.

1. Turbidimetric methods

2. Direct spectrophotometric methods

3. Biuret method

4. Lowry-Folin-Ciocalteu method

5. Bradford dye-binding assay

6. BCA protein assay

These methods employ different principles and seldom give identical results. None is entirely specific for protein, and all are subject to interference by certain salts and organic solvents. The method of choice depends on the nature of the problem since each method has unique and useful features and specific limitations.

Although we will use only the Bradford dye-binding assay, the procedure and qualifications of all five methods will be discussed.

1. The turbidimetric method[3,4] involves the controlled precipitation of protein with trichloroacetic acid. A standard curve using serum albumin is prepared by mixing aliquots of a stock protein solution with water to give solutions containing 0.1 to 0.9 mg/mL protein in a volume of 2.6 mL. Ammonium sulfate, 0.3 mL of 0.6 M, is added and the absorbance A_1 at 340 nm is measured (vs. water). Trichloroacetic acid, 100 μL, is added and after exactly 2 minutes the absorbance A_2 at 340 nm is redetermined. A plot of A_2-A_1 against protein concentration is used as a standard curve. This method works only for proteins with solubilities in highly acidic solutions akin to that of serum albumin; proteins that remain in solution or proteins that rapidly precipitate cannot be assayed turbidimetrically using albumin standards.

2. The direct spectrophotometric method[5-7] depends on the ultraviolet absorption of aromatic amino acid residues in proteins. The absorbances at 260 and 280 nm of a series of solutions of serum albumin are plotted as a function of protein concentration to give a standard curve. Proteins deficient in aromatic amino acids (e.g., histones) can be quantitated by absorbance at 230 nm. A method employing only one absorption wavelength cannot be used for assaying protein levels in the presence of nucleic acids but a direct spectrophotometric method that uses absorbances at 230 nm and 260 nm to determine both protein and nucleic acids has been developed by Kalb and Bernlohr.[5] Direct photometric methods can be used in the presence of contaminants that strongly absorb at the observation wavelength.

3. The biuret[8,9] method depends on the purple square planar complex formed between Cu^{2+} and the four amide nitrogen atoms of two parallel running peptide chains in proteins. The copper ion lies between the peptide chains. The chromophoric properties of this complex depends on the delocalized electrons in the coplanar amide linkages which interact with the copper as bidentate ligands. Crystal structures of related copper complexes have been worked out.

The colorimetric reagent is prepared by dissolving 1.5 g of $CuSO_4 \cdot 5H_2O$ and 6 g of sodium potassium tartrate in 500 mL water. This is thoroughly mixed with 300 ml 2.5 M NaOH (carbonate free). When all suspended solid has dissolved, 1 g of KI is added. The reagent is diluted to 1000 mL and stored in a plastic bottle. A standard curve is prepared from bovine serum albumin aliquots ranging from 1 to 8 mg/mL. One milliliter of protein solution is combined with 4 mL of reagent and allowed to stand at room temperature for 30 minutes; then the absorbance at 540 to 560 nm is measured. A violet or pink coloration is a positive test for proteins. Ammonium salts interfere with this test. The biuret test is one of the most reliable total protein assays in terms of interfering substances; however, it is the least sensitive.

Biuret Complex

4. The Lowry-Folin-Ciocalteu method[10,11] is based on two color reactions: the biuret reaction (see above) between Cu^{2+} and two peptide chains and the reaction of a complex salt of phosphomolybdotungstate with the Cu^{2+} biuret complexes of tyrosine and tryptophane. Hence the color intensity is sensitive to the presence of tyrosine and tryptophane in proteins. A reagent is prepared from the following stock solutions:

A. 2% (w/w) Na_2CO_3 in 0.1 M NaOH

B. 0.5% $CuSO_4 \cdot 5H_2O$ in 1% sodium tartarate.

C. Mix 50 mL of A with 1 mL of B. Must be made fresh each day.

D. Commercial 2.0 N Folin-Ciocalteu reagent diluted 1:1 with water (to 1.0 N).

Folin-Ciocalteu reagent contains sodium tungstate and sodium molybidate acidified with HCl and phosphoric acid and oxidized with bromine. Commercial reagent is preferred in a teaching laboratory because the process of preparation is very time consuming. The reagent should be stored in the refrigerator and the diluted solution stored in a brown bottle. The reagent solution should be a bright yellow color without a greenish tint. The protein solution (0.5 mL) is added to 5.0 mL of C and the solution allowed to stand for ten minutes. Then 0.5 mL of D is rapidly added with vigorous shaking. The solution is allowed to stand for *exactly* 30 minutes, and the absorbance is measured at 600 nm. A standard curve is prepared by diluting a stock protein solution containing 1 mg/mL serum albumin in water to yield a number of 0.5-mL aliquots containing 20 to 400 µg/mL protein. The Lowry method is about 100 times more sensitive than the biuret method but the readings are sensitive to the reaction times. The method is sensitive to interference by carbohydrates, TRIS, ammonium cations, and reducing agents. The instability of the reagent and the rigid timing required for reproducible results are the most serious limitations to this method in a teaching laboratory. To give the reader a feel about the importance of this method, it is important to note that the Lowry assay paper is the most widely cited scientific paper in all science.

5. The Bradford[12,13] assay is based on the color change (red to blue) of the dye Coomassie Brilliant Blue G (see Figure 4-4) upon binding to added protein. This method is relatively free from interference from most buffers, but the dye does bind to detergents. Rio-Rad offers a dye reagent concentrate that can be diluted with 4 parts of water and filtered to give a stock reagent that is stable for about 2 weeks. Alternatively, 100 mg of dye can be dissolved in 50 mL of 95% ethanol and combined with 100 mL of 85% (w/v) phosphoric acid and diluted to 1000 mL with water. The Bradford assay paper is the second most widely cited paper in science. The Bradford protein assay is used in a number of experiments described in this book because of its ease and relative sensitivity, however it must be kept in mind that it is most reliable when used when standardized against a protein closely like the unknown protein being determined.

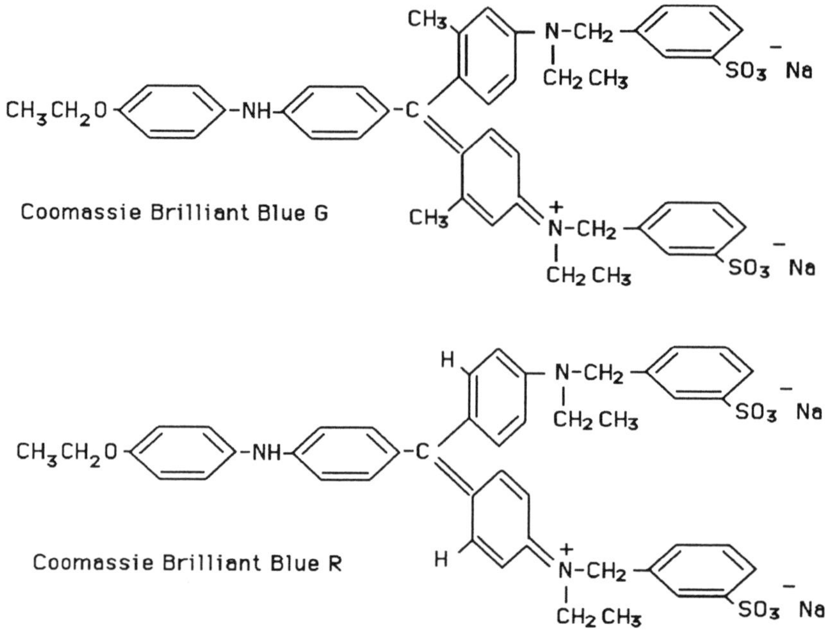

Figure 4-4 Coomassie Brilliant Blue dyes

The standard procedure is applicable to 20 to 140 µg protein in 0.1-mL solution aliquots (0.2 to 1.4 mg protein per milliliter of solution). Five milliliter of the dye reagent is combined with 0.1 mL of protein solution and shaken vigorously. After 5 minutes the sample is read against a reagent blank at 595 nm. The dye-protein complex is stable for at least an hour. The dye differs in structure from Coomassie Brilliant Blue R, which is the protein stain for gel electrophoresis.

One must be sure all cuvettes are clean when using this reagent since the reagent stains dirty glass or plastic. Detergents such as sodium dodecyl sulfate and Triton X-100 interfere but the assay can be run in the presence of amino acids, peptides, EDTA, and reducing agents such as mercaptoethanol and sugars. Because this method involves dye binding to proteins by hydrophobic and charge neutralization interactions, the nature of the complex reflects the amino acid composition of the protein. It was originally asserted that most proteins produce nearly identical assay response curves; however, it is now understood that this method is often unreliable for some proteins unless based on a standard curve prepared with that specific protein.[14] For many proteins a standard curve based on gamma globulin is suggested.

6. The BCA Protein assay[15] consists of a bicinchinic acid (BCA) solution and copper sulfate solution. When mixed together, an apple green working reagent is produced. This reagent is stable for at least a week on the shelf - a marked improvement over the Lowry method. Protein reduces the Cu(II) ions in the reagent to Cu(I) ions which form an intense purple color at 562 nm with the BCA ligand as shown in Figure 4-5. High levels of salts (1 M) and 40% non-reducing carbohydrates, 3 M urea, and 1% detergents will not interfere with this reagent. A reagent stock solution for determining protein concentrations in the range 100 to 1200 µg/mL is available from Pierce Chemical Company. Because the BCA reaction depends on the Cu(II) $\rightarrow$ Cu(I) reduction, this method is susceptible to reducing compounds (e.g., sugars).

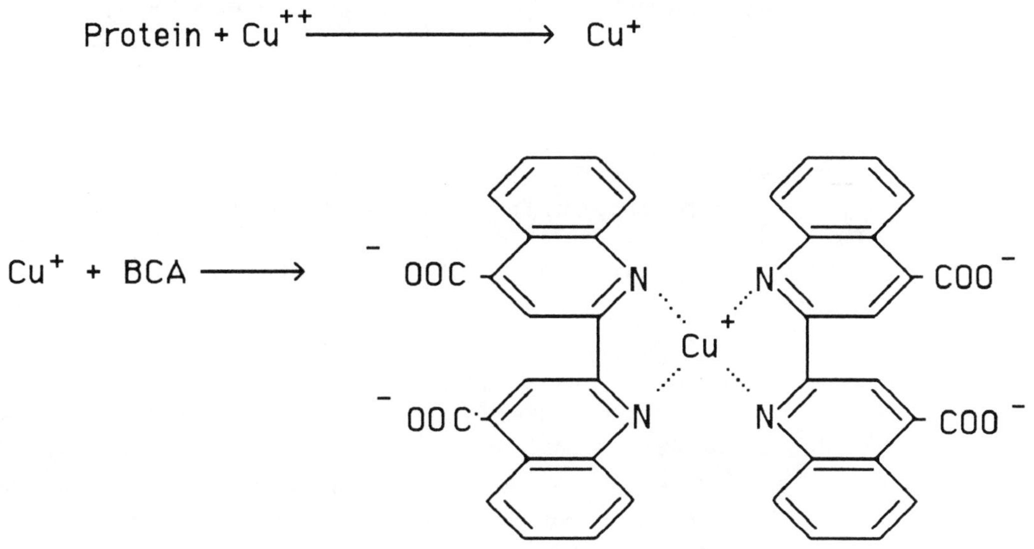

$$\text{Protein} + \text{Cu}^{++} \longrightarrow \text{Cu}^{+}$$

$$\text{Cu}^{+} + \text{BCA} \longrightarrow$$

Figure 4-5 Bicinchinic acid Cu(I) complex

EXPERIMENTAL SECTION

Students may work in pairs during this experiment but each must turn in a separate laboratory report. After consultation with the instructor, each pair will prepare a 100-mL volume of ice-cold 5 mM NaCl. The pH must be adjusted to near neutrality (7.0±0.5). Fresh mung bean sprouts (10 to 20 g) are suspended in the cold salt solution in a blender and pulverized at a high-speed setting for about a minute. The resulting "soup" is clarified by first pouring it through a mat of glass wool or through several layers of cheese cloth, then by 5 to 6 minutes of low-speed centrifugation (2000 x g) with a tabletop centrifuge. The resulting straw colored liquid should be clear or only slightly cloudy. The protein concentration is then determined by the Bradford protein assay. Protein concentrations should be expressed as mg protein/ mL and ammonium sulfate concentrations as %-saturation $(NH_4)_2SO_4$. Mung bean sprout "soup" very quickly spoils and there is very little advantage in preparing the crude protein extract in advance. It is best to blend sprouts on the day of the protein assays.

Preparation of a Standard Curve

The following steps are required to obtain a standard curve for converting measured absorbances into concentration values.

1. A 10-mL protein standard solution (gamma globulin) containing 1.0 mg/mL protein is prepared for the entire class either by diluting a Bio-Rad protein standard sample or by weighing out and diluting the solid protein. One can be assured that there is no undissolved solid by centrifugation at 5000 x g for 10 minutes.

2. Each pair of students prepares five dilutions of the standard solution, ranging from 0.2 to 1.0 mg/mL; each solution is made up to 100 μL total volume in a small (12 x 100 mm) test tube.

3. To each test tube is added 2 mL of Bradford reagent and the mixture is shaken by gentle "finger flicking." Your instructor will demonstrate the appropriate technique.

4. The absorbance at 595 nm is measured and a plot of protein concentration versus absorbance is prepared. If the blank was made by combining 100 µL of 5 M NaCl with 2 mL of Bradford reagent, the standard plot should extrapolate through the origin. Over this range of concentrations the plot should be linear. At higher protein concentrations, the curve will bend down (i.e., the slope will decrease).

Assay of Total Protein in Mixture

1. Portions of the ice-cold sprout extract are diluted by *known* volumes of 5 mM NaCl solution to give five 100-µL samples of diluted extract, which are then combined with 2-mL aliquots of reagent. Dilutions of 10, 25, 50, and 75 µL of extract diluted up to 100 µL with salt solution, as well as one undiluted extract solution are recommended.

2. After 5 minutes, the absorbance at 595 nm of each sample is measured relative to a blank containing 2 mL of reagent and 100 µL 5 mM NaCl. The concentration of the total mung bean protein in each sample is estimated from the standardization plot of absorbance vs. protein concentration, prepared for gamma globulin.

3. In the likely event that the absorbances for one or more mung bean samples are too high (>1.0), other more dilute samples of the original extract can be prepared so that five samples give reliable estimates of the protein concentration in the diluted sample. The concentration of protein in the undiluted extract is then calculated from the dilution factor. (This calculation is the most frequent source of student error; one must think through the calculation very carefully.)

4. The average protein concentration calculated for the undiluted extract based on the five determinations must be reported with the standard deviation $\pm \sigma$.

Fractionation of the Protein Mixture

There are two ways a sample of protein can be fractionated with added salt. Several successive additions of salt to the same sample, with a separate centrifugation step to remove the precipitate after each addition, will give fractions containing different proteins in each fraction. Alternatively, dividing the whole sample into a number of aliquots and adding increasing amounts of salt to each successive aliquot will give fractions with increasing amount of protein precipitated as the salt is increased. Each precipitate will consist of all those proteins salted out at or below the selected salt concentration. so that high salt precipitates will contain proteins also precipitated at lower salt concentrations. This second method is faster because all the fractions can be centrifuged simultaneously. and it has the advantage that the combined protein concentrations in the supernatant and redissolved precipitate, for each fraction, must equal the original protein concentration in the unfractionated mixture thus providing a check on the recovery of each precipitate. The second method is the method demonstrated in this experiment.

The cold protein extract is divided into five 10-mL portions with the remainder reserved for later use (in case of spills, etc.). To each portion is added either a volume of saturated $(NH_4)_2SO_4$ solution or solid salt so that the final ionic strength of each sample is different and the five samples span the range between 20 and 100% saturation.

Solid $(NH_4)_2SO_4$ will have to be added to the most concentrated sample to saturate it completely. *Note*: adding $(NH_4)_2SO_4$ solution necessarily dilutes the protein concentration, so solid salt is preferable when preparing the higher saturation levels. The salt must be added slowly to the cold solution with very gentle shaking to avoid irreversible denaturation. Any precipitate that forms is collected by centrifugation at 9000 x g for 20 minutes (or as long as it takes to clarify the solution). The precipitate should be redissolved in 10 mL of cold 5 mM NaCl and the protein concentration determined in duplicate. Vortex mixing may be required. The protein concentration in the clarified supernatants from the original extract samples must also be determined. The protein concentrations of precipitate and supernatant should sum to the original concentration of the extract. The results should be reported as a plot of protein concentration (mg/mL) vs. percent saturation $(NH_4)_2SO_4$.

Operation of a Refrigerated, High-speed Centrifuge (a Checklist)

1. Fill the appropriate centrifuge tubes in pairs on a double-pan balance. Make all adjustments in volume so that pairs of tubes weigh within 0.1 g of each other. Do not depend on equal levels of liquid. Transfer the tubes, without spilling, to the rotor. Be sure that equal weight tubes are exactly opposite each other in the rotor and that the outsides of the tubes are dry. Clean up drips or spills at once.

2. Open the chamber door of the centrifuge carefully and slowly. Observe the quality of the seal and check for foreign material along the seal or in the rotor chamber.

3. Apply a light coat of silicon grease to the top taper and bottom skirt of the drive hub before installing the rotor. Set the rotor down gently on the drive shaft and tighten the knob on the top of the rotor. Rotors should never be left in the centrifuge since they will become difficult to remove after a day or so.

4. After the rotor is installed, the door is closed carefully to avoid damaging the o-ring seal around the chamber in some models and preventing a good vacuum seal.

5. Unless the centrifuge has been precooled, the power switch must be turned on now.

6. Set the desired sample temperature with the temperature "control." For some models temperature compensation for your rotor type is set with the "compensate" control. The compensation setting can be taken from the table associated with that centrifuge. The centrifuge will not operate if the set temperature is too far below the actual rotor temperature- in such cases set the temperature to 25°C and slowly lower it over 10 minutes of centrifugation time. Cooling the rotor ahead of time is necessary if this cooling down sequence is inappropriate.

7. For runs of 3 hours or less, adjust the time selection controller on the control panel to the appropriate number of minutes. For long runs, the controller should be set to hold. For older Beckman models, if the run is 30 minutes or less, turn the timer clockwise past 30, then back to the desired time.

8. Turn the speed selector to the desired speed and push the start button. The rotor should begin to accelerate. Always remain with the centrifuge until it reaches the set speed. Any load banging or rattle indicates a loose or unbalanced rotor. **Turn the time and speed to zero and set the brake immediately.**

CAUTION: DO NOT ATTEMPT TO STOP A SPINNING
ROTOR WITH YOUR HANDS!

9. The brake does not stop the rotor immediately. The brake toggle may be set or changed any time during the run. If the brake is turned off, the rotor will coast to a stop. With the brake on, the rotor will be smoothly slowed and will come to rest in approximately half the coasting time. Wait for the indicator light to go out, indicating full rotor stop, before opening the door.

10. The tachometer on the control panel indicates the actual rpm in thousands.

11. Adequately clean the rotor and all the tubes after each run with a liquid soap solution and soft sponge. Never use abrasive detergents or scouring powders or brushes when cleaning the rotor. Aluminum rotors are especially sensitive to alkali. Be sure to remove all frost from the centrifuge chamber with a clean dry towel.

DISCUSSION QUESTIONS

1. Why was 5 mM NaCl used as the extraction solvent rather than distilled water?

2. Why is it better to use solid ammonium sulfate rather than 100% saturated when preparing the high salt (>60%) fractions?

3. Why is it essential that one record the volume of solvent into which precipitated protein is redissolved?

PROJECT EXTENSIONS

1. The fractions can be assayed for specific enzyme activities using the assay procedures outlined in this book; e.g., acid phosphatase, esterase, galacturonidase, and ribonuclease activity.

2. SDS gel electrophoresis of the protein fractions will permit an estimate of the number of proteins in each fraction. Students can look for prominent bands that show up in more than one fraction. The procedure for SDS gel electrophoresis is given in Chapter 9. The method of successive fractionation, by adding increasing amounts of salt to the same sample, can also be used and the SDS gel protein pattern of these fractions compared to the pattern seen with the above method.

REFERENCES

1. W. P. Jencks, *Catalysis in Chemistry and Enzymology*, Chap.7, New York: McGraw-Hill, N. Y. 1969.

2. D. Rickwood, ed., *Centrifugation*, Washington D. C: IRL Press, 1984.

3. P. L. Kirk, *Adv. Protein Chem.*, **1947**, *3*, 139.

4. T. Bucher, *Biochem. Biophys. Acta*, **1947**, *1*, 292.

5. V. F. Kalb and R. W. Bernlohr, *Anal. Biochem.*, **1977**, *82*, 362.

6. W. J. Waddell, *J. Lab. Clin. Med.*, **1956**, *48*, 311.

7. J. B. Murphy and M. W. Kies, *Biochem. Biophys. Acta*, **1960**, *45*, 382.

8. G. Toennies and F. Feng, *Anal. Biochem.*, **1965**, *11*, 411.

9. E. F. Hartree, *Anal. Biochem.*, **1972**, *48*, 442.

10. O. H. Lowry, N. J. Rosebrough, A. Farr, and R. J. Randall, *J. Biol. Chem.*, **1951**, *193*, 256.

11. S.-C. Chou and A. Goldstein, *Biochem. J.*, **1960**, *75*, 109.

12. M. Bradford, *Anal. Biochem.*, **1976**,*72*, 248 ; **1977**, *79*, 544.

13. S. J. Compton and C. G. Jones, *Anal. Biochem.*, **1985**, *151*, 369.

14. H. B. Pollard et al, *Anal. Biochem.*, **1978**, *105*, 202.

15. P. K. Smith, *Anal. Biochem.*, **1985**, *150* , 76.

Chapter 5

Ion-Exchange Chromatography of Amino Acids

In this experiment a protein is hydrolyzed in boiling hydrochloric acid. The resulting amino acid mixture can be fractionated on a strong acid cation exchange resin at 50°C. The acidic amino acids are separated by elution with 0.2 N sodium citrate (pH 3.49) and the remaining neutral and basic amino acids flushed from the column as a group with 0.2 N NaOH. Alternatively, if laboratory time permits, all the amino acids can be separated by elution with a series of buffers of successively higher pH and ionic strength, followed by 0.2 N NaOH, which serves to regenerate the column. Fractions of the effluent are collected on an automatic fraction collector and their amino acid content quantitated with ninhydrin by a modification of the method of Stein and Moore.

KEY TERMS

Amino acid elution order   *Anion exchangers*
Cation exchangers   *Column regeneration*
Ion-exchange chromatography   *Ninhydrin derivativization*
Protein hydrolysis

BACKGROUND

Proteins can be hydrolyzed into their constituent amino acids in either concentrated acid or alkali. Under either condition some amino acids are destroyed. Under acidic conditions, serine, threonine, cysteine and especially tryptophan are partially destroyed. The destruction of other amino acids is minimized by carrying out the reaction in evacuated sealed tubes. The amino acid hydrochlorides can be separated on an ion-exchange column. As the eluting solvent is progressively changed from acidic to basic pH values, the amino acids come off the column one at a time. Stein and Moore developed the original strategy of using cation-exchange columns for the separation of amino acid hydrochlorides and the use of ninhydrin to quantitatively determine the eluting amino acids.[1-5] Using their method the amino acid composition of a protein could be determined after only several days of hard work. Several large columns were used and considerable skill was required in changing buffers, adjusting column temperatures, purifying solvents (ammonia was the main problem) and applying samples so that unambiguous separations could be achieved. The only way that specific amino acids could be identified was by elution order, so that overlapping or missing peaks would put the identity of all later eluting amino acids in limbo. Despite these experimental hardships, the method led to major advances in protein chemistry in the 1950s and, within a few years, "automatic amino acid analyzers," adapted from the method of Stein and Moore, became commonplace. Stein and Moore received a Nobel Prize for their contributions to protein chemistry. Today amino acid analysis employing ion-exchange High-performance liquid chromatography (HPLC) and postchromatographic derivatization with o-phthaldehyde and fluorescent detection permits the quantitative amino acid analysis of picomolar quantities of protein hydrolysate within hours.[6] Precolumn derivatization with phenylisothiocyanate (PITC) permits routine amino acid analysis of even smaller samples.[7] The experiment described here will give a taste of the "good old days" of column ion-exchange chromatography. HPLC is introduced as an optional section in Chapter 6.

Ion-Exchange Chromatography

An ion-exchange resin consists of an insoluble matrix (usually, polystyrene) to which charged groups are covalently bound. Associated with these charged groups are mobile counterions which may be reversibly exchanged with other ions of the same charge. Positively charged ion-exchange resins have associated negative ions, hence they are called **anion exchangers**. Similarly cations are reversibly bound to **cation exchangers**, which are anionic resins. There are a large number of ion-exchange materials available to the biochemist and the choice depends on the ionic species to be purified.[8-10] Amino acids at pH values below their pI values are cations and bind reversibly to cation-exchange resins.

There are two general categories of cation-exchangers:

1. Weakly acidic exchangers containing carboxylic acid exchange groups ($-CO_2^-$). Two common varieties are marketed by Bio-Rad (Bio-Rex 70) and Rohm & Haas Co. (IRC-50). Weakly acidic resins have pK_a values below the pI values of many amino acids, so that the resin is not significantly charged at pH levels above the pK_a of the resin. Consequently there will only be weak binding between resin and amino acid in the pH range.

$$pK_{a,resin} < pH < pI_{amino\ acid}$$

2. Strongly acidic exchangers composed of sulfonic acid functional groups ($-SO_3^-$). Bio-Rad offers a sulfonic acid resin attached to a styrene divinylbenzene resin (AG 50W); Dow and Rohm & Haas market the related resins Dowex 50W and IR-120, respectively.

Bio-Rad AG 50W-X8 with 8% cross-linking is an excellent material for the separation of amino acids by conventional column chromatography. There are also Dionex resins, developed to permit separations of amino acids within a matter of hours at the high flow rates and operating pressures of an HPLC column. Like AG 50W resins, the Dionex resins are sulfonated divinylbenzene-styrene copolymers but they are very carefully sized so that the size distribution of the resin beads is very homogeneous. This permits high flow rates under pressures that would tightly pack more heterogeneous resins. The high operating pressures (300 to 500 psi) required for their optimal use and the high cost of such resins (up to $250 per gram) preclude the use of such material in an introductory experiment in ion-exchange chromatography. The procedure described in the Experimental Section will use AG-50W.

The movement of an amino acid down an ion-exchange column depends on the distribution of the amino acid between the mobile aqueous buffer and the stationary ion-exchange resin. Amino acids that bind tightly to the resin come through the column more slowly than amino acids more weakly bound. The affinity for the resin is a function of the electrostatic interaction between the amino acid ammonium group and the SO_3^- groups of the resin, as well as the hydrophobic interaction between the amino acid hyrophobic (aromatic) side chain and the nonpolar portion of the resin. Buffer pH and electrolyte concentration alter the ionization state and net charge of the amino acid and therefore the use of stepwise pH and ionic strength gradients permit the resolution of amino acids with quite closely related pK_a values.

The original method of Stein and Moore employed a Dowex-type cation-exchange resin with a structure not unlike AG-50W, but the higher homogeneity of a modern resin permits reasonable separations on a single short (1.5 x 50 cm) column in a much shorter time if a carefully selected set of buffers with gradiated pH and electrolyte concentration are used. In our procedure the amino acids are loaded onto the column in pH 2.2 buffer so that all amino acids will have a significant net positive charge. (An oxidation product of cysteine, cysteic acid will have a net negative charge and will be eluted rapidly from the column.) The resin should be in the sodium form at pH 2.2 and the column should be thermally equilibrated to $53.5 \pm 0.5°C$ by use of a circulating warm-water bath and water jacket around the column. The water bath must be brought up to temperature at least 1 hour before the sample is applied. The acidic and neutral amino acids can be removed from the column with pH 3.49 (0.2 N) and pH 4.12 (0.4) sodium citrate buffer. The first buffer change should occur between the alanine and glycine peaks. The temperature should be around 53.5°C but can be adjusted down 3 or 4°C to improve the threonine-serine resolution. A second buffer change (pH 4.12 to 6.40), permitting separation of the basic amino acids, should be made well before the tyrosine-phenylalanine pair of peaks. The basic amino acids can also be quickly eluted from the column as a group with 0.2 N NaOH (which also serves to regenerate the column). The elution conditions and order of the amino acids are given in Table 5-1.

The resin bed shrinks with the third, high-normality (1.0 N) buffer. This is a general phenomenon observed in ion-exchange chromatography; higher counterion concentrations shield more effectively the charged groups on the resin, permitting their closer approach and a smaller resin volume. Conversely, resins swell at low electrolyte concentrations.

The amino acid concentration in the effluent fractions can be determined colorimetrically by the addition of a strongly buffered ninhydrin solution containing some reduced ninhydrin (hydrindantin) which prevents interference from oxygen (Figure 5-1). The yield of blue color in the reaction of ninhydrin with amino acids is nearly 100% of theory if the reaction is allowed to run 20 minutes or so at elevated temperatures. Because buffer changes must be made during interludes between peaks, the ninhydrin reaction must be run on all fractions as soon as they are collected. It is not necessary to wait the full 20 minutes to estimate roughly whether a fraction contains any amino acid. Prolonged heating in a steam bath and spectrophotometric absorbance measurements at 570 nm are necessary to get quantitative results that can be plotted as an elution profile.

Table 5-1
ELUTION ORDER OF SOME AMINO ACIDS

Amino Acid	Buffer (conc., pH)	Approx. Elution Volume (mL)
Acidic and neutral amino acids		
Aspartic acid		20
Methionine sulphone	0.2 N Na citrate,	
Threonine	pH 3.49	
Serine		
Glutamic acid		60
Cysteine		
Proline		
Glycine		80
Alanine		
Valine	0.4 Na citrate,	
Methionine	pH4.12	
Isoleucine		
Leucine		120
Tyrosine		
Phenylalanine		150
Basic amino acids		
Lysine	1.0 Na citrate,	155
Histidine	pH 6.40	
Arginine		210

The ninhydrin reaction involves the formation of a colored species, Ruhemann's purple, by the decarboxylative deamination of all the amino acids but proline. The intermediate iminium ion is hydrolyzed to form an amine that condenses with a second molecule of ninhydrin. The colored product has a λ_{max} value of 570 nm and a molar extinction coefficient of 2×10^4.

Hydrindantin

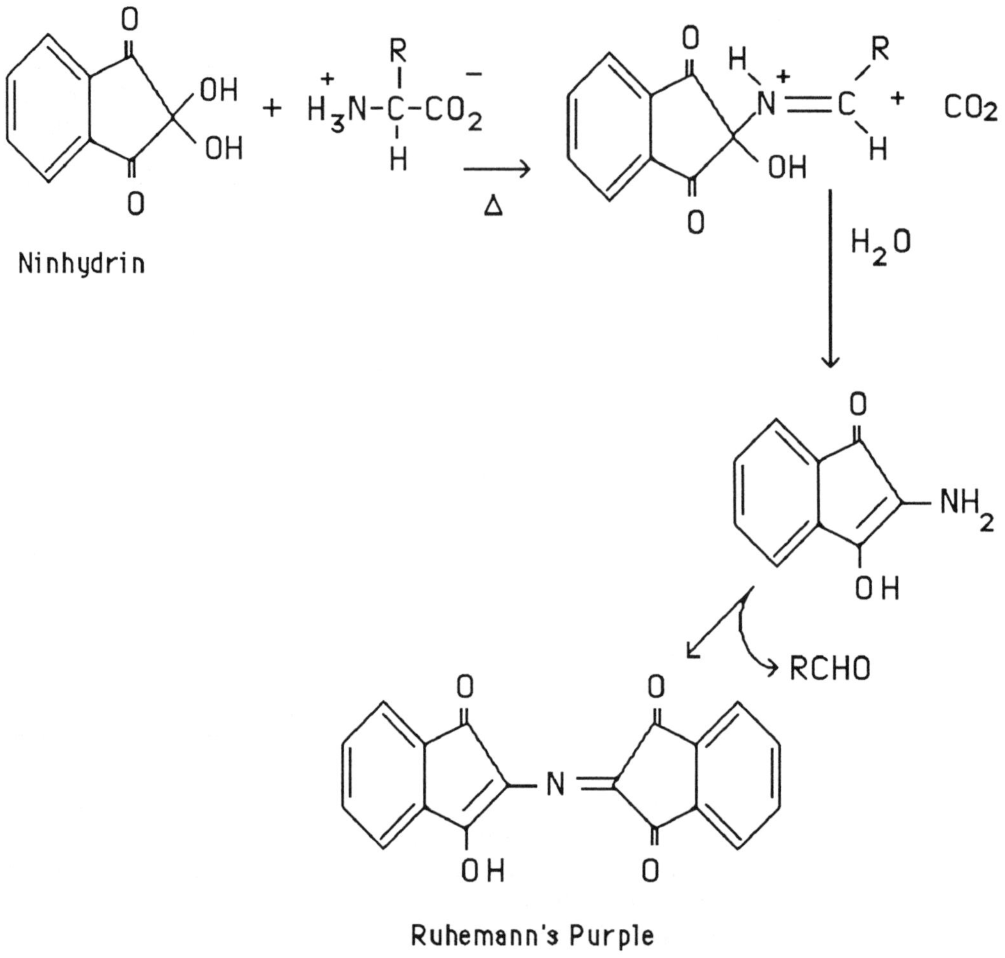

Figure 5-1 Ninhydrin color reaction scheme

EXPERIMENTAL SECTION

Students should work in groups of three to six, dividing the chores as they see fit. This experiment requires that several steps be done simultaneously to minimize the total time required. For example, the hydrolysis of protein and the preparation of buffers can be done on the same day.

Protein Hydrolysis

A protein sample (10 mg) can be hydrolyzed by heating in 3 mL of 6 N HCl in an evacuated sealed ampule at 100°C for 22 hours. The contents of the ampule are cooled in an ice bath; then the open neck of the ampule is connected to a vacuum line. Brief evacuation followed by application of a torch flame **all the way around the neck of the ampule** will cause the ampule to seal.

CAUTION: WEAR SAFETY GLASSES AND GLOVES.

The sealed ampule neck is heated to softness while the ampule is pulled away from the neck stub, which remains attached to the vacuum line. The stub must be cooled down before it is removed from the line.

The laboratory instructor may elect to seal the ampules. The tubes must be wrapped in a wire screen to prevent flying glass in case of an explosion in the oven. Once the tubes have been removed from the oven, cooled to room temperature, and opened, the contents are concentrated on a rotary evaporator (avoid contact with metal) and reduced to dryness with a little added benzene to form an azeotrope with the water.

CAUTION: BENZENE IS A KNOWN CARCINOGEN;
AVOID DIRECT CONTACT WITH SKIN OR INHALATION.

The solid residue is dissolved in 5 mL of Na citrate buffer (pH 2.2) and applied to a 1.5 x 50 cm column of AG-50W-X8 ion-exchange resin (see below.)

Buffers for Amino Acid Analysis

Sodium citrate buffers are prepared according to Table 5-2. Citrate buffers are quite susceptible to the growth of microorganisms and must be stored in the refrigerator until just before use. (Octanoic acid is a mold inhibitor, and thiodiglycol inhibits oxidation of methionine; both can be omitted if these buffers are prepared fresh the week of the experiment and stored in the refrigerator.) The final pH may be adjusted with small amounts of concentrated HCl or 50% NaOH solution. The volumes indicated are sufficient for a number of columns and can be used by several groups.

Table 5-2
SODIUM CITRATE BUFFERS FOR AMINO ACID ANALYSIS

pH	Na+ conc. (N)	Sodium Citrate[a] (g)	Sodium Chloride (g)	Conc. HCl (mL)	Thio- diglycol (mL)	Octanoic Acid (mL)	Final Volume (mL)
2.20	0.2	19.6	-	16.5	5	0.1	1000
3.49	0.2	78.4	-	46.8	10	0.4	4000
4.12	0.2	78.4	46.8	33.5	10	0.4	4000
6.40	1.0	78.4	187.2	1.0	-	0.4	4000

[a] Monohydrate

Resin Treatment

To save laboratory time and because the AG-50W-X8 ion-exchange resin used in this experiment is expensive, the column will be poured by the teaching staff before the second laboratory session. The column must be thermally equilibrated at 50°C for 1 hour before use. The resin can be reused after regeneration with 0.2 N NaOH followed by pH 2.20 buffer. The resin need not be removed from the column for routine regeneration. If the resin must be removed from the column, the pump and circulating water bath must be turned off and disconnected. The upper inlet of the column water jacket should be disconnected first to allow the jacket to drain. To remove resin from the column, the column should be inverted and the resin pumped out. Care must be taken to avoid damage to the resin support screen. The resin material can be freed from contaminates by stirring with 2 N NaOH containing 1 g per liter of EDTA, followed by repeated washings with distilled water, then 4 N HCl (at 70°C). The washed resin is decanted free from acid and then suspended in pH 2.20 buffer and repacked into the column as a slurry. Care must be taken to avoid uneven packing and air pockets. Columns should be stored wet.

Automatic   Fraction   Collection

The outlet from the column is connected to a fraction collector. Figure 5-2 illustrates an ISCO model 1200 fraction collector but a number of different types are available. The cheapest is a pair of steady hands but this chore requires a disposition impervious to acute boredom. More likely you will use a rotary fraction collector with tubes mounted in a carousel or a square fraction collector such as the ISCO with tubes mounted in "trays" that slide in tracks. Both types of fraction collectors must be adjusted to ensure uniform fraction volumes. The flow rate from the column, the buffer pumping rate, and the use of the fraction collector must be worked out while the column is being equilibrated with pH 2.2 buffer.

Sample   Addition

The 5 mL of protein hydrolysate in citrate buffer (pH 2.20) is applied evenly over the surface of the resin bed with a pipet. The flow rate from the column is adjusted so that the sample is allowed to seep into the top 1 to 2 cm of the resin bed. As soon as the sample layer has entered the top of the bed, the bed is overlayered with three 2-mL portions of the pH 2.20 buffer, which will ensure that all the amino acids are bound firmly to the column. Great care must be taken to ensure that the sample is applied evenly across the top of the column. If the column is mounted on a laboratory bench, the top of the column will be well above normal eye level. If this is the case, then a student should stand on a short step-ladder or some other **safe** platform so that the top of the column can be easily viewed. The student must have something firm, other than the glass column, to grasp to maintain balance when leaning over the bench top.

DO NOT LEAN OVER A COLUMN

WITHOUT A GOOD HANDHOLD.

A preferable option would be to mount the column on the floor or on a short platform so that the column top is at eye level. Care must be taken to place the column where students will not trip into it.

Buffer   Addition

The elution buffers are pumped into the column using the flow rate adjusted earlier. It is essential that the flow rate be moderately slow (≤ 0.5 mL/min) and steady. (At this flow rate the complete amino analysis will require an overnight run.) It is also essential that all the solvent lines to and from the pump are clean, free from kinks and that all connections are wired tight. Spilled buffer must be cleaned up immediately and should never be allowed to accumulate under the pump or fraction collector. A piston-driven pump must be washed by pumping several hundred milliliters of distilled water after this experiment.

DO NOT LEAVE BUFFER SOLUTION

STANDING IN THE PUMP.

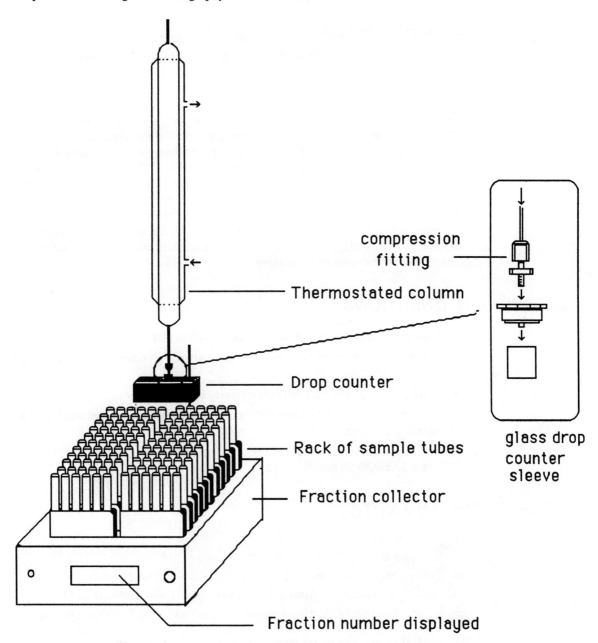

Figure 5-2 Column chromatography with an ISCO fraction collector

Ninhydrin Reaction

A stock solution of ninhydrin reagent is prepared from 2 g of ninhydrin and 0.3 g of hydrindantin in 75 mL of 2-methoxyethanol followed by dilution with 25 mL of 4 N Na acetate (pH 5.5). The resulting red solution is stored under nitrogen in the dark. If long- term storage is anticipated, the bottle should be equipped with a nitrogen syphon. Care must be taken to ensure that the nitrogen pressure is just high enough to assure efficient siphoning. The stopper and siphon will pop off if the nitrogen pressure is too high. For analysis, 2 mL of effluent from the ion exchange column is added to 1 mL of ninhydrin reagent. No adjustment of pH is necessary. The tubes are capped with glass marbles, shaken by hand (use gloves), and heated for 30 minutes in a boiling-water bath. An entire rack of tubes can be heated at a time. The absorbance of the contents of each tube at 570 nm is measured with a colorimeter.

Each student should prepare a plot of absorbance vs. fraction number. Care must be taken to observe overlapping elution peaks since the amino acids are identified by their relative order of retention on the column.

DISCUSSION QUESTIONS

1. Explain why aspartate comes off the column before glutamate.

2. Why does this experiment fail to distinguish between glutamate and glutamine or aspartate and asparagine?

3. What is the function of the elevated column temperatures?

4. Column flow rates can be faster if the column packing material is of uniform size particles. Explain.

5. Ninhydrin is not very suitable as a postcolumn derivatization method for HPLC amino acid analysis methods. Why?

REFERENCES AND NOTE

1. S. Moore and W. H. Stein, *J. Biol. Chem.*, **1951**, *192*, 663.

2. S. Moore and W. H. Stein, *J. Biol. Chem.*, **1954**, *211*, 907.

3. H. Hirs, W. H. Stein, and S. Moore, *J. Biol. Chem.*, **1954**, *211*, 941.

4. D. H. Spackman, W. H. Stein and S. Moore, *Anal. Chem.*, **1958**, *30*, 1190.

5. S. Blackburn, *Amino Acid Determination*, New York: Marcel Dekker, **1968**, pp. 1-137.

6. G. R. Barbarash and R. H. Quarles, *Anal. Biochem.*, **1982**, *119*, 177.

7. R. R. Granberg, *Liquid Chromatog. LC*, **1984**, *2*, 776.

8. J. Khym, *Analytical Ion-Exchange Procedures in Chemistry and Biology*, Englewood Cliffs, N. J.:Prentice-Hall, **1974**.

9. C. J. O. R. Morris and P. Morris, *Separation Methods in Biochemistry*, (2nd ed.) New York: Interscience Publishers, **1976**.

10. See also chromatographic equipment supplier technical literature from the following laboratories: Bio-Rad Laboratories, Pharmacia Fine Chemicals, Pierce Chemical Company, Waters Associates, Varian Instruments, and Rainin Instruments.

Chapter 6

Gel Filtration Chromatography

An unknown protein and a mixture of known proteins are chromatographed on a Sephadex G-75 column. The partition coefficients K_D for each protein is determined from its elution volume. The protein concentration in eluted fractions is determined by the Bradford protein assay. The K_D values for the known proteins are used to construct a linear plot of K_D vs. log (MW) and the molecular weight of the unknown protein is estimated from the position of its K_D on this line. In a second portion of this experiment, cytochrome c is cross-linked to give a family of "*n*-mers" which can be partially resolved on a G-75 column. The theory of HPLC size-exclusion chromatography is introduced to permit a possible extension of this experiment.

KEY TERMS

Column capacity factor	*Eution volume*
Gel Filtration	*HPLC*
Interstitial or void volume	*Partition coefficient*
Peak resolution	*Pore volume*
Radius of gyration	*Retention constant*
Sephadex and Bio-Gel media	*Theoretical plates*
Total permeation volume	

BACKGROUND

Gel filtration is a chromatographic technique that discriminates on the basis of molecular volume.[1-6] Large molecules elute from a gel filtration column before small molecules. The technique is also known as gel permeation chromatography or size-exclusion chromatography. A gel filtration column consists of a stationary uncharged gel phase, comprised of porous beads, and a mobile aqueous phase known as the eluent. The eluent percolates through the column and passes slowly out the exit port at the bottom. Solvent passing from the column is replaced by an equivalent flow of fresh solvent. Each type of solute molecule is partitioned between the interior of the stationary porous gel beads and the mobile phase and is eluted from the column by a volume of solvent equal to the volume accessible to that solute molecule within the column. The beads have a distribution of pore sizes and molecules only can be taken into pores large enough to hold them. Very large molecules cannot penetrate any of the the bead pores and they are eluted by a solvent volume V_o known as the *void volume*, which is equal only to the *interstitial volume* between gel beads. Smaller molecules can penetrate the beads and emerge from the column at an *elution volume* V_e, which is greater than V_o. Each column has a *total permeation volume* V_T, which is equal to the sum of the void volume V_o and the *pore volume* V_p (i.e., the volume accessible to solvent within the pores):

$$V_e \leq V_T = V_o + V_p \qquad\qquad (6\text{-}1)$$

These volume relationships are illustrated in Figure 6-1.

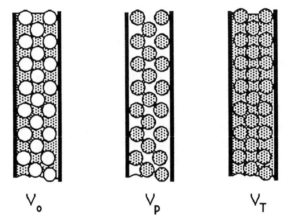

Figure 6-1   Volume relationships in gel-filtration

The partition coefficient of the solute between the stationary and mobile phase is described by the partition coefficient K_D.

$$K_D = \frac{V_e - V_o}{V_p} \qquad\qquad (6\text{-}2)$$

Species that are too large to permeate the pores of the gel packing elute at V_o and solutes of low molecular weight that can penetrate all the pores elute at V_T. The separation of species of intermediate size depends on their relative extent of penetration into the pore volume. The preceding equation can be rearranged as

$$V_e = V_o + K_D V_p \qquad\qquad (6\text{-}3)$$

The constant K_D can range from zero for a solute too large to diffuse into any of the pores to unity for a small solute that can enter all of them. In the absence of specific chemical interactions between solutes and chromatographic packing, the entire sample elutes within a K_D range of 0 to 1. The retention mechanism involves only entropic rather than enthalpic interactions with the packing so that the separation depends only on the dimensions of solute and the geometry of the pores within the packing.[1,2] In an ideal case of size-exclusion chromatography:

$$K_D = \exp(\Delta S/R) \qquad (6\text{-}4)$$

where ΔS is the entropy change upon transfer of solute from the mobile phase to the pores.[2,3] For large solutes ΔS is negative and $K_D < 1$.

The elution volume cannot be greater than V_T unless some additional retention mechanism is operating. Effects caused by specific absorption mechanisms such as hydrogen bonding or electrostatic interactions can lead to $K_D > 1$ but such effects can usually be avoided by operating at pH values and with ionically neutral packing, where such effects are minimized. (Polysaccharide packing material can hold onto cationic proteins due to cation-exchange interactions with anionic carboxylate groups produced by partial oxidative degradation of the packing.)

Protein Detection

The elution volume for a colorless protein can only be obtained after the concentration of protein in each fraction is determined by UV absorption (280 nm) or by a protein assay such as the Bradford dye-binding assay. Heme proteins such as cytochrome c are colored and their progress through a column can be seen immediately as a light red or reddish-brown band. Cytochrome c with a molecular weight of 12,000 is near the lower limit of the protein fractionation range of Sephadex G-75; however, higher oligomers of cytochrome c prepared by cross-linking this protein with glutaraldehyde[7,8] or other cross-linking reagents, at high protein concentrations, come off the column more quickly. The dimer, trimer, and tetramer can be separated from the monomer by virtue of their molecular weight differences [24,000, 32,000, and 44,000 daltons, respectively] and each can be seen as a separate maxima on a broad band eluting from a well-packed column of 50 cm or more. Cytochrome c exists in both Fe(II) and Fe(III) forms; therefore, the protein must all be oxidized to the Fe(III) form to give a sharper elution pattern. This is done by adding *a small excess* (5%) of ferricyanide. Cytochrome c has an $E^{1\%} = 21.9$. If higher levels of ferricyanide are used, unconsumed ferricyanide is retained by the column and appears as a yellow band eluting at V_T.

Molecular Weight Determination

The partition coefficient K_D can be related with reasonable confidence to the molecular size of the solute and to parameters related to size, such as molecular weight (MW). For a series of spherical proteins of increasing radii, K_D decreases as a function of molecular weight and a plot of K_D vs. log(MWt) is linear throughout most of the fractionation range. The real parameter determining K_D is the radius of gyration R_G of the protein molecule which is a function of its shape.[1] R_G is defined by the distribution of all atoms of mass m_i at distances R_i from the center of mass of the molecule:

$$R_G = \sqrt{\frac{\sum_i m_i R_i^2}{\sum_i m_i}} \qquad (6\text{-}5)$$

The relationship between R_G and MW for sherical, rod-like and random coil molecules has been developed.[1,2]

R_G (spheres)   is proportional to $MW^{1/3}$

R_G (rods)   is proportional to M

R_G (random coils) is proportional to $MW^{1/2}$

Rod-shaped molecules will elute earlier than spherical ones of equivalent mass and their points will fall below the line defined by the more spherical molecules. However since most proteins are "roughly" spherical and a plot of K_D vs. log(MW) is linear over two or three decades of molecular weight, and because of the nondestructive nature of gel filtration chromatography, this method remains a useful technique for estimating molecular weights. It is convenient, fast and nondestructive.

Accurate determinations of K_D require well separated, narrow, symmetrical elution curves. This is achieved only if the sample is small and applied evenly, the column is uniformly packed and long enough for good resolution, and the solvent flow rate is constant. It is also essential to minimize the dead volume below the bed, to exclude contaminates and denaturing agents, and to protect the bed surface during application. The procedure suggested below satisfies these requirements with a minimum investment of time and materials.

One of the principal problems encountered in a teaching laboratory is faulty or uneven application of the sample to the column. The sample must be applied in a gentle flow distributed evenly over the surface. Sample application to only one spot may create a pit in the packing material, so that part of the sample starts ahead of the rest. Solid debris in the sample may coat the packing and lead to uneven takeup of the sample.

Peak Resolution and Estimating V_e

When one attempts to separate two molecules of closely related molecular weight on a gel-filtration column, the two components may be eluted together and appear as one peak. It is also possible that the peak maxima may be well separated but with the peaks so wide that they seriously overlap. Both are examples of poor resolution. The resolution of two components of a mixture R_S is defined as the ratio of the peak-to-peak separation (in volume units) to the mean peak width, also in volume units (see Figure 6-2).

For two peaks emerging at V_{e1} and V_{e2} with widths w_1 and w_2 :

$$R_S = \frac{V_{e1} - V_{e2}}{\frac{(w_1 + w_2)}{2}} \qquad (6\text{-}6)$$

The resolution of two separated peaks in gel-filtration increases as the square root of the column length. Peaks also spread gradually as they move down a chromatographic column but the rate of broadening is less rapid than the rate of separation of the peaks. In general resolution can be increased by reducing the elution rate or by lengthening the column but this increases the time required and the protein emerging from the column from a broad band will be more dilute. One must make a judicious choice between resolution, speed, and bandwidth. A protein mixture is cleanly separated if the resolution is sufficient to give baseline resolution of the components of interest. This corresponds to $R_S > 1.2$.

A sample volume of 5% of the total column volume V_{tot} is recommended; samples smaller than this do not lead to improved resolution. Optimum resolution is observed with slow flow rates (2 mL/cm.^{2} h), but flow rates up to five times faster can be used with reasonable resolution. Very rapid flow leads to *channeling*, in which the rapidly flowing solvent causes the formation of channels or pockets within the column. Solution mixing within channels destroys resolution.

The elution volume for a narrow but unsymmetrical elution peak is taken as the solvent volume corresponding to the high point (maximum) of this elution peak. For broad unsymmetrical peaks, the leading edge is used to define V_e. Broad unsymmetrical elution profiles can be due to overlapping peaks from several proteins eluting close together. Overloaded columns also give characteristic unsymmetrical profiles with a sharply rising leading edge but a long trailing edge.

Column Capacity Factors

The retention of a sample component is best expressed in terms of elution volume with respect to a non-retained component. If the elution volume of the nonretained component is V_0 and that of the retained component is V_e, the *column capacity factor* (k') is expressed as the ratio

$$k' = \frac{V_e - V_0}{V_0} \qquad (6-7)$$

The capacity factor is a complex function of the relative partitioning of the solute between the mobile and stationary phases. It depends on the relative affinity of the solute for these two phases and the column volume occupied by the mobile phase and the surface area of the stationary phase. It characterizes the column and the solute. The *selectivity* of the column for two different solutes is given by α. Two components must have differing capacity factors in order to be separated.

$$\alpha = k'(2)/k'(1) \qquad (6-8)$$

Efficiency is the ability of a column to produce sharp peaks. The efficiency of a chromatographic system is measured as the number of theoretical plates (N). Band broadening results from diffusion and nonuniform flow patterns induced by all parts of the system including turbulence when the sample is injected onto the column or into the solvent stream immediately above the packed bed in a pumped solvent system. For a symmetrical Gaussian (bell-shaped) peak emerging from the column the number of theoretical plates can be calculated from the expression.

$$N = 16 \frac{V^2}{w^2} \qquad (6-9)$$

where w is the width of the peak at half height. The resolution of a column is the amount of separation between two peaks and is a function of column efficiency, selectivity, and k', the average capacity factor..

$$R_s = (0.25) \frac{\alpha - 1}{\alpha} \sqrt{N (k'/1 + k')} \qquad (6-10)$$

These relationships are illustrated in Figure 6-2.

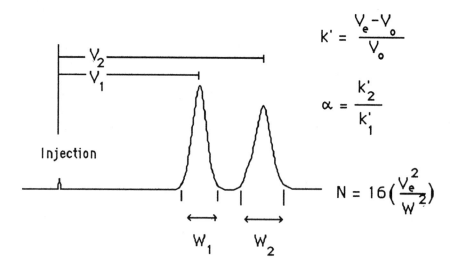

Figure 6-2   Resolution defined in terms of peak widths and elution volumes.

Characterizing a Column

The column parameters V_0 and V_T can be determined by measuring the elution volumes of very large and very small colored molecules, respectively. Blue dextran is the traditionally used large molecule; it is a dye cross-linked to a high-molecular-weight carbohydrate, and the colored complex elutes from the column when $V_e = V_0$. There have been reports that although Blue Dextran is satisfactory for teaching purposes, it is heterogeneous and that colored proteins such as ferritin (470,000 MW) should be used in research applications, where precise molecular weight determinations are required.[4] A low-molecular-weight dye or colored inorganic salt that penetrates all the pores in the gel phase is used to determine the total volume within the gel bed accessible to a low-molecular-weight solute. The maximum volume of solvent necessary to elute any protein, that is bound to the column by entropic interactions alone, is equivalent to V_T. Molecules that bind by some sort of specific interaction (hydrogen-bonding) may require elution volumes greater than V_T. For spherical beaded packing V_0 is equal to about 30 to 35% of V_T so that the useful resolving range ($V_T - V_0$) is about 70 to 80% of V_T. Both V_0 and V_T must be determined to fully characterize a gel-filtration column.

Gel Filtration Media

Two common types of widely used gel-filtration media are polysaccharide-based *Sephadex G* (Pharmacia) and polyacrylamide-based *Bio-Gel P* (Bio-Rad). Each is a water-insoluble, spherical beaded gel, homogeneously sized, with a narrow pore size distribution for high-resolution gel-filtration. Sephadex is prepared by cross-linking dextran with epichlorohydrin. The large number of hydroxyl groups affords a very hydrophilic gel. The G types differ in their degree of swelling and the fractionation range (Figure 6-3). Table 6-1 gives the different types of Sephadex and their physical properties. Sephadex is stable in water, salt solutions, alkaline and weakly acidic solutions, and can be exposed to 0.1 M HCl for 1 to 2 hours or in 0.02 M HCl for 6 months, without significant hydrolysis of the glycosidic linkages.

The Bio-Gel P series consists of porous polyacrylamide beads. The gels are highly hydrophilic and free of charge. At room temperature the recommended operating pH is 2 to 10, with significant hydrolysis of amide groups at higher or lower pH. The properties of Bio-Gel P types are given in Table 6-1.

Both Sephadex and Bio-Gel beads are soft and deform at pressures produced by high flow rates or osmotic pressures. Extra cross-linking of dextran with acrylamide affords more rigid beads of Sephacryls that can be operated at faster flow rates. Pharmacia also offers rigid cross-linked agarose beads of CL-Sepharose which are often used for the purification of high-molecular-weight biomolecules. These media are not suitable for high-performance liquid chromatography, which requires higher column pressures.

Table 6-1
TECHNICAL DATA FOR GEL-FILTRATION MEDIA

Gel Type	Protein Fractionation Range	Water Regain (mL/g dry gel)	Bed Volume (mL/g dry gel)
Sephadex			
G-25	1000-5000	2.5	4-6
G-50	1500-30,000	5.0	9-11
G-75	3000-70000	7.5	12-15[a]
G-100	4000-150,000	10.0	15-20[a]
G-150	5000-300,000	15.0	20-30[a]
G-200	5000-600,000	20.0	30-40[a]
Bio-Gel			
P-2	100-1800	1.5	4
P-4	800-4000	2.4	5
P-6	1000-6000	3.7	7
P-10	5000-20,000	4.5	9
P-30	2500-40,000	5.7	11[b]
P-60	3000-60,000	7.2	14[b]
P-100	5000-100,000	7.5	15[b]
P-150	15,000-150,000	9.2	18[b]
Sephacryl			
S-200	5,000-250,000	-[c]	
S-300	10,000-1.5 $\times 10^6$	-[c]	
Sepharose			
6B/CL-6B	10,000-4 $\times 10^6$	-[c]	
4B/CL-4B	60,000-2 $\times 10^7$	-[c]	

[a] After more than 5 hours on a boiling-water bath. [b] After more than 10 hours at room temperature. [c]Supplied pre-swollen.

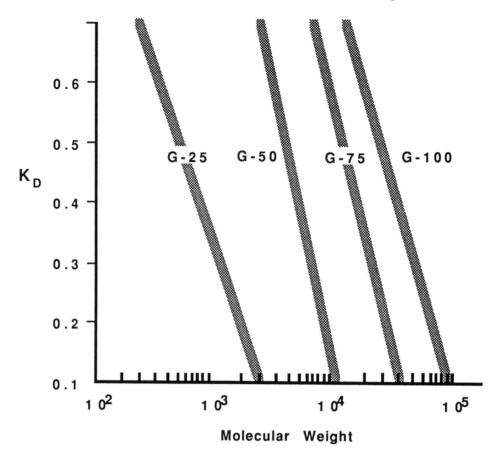

Figure 6-3 Effective fractionation ranges for globular proteins with Sephadex media

High Performance Liquid Chromatography

High-performance liquid chromatography (HPLC) has become one of the most powerful analytical and preparative techniques available to the modern biochemist. The method offers speed, sensitivity, and the benefits of automatic operation which are more than sufficient return for the initial investment in instrumentation. Although HPLC is often not employed in an undergraduate biochemistry teaching laboratory because of the expense and special supervision that such instrumentation requires, the following is a brief introduction to the theory and practice of the HPLC method, which builds on the column chromatography theory introduced earlier in this chapter.

An HPLC system requires a solvent delivery system in which the mobile phase is pumped through a column containing packing material of smaller and more uniform particle size and greater mechanical rigidity than in conventional liquid chromatography systems (Figure 6-4). The separated solutes in the solvent stream, emerging from the column, are detected by a nondestructive detection system such as a UV/visible monitor, and the resulting elution chromatogram is automatically plotted on a recorder or stored within a microcomputer. The term *high performance* in the name arises because the high operating pressure (100 to 300 psi), the packing material, the sample injection system, and a detection system that does not impede the flow from the column, all permit separations with much less band spreading than in conventional glass columns.

An understanding of the physical and chemical interactions whereby an HPLC column packing material discriminates between solutes is necessary for selecting the best column and mobile phase conditions. There are a number of mechanisms by which solutes are retained by the packing material and a large number of column packing materials that discriminate between biochemically important solutes on the basis of solute size, charge, polarity or special chemical affinities. Most work on biomolecules involves ion exchange, gel filtration, and reversed phase absorption chromatography (RPC). A new method of hydrophobic-interaction chromatography (HIC) is also becoming popular.[5] Two basic types of packings are used in HPLC : (1) silica packing, which can be either totally porous particles, partially porous or pellicular particles, and (2) polymeric packing, which can have all sorts of specially designed chemical properties. The totally porous particles have the highest surface area (200 to 500 m^2/g) and consequently the largest sample capacity; they can be very small (3 to 10 μm). Pellicular particles, glass beads coated with silica, are much larger (20 to 26 μm) and have a much lower surface area per weight.

In many cases the silica particles have chemically bonded to them a coating that imparts special selective features, such as ionic groups, imparting ion-exchange capacity, or organic moieties that make the particles hydrophobic. Some HPLC packings are based on synthetic resins rather than silica. There are a very large number of types of column packings available and more are being developed every year. The partial list in Table 6-2 will impress upon the beginner the number of HPLC separation options open to the biochemist. Both silica and organic gels are used in size-exclusion or gel filtration chromatography. Generally speaking, silica-based gel filtration materials are more suitable for protein separations because they provide higher resolution than do resin-based materials. Toya Soda's TSK-type SW gels are among the most commonly used and are available in three pore sizes corresponding to the various separation ranges (in parentheses): G2000SW (5,000 to 100,000 D), G3000SW (10,000 to 500,000 D), and G4000SW (20,000 to 7,000,000 D). These packings are available from Bio-Rad as Bio-Sil TSK and from LKB as LKB Ultrapac TSK. Aquapore OH is a glyceropropylsilane-modified silica produced by Brownlee labs and marketed by Rainin. Aquapore OH-100 will separate proteins from 10,000-100,000 D in minutes. Waters Associates markets a protein separation system based on their I-125 size-exclusion column. [Size-exclusion columns of highly cross-linked styrene divinylbenzene polymers (e. g., Waters' Styrogel) are usually used for fractionating very hydrophobic synthetic polymers.]

Both RPC and HIC employ silica-based packing materials that have been chemically modified to render the silica material hydrophobic. If a high percentage of the chemically bonded phase is hydrophobic C_8 or C_{18} aliphatic chains, then the packing material can discriminate between solutes on the basis of their partition between a polar (aqueous methanol or acetonitrile) mobile phase and a hydrophobic stationary phase. All remaining silica sites are capped with a smaller hydrophobic reside such as trimethyl silane to minimize ion exchange effects. HPLC employing such packing is known as reverse phase chromatography and is widely used for protein separations. However RPC can result in denaturation of proteins and large peptides.[6] Altex ultrasphere and ultrasil reverse phase packings are widely used for peptide and protein work. Bio-Rad offers a large pore reversed-phase column (Hi-Pore) for separating macromolecules. Waters' μ-Bondapak C_{18} reversed-phase columns are useful for low-molecular-weight organic compounds. Columns with intermediate polarity packing including cyanopropyl-bonded phases, have found some applications in reversed-phase protein purifications especially with solvent gradients.

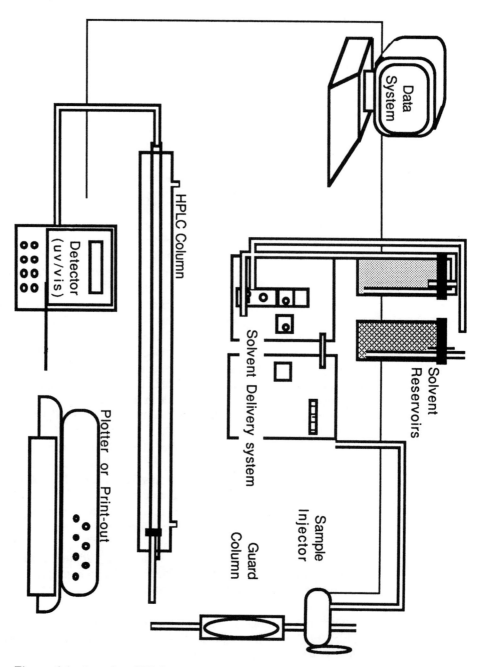

Figure 6-4 A modern HPLC system

Amino acid analysis (see Chapter 5) can be accomplished on ion-exchange HPLC systems operating with sample loads in the range of 100 pmol to 10 nmol and with less than 50-μL sample volumes. Postcolumn derivatization with ninhydrin is the principal means of detection. Analysis times of less than 1 hour are common. Recent methods employing precolumn derivatization combined with reversed-phase HPLC can increase the detection limits by as much as 1000-fold. Precolumn derivatization with *ortho*-phaldialdehyde (OPA) can resolve amino acids and other metabolites in physiological fluids on the order of 5 to 10 fmole amino acid[7] and amino acid detections (A *femtomole* is 10^{-15} mole). Ion exchange can be done on conventional cross-linked resins or bonded phase silica packings. Anion exchangers contain diethylammonium groups (Bio-Gel TSK DEAE and Bio-Sil DEAE packings). Cation exchange usually involves carboxy and sulfonate groups (Bio-Sil CM and Bio-Gel SP packings, respectively). The fact that all peptides are ionic can be exploited chromatographically by using polymeric high-performance ion-exchange columns.[8] Silica materials have not been entirely useful in separating acidic peptides and have been markedly unsuccessful in separating basic peptides, which are not retained on silica packings at all. Polymeric materials can overcome these problems.

Very often the mixed aqueous-organic solvent composition is varied by a gradient former in the solvent delivery system so that the solvent polarity decreases (with increasing percentage methanol or acetonitrile) during the course of the chromatographic run.[9] This **gradient elution** facilitates separations by introducing subtle variations in the partitioning of solute between the two phases. Modern HPLC instruments use a computer-controlled solvent delivery system that can be programmed to control the solvent composition; the solvent component ratio can be varied over the course of the chromatographic run, either in step increments or as a continuous function of the running time.

Table 6-2
COMMON HPLC COLUMNS AVAILABLE TO THE BIOCHEMIST

Column Type/ Specific Examples	Source or Supplier
Gel-filtration	
TSK SW	Toya Soda /Kratos
Spherogel TSK SW	Beckman
Ultrapac	LK
Bio-Sil TSK	Bio-Rad
Aquapore OH	Brownlee /Rainin
Protein PAK 60 or 125	Waters
Reverse Phase	
Hi-Pore RP	Bio-Rad
Bio-Sil ODS	Bio-Rad
Ultrapore RPSC (C_3)	Beckman
Spherogel TSK Phenyl-5PW	Beckman
LiChrosorb RP-2, -8,-18	Merck/Pierce
Ultrasphere ODS, or octyl	Altex /Rainin
Ultrasil ODS or octyl	Altex /Rainin
Aquapore RP	Brownlee /Rainin
μ-Bondapak C_{18}	Waters
μ-Bondapak CN	Waters
Ion Exchange	
Bio-Sil TSK DEAE (anion exchanger)	Bio-Rad
Bio-Gel TSK DEAE (anion exchanger)	Bio-Rad
Bio-Sil TSK CM (cation exchanger)	Bio-Rad
Bio-Gel SP (cation exchanger)	Bio-Rad
Aquapore AX 300 (anion exchanger)	Brownlee /Rainin
Aquapore CX 300 (cation exchanger)	Brownlee /Rainin

EXPERIMENTAL SECTION

Students are expected to work in pairs or groups of three, but each is responsible for a laboratory report. The column setup can be done during any free time in the previous laboratory experiment and the column stored in the cold room until needed. The cross-linking experiment should be done as near in time to the chromatography step as possible.

Column Preparation

Sephadex G-75 is purchased as a dry powder and must be allowed to swell overnight in 0.02 M phosphate buffer (pH 7), then heated to near boiling just prior to use. This heating ensures maximum swelling and degases the slurry. The gel is allowed to cool and settle and the bulk of the supernatant and any floating fine particles are removed by aspiration. Just enough supernatant liquid is left to yield a pourable suspension; the slurry should not be so thick as to retain air bubbles. The slurry is poured along a glass rod into an upright glass column until there is less than 1 to 2 cm of space above the resin bed. All the gel should be added at one time. The gel surface may be covered with a mesh protection device after the bed has settled due to gravity. The column should not be filled while the solvent is flowing through it; beds settled under continuous downward solvent flow often give very slow flow rates.

The column can be used in the conventional downward flowing configuration with the solvent flow regulated with a Mariotte flask or similar device (Figure 6-5). This is the most common configuration used in a teaching lab that must serve a number of students with inexpensive gel-filtration setups. If money permits and if the resolution and time requirements are strenuous, the upward-flowing configuration should be used. The solvent is pumped into the column from the bottom and the column effluent will come out the top. This inverted configuration permits a higher flow rate, especially with longer columns, because the packing action of solvent flow is reduced rather than augmented by gravity.

The column must be filled entirely with gel because solvent-filled gaps can lead to mixing and diffusion, resulting in very broad bands. For this reason one must never use too fast a flow rate when pumping in either direction, since this will compact the column bed, leaving a space in the column at the solvent inlet side filled only with solvent. [A large column ($V_T = 200$ mL or more) should be eluted with solvent for several hours, to ensure that the bed will not compact, before the sample is applied.]

If the downward-flowing configuration is used, the sample is applied directly to the drained surface of the bed. The solvent must be drained just down to the surface of the bed and the sample gently overlayered onto the bed and allowed to enter the bed surface just as the solvent is drained from it. Great care must be taken to ensure that the sample is applied in a thin, even band. Care must also be taken to prevent any of the column from running dry and cracking or shrinking. After the sample has been applied, additional solvent is overlayered and allowed to drain into the column to wash all the sample into the bed. Then the top of the column is filled with solvent and the flow rate adjusted to the desired level. There should not be an air space above the column as entering eluent will drop onto the bed surface. Irregular bed surfaces lead to broad bands.

In the upward-flowing or inverted configuration the sample is injected into a flowing solvent stream either with a syringe through a septum cap on a T-tube or, preferentially, with a sample injection loop that enables the sample to be incorporated into the solvent stream entering the column. The injection loop volume should never be larger than 2% of the column bed volume. An injection loop facilitates the reproducible application of samples of a fixed volume.

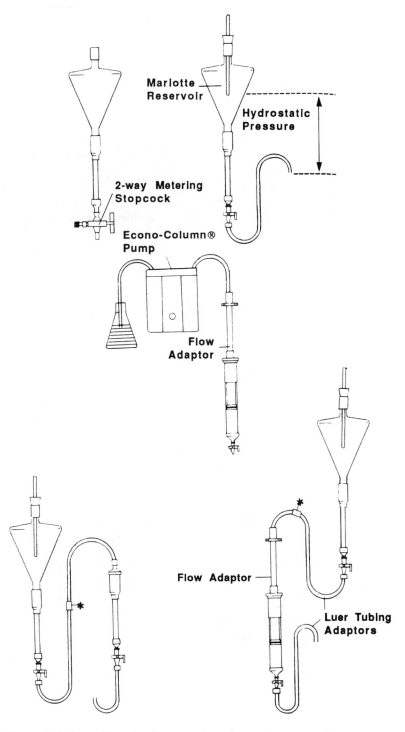

Figure 6-5 Controlling the flow rate in column chromatography

Applying the Sample

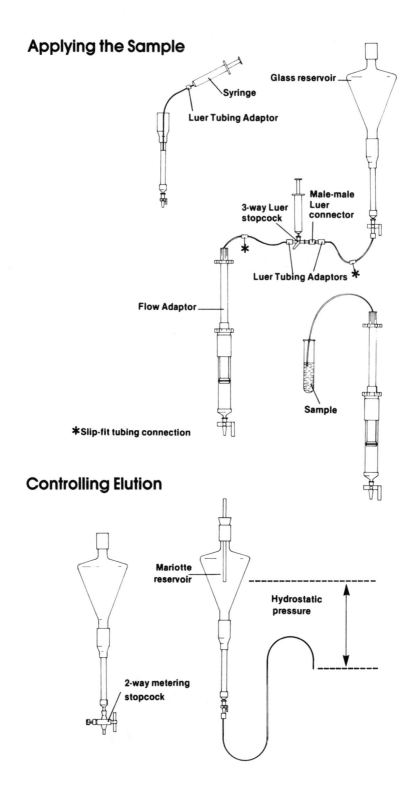

Controlling Elution

Figure 6-6 Applying a sample to a chromatography column

Molecular Weight Determination

To determine the molecular weight of an unknown, three separate experiments must be performed:

1. Determination of V_O, from the elution volume of the large protein ferritin or with Dextran Blue.

2. Determination of V_e and K_D values for the unknown protein.

3. Determination of V_e and K_D for three known proteins spanning the fractionation range. Cytochrome c, carbonic anhydrase, and hemoglobin can be used. The two heme proteins do not require a separate protein assay step but carbonic anhydrase does. The method suggested in Chapter 4 is recommended.

In the third experiment, care should be taken to avoid overlapping peaks. A linear plot of V_e vs. log(MW) for the known proteins will be used to interpolate a reasonable estimate of the molecular weight of the unknown protein from its observed V_e value.

Because the protein samples are colorless, the elution volumes can be determined only after the protein concentration of each recovered aliquot is determined by a Bradford protein assay. Optimum resolution of closely associated peaks will require very small aliquot volumes (e.g., 0.5 mL). This implies the collection of 100 fractions for a column with a V_T of 50 mL. Less resolution is required if peaks are nicely separated.

The molecular weight standards commonly used in gel-filtration chromatography are given in Table 6-3.

Table 6-3
PROTEIN MOLECULAR WEIGHT STANDARDS FOR GEL FILTRATION

Molecular Weight Marker	Molecular Weight (approximate)
Glutathione, reduced	300
Glutathione, dimer	600
Bacitracin	1,400
ACTH	3,500
Cytochrome c	13,000
Ribonuclease A[a]	14,000
Myoglobin[b]	16,800
Chymotrypsinogen A[a]	25,000
Carbonic anhydrase	31,000
Ovalbumin[a,b]	44,000
Hemoglobin	64,500
Bovine serum albumin	68,000
IgG[b]	150,000
Aldolase[a]	158,000
Catalase[a]	210,000
Fibrinogen	341,000
Ferritin[a]	470,000
Thyroglobulin[a,b]	670,000

[a] Premixed in one of the Pharmacia gel-filtration calibration kits. [b] Premixed in Bio-Rad gel-filtration standard

Protein Cross-Linking (an Optional Experiment)

Covalently cross-linked oligomers of cytochrome c, or other conveniently available small protein, can be prepared by reaction with Traut's reagent (2-iminothiolane-HCl),[10] which reacts with the epsilon amino group of lysines to introduce sulfhydryl groups into proteins. This modified protein can then be cross-linked to another sulfhydryl containing group. This treatment affords dissulfide cross-links between modified protein subunits in the presence of low levels of the oxidizing agent, hydrogen peroxide. Cytochrome c is no-longer strongly colored after peroxide treatment and an alternative method of protein detection such as the Bradford assay must be used.

$$\text{(2-iminothiolane ring with } \overset{+}{NH_2}) + R-NH_2 \xrightarrow{\text{pH 7-10}} R-NH\overset{\overset{+}{NH_2}}{\underset{\parallel}{C}}CH_2CH_2CH_2SH$$

Traut's reagent stock solution is prepared by dissolving 68.5 mg of 2-iminothiolane-HCl in 1 mL of 1 M Triethanolamine-HCl buffer (TEA buffer, pH 8). The cross-linking is accomplished in a microfuge tube by adding 250 µL of this reagent to 500 µL of an ice-cold solution of 3mg/mL cytochrome c in 25 mM TEA buffer with 25 mM KCl. The reaction mixture is incubated at 4°C for 30 minutes then 100 µL of 30% hydrogen peroxide is added. The solution goes from reddish pink to colorless. After ten minutes excess hydrogen peroxide is destroyed by adding 5 to 10 µg catalase to the reaction mixture. Care must be taken when catalase is added because foaming can occur and enough gas is released to blow the lid off a microfuge tube.

Students are encouraged to vary the parameters (protein to reagent concentration ratios, temperatures, etc.) of the cross-linking experiment to improve the resolution of the oligomers in the gel-filtration step. Not all dimers will elute at the same V_e if their radii of gyration are different and longer oligomers will behave like rods rather than spheres. There is a tendency for cross-linked proteins to migrate slightly faster on SDS electrophoresis gels than standard proteins of the same molecular weight[11] and oligomeric molecular weight standards can lead to an overestimation of molecular weight by 5-15%.

DISCUSSION QUESTIONS

1. The total permeation volume V_T is slightly less than the total bed volume V_{tot} calculated from the bed height and diameter because of the volume occupied by the gel material V_{gel} and can only be obtained by measuring V_o and V_p or by calculating V_{gel} from the weight of dry gel and the partial specific volume of the gel matrix when fully hydrated (see Table 6-1). Calculate V_{gel} for a column made up with 10 g of G-25. What volume of glass column would one require for this much resin?

$$V_{tot} = \pi(\text{height})(\text{diameter})^2 = V_T + V_{gel} \qquad (6\text{-}11)$$

2. To characterize the elution properties of the solute in a way independent of the geometry of the column, the retention constant $\mathcal{R} = V_0/V_e$ is used. $\mathcal{R}$ is equal to the fraction of solute in the void volume V_0 and $(1-\mathcal{R})$ is the fraction of solute in the pore volume V_p during the chromatographic run. Show that $[(1-\mathcal{R})/\mathcal{R}] = k'$.

3. Given the $E^{1\%} = 21.9$ for cytochrome c, how much potassium ferricyanide is required to react with 10 mL of a cytochrome c solution with an absorbance of 2.0 for a 1 cm path length?

REFERENCES

1. W. W. Yau, J. J. Kirkland and D. D. Bly, *Modern Size Exclusion Chromatography*, New York: John Wiley & Sons, **1979**.

2. H. G. Barth, *Liquid Chromatog. ,LC*, **1984**, *2*, 25.

3. E. F. Cassasa and Y. Tagami, *Macromolecules*, **1969**, *2*, 14.

4. J. J. Marshall, *J. Chromatog.*, **1970**, *53*, 379.

5. A. J. Albert, *Bio-Chromatography*, **1987**, *2*, 131.

6. S. C. Goheen and A. Stevens, *Bio-Techniques*, (Jan-Feb), **1984**, 48-54.

7. G. Ogden and P. Foldi, *Liquid Chromatog-Gas Chromatog.*, LC-GC, **1987**, *5*, 28-40.

8. P. Mychack and J. R. Bensen, *Liquid Chromatog-Gas Chromatog.*, LC-GC, **1986**, *4*, 462-68.

9. A. Pryde and M. T. Gilbert, *Applications of High-Performance Liquid Chromatography*, London: Chapman & Hall, **1979**.

10. R. Jue and R. Traut, *Biochemistry*, **1978**, *17*, 5399.

12. J. C. H. Steele and T. B. Nielsen, *Anal. Biochem.*, **1978**, *84*, 218.

Chapter 7

Radioisotopes and Liquid Scintillation Counting

The general theory of radioactive decay and the measurement of radioactivity by liquid scintillation spectroscopy are discussed. The reasonable requirements for safe handling of radioisotopes are emphasized. Five detailed experiments illustrating liquid scintillation counting are offered: the determination of the β-spectra of unquenched (standardized) ^{14}C samples, quench correction by automatic external standardization, the determination of the binding constant for ^{14}C-labeled IPTG to crude *lac* repressor, an *invitro* protein synthesis project employing commercially available rabbit reticulocyte lysate, and the reductive methylation labeling of cytochrome *c* with [^{3}H]formaldehyde and $NaBCNH_3$, followed by purification on a G-75 column.

KEY TERMS

Autoradiography	*Channel ratio method*
Coincidence counting	*Counting efficiency*
Disintegration constant	*DPM and CPM*
External quench correction	*Half-life*
Isotopic labeling	*Liquid scintillation counting*
Quenching	*Radioisotopes*
Rads and Curies	*Roentgens*
Scatchard plots	*Scintillation cocktails*
Specific activity	*TCA precipitable counts*
Tracers	*β–Decay*
β–Spectra	*γ–Emission*

BACKGROUND

The chemical behavior of an atom is determined by its atomic number not its atomic weight. *Isotopes* are atoms that contain the same number of protons (have same atomic number) but different numbers of neutrons and consequently different atomic weights. These atoms have very nearly identical chemical properties. For example, the three principal isotopes of carbon (^{12}C, ^{13}C, and ^{14}C) all contain six protons but six, seven, and eight neutrons, respectively. The carbon isotopes all have the same valences and bonding geometries so that biomolecules made from them are chemically identical. However, isotopes differ in their nuclear stabilities and unstable isotopes (radionuclides) break down spontaneously so that a small sample of a biomolecule containing an unstable isotope such as a ^{14}C atom is uniquely labeled by the fact that some of these atoms are disintegrating at any time. The products of the disintegration are high-energy subatomic particles, *radiation*, that can be easily detected. By monitoring this *radioactivity*, the passage of that labeled biomolecule and its chemical reaction products can be traced in a biological system; radioactive isotopes are frequently called *tracers*.

The two stable isotopes of carbon ^{12}C and ^{13}C make up 98.89 and 1.11% of naturally occuring carbon respectively. The unstable isotope ^{14}C has a very low natural abundance and would have long since decayed completely away were it not replenished by transmutation of nitrogen upon collision with cosmic ray neutrons in the upper atmosphere. All radioactive isotopes occur in very low natural abundance and most are the human-made by-products of specialized nuclear reactors. There are a large number of nuclear transformation mechanisms whereby one type of nucleus is transformed into another; however only two mechanisms are important in biochemistry. The emission of high-energy electrons or β-particles is the nuclear transformation mechanism for the principal isotopes used in biochemistry; ^{3}H, ^{14}C, ^{35}S, and ^{32}P are all β-emitters. A high-energy electron is ejected when a neutron is converted to a nuclear proton and an antineutrino; the product is a stable isotope with an atomic number one higher than the parent atom.

$$^{3}H \longrightarrow \beta + \, ^{3}He + \tilde{\nu}$$

$$^{14}C \longrightarrow \beta + \, ^{14}N + \tilde{\nu}$$

$$^{32}P \longrightarrow \beta + \, ^{32}S + \tilde{\nu}$$

$$^{35}S \longrightarrow \beta + \, ^{35}Cl + \tilde{\nu}$$

In all four cases the atomic mass number of the product is the same as that of the parent isotope, but the atomic number has been increased by one so that the product is the element to the left of the parent on the periodic table.

Another mode of nuclear transformation is the emission of gamma particles or rays. This process may be very complicated and involve a number of steps including more than one γ (gamma)-particle. The most important gamma emitters used in biochemistry are ^{125}I and ^{131}I. The latter nucleus decays to produce both β and γ radiation.

Any nuclear transformation is characterized by two parameters which are not altered by ordinary chemical or physical processes: the nuclear decay rate constant and the decay energy. The rate law governing nuclear decay is a first-order rate law so that the decay rate can be specified by a single constant, λ, the *disintegration constant*. The first-order rate law can be written either as the differential equation or its integral solution :

$$-\frac{dN}{dt} = \lambda N \quad \text{or} \quad \ln \frac{N_0}{N_t} = -\lambda t \qquad (7\text{-}1)$$

The integrated form of the disintegration rate law relates the number of radioisotopic atoms N_t at time t to the number N_0 at time $t = 0$, with λ being the disintegration rate constant. This can be seen in Figure 7-1.

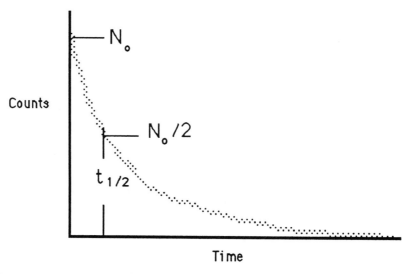

Figure 7-1   First-order decay curve

The easiest way to grasp the physical meaning of the disintegration constant is in terms of the time required for N_t to reach one-half of its original value, $N_0/2$; this time is a constant $t_{1/2}$ for any particular first-order disintegration process and is known as the *half-life*. The half-life is inversely proportional to the disintegration constant:

$$t_{1/2} = \frac{\ln 2}{\lambda} = \frac{0.693}{\lambda} \qquad (7\text{-}2)$$

Decay Energies

The decay energy of a nuclear transformation is a unique value for any pair of parent and product atoms and represents the upper limit of the energy of the particles emitted during that particular kind of atomic disintegration. However, the energy of individual beta particles emitted in that disintegration can be quite variable. The kinetic energy released in the disintegration is shared randomly between the β particle and the neutrino, which is also liberated in a β-decay process, so that the β-particle energies are best expressed as an energy profile (i.e., a plot of particle abundance vs. particle energy). The maximum energies (endpoint energies) and the average β-particle energies differ from isotope to isotope and they are important in discriminating between two separate radioisotopes in double-label experiments where two isotopes are assayed simultaneously. The characteristics of the most common radioisotopes used in biochemistry are given in Table 7-1.

Table 7-1

CHARACTERISTICS OF BIOCHEMICALLY IMPORTANT ISOTOPES[a]

Isotope	Half-Life	Decay Energy (MeV)	Average. β- Particle Energy (MeV)
^3H	12.43 years	0.0186	0.0057
^4C	5730 years	0.156	0.049
^{32}P	14.3 days	1.709	0.69
^{35}S	87.4 days	0.167	0.049
^{125}I	60 days	0.035	γ only
^{31}I	8.04 days	0.806	0.180[b]

[a]Data from Amersham Corporation. [b] Average for four β–particles; also emits five types of γ-particles.

Quantitative Measurement of Radioactivity

Radioactivity can be quantitated in terms of the number of nuclear disintegrations per minute (dpm), the energy imparted to a unit mass of matter by ionizing radiation (rad = 100 ergs/g) or the number of β-particles detected per minute by a detecting device (cpm). Conceptionally, the three commonly used units of radioactivity, *dpm, rad* , and *cpm*, are not equivalent and cannot be interconverted without additional information specific to the radionuclide in question. Rads and the related unit *Roentgens* (r = 0.87 rad) are used primarily in the health and safety fields for measuring human and animal exposure to ionizing radiation, X-rays,and so on, and are not used in the sort of biochemistry discussed in this chapter. Disintegrations per minute (dpm) and counts per minute (cpm) are related to the efficiency of detection by the relationship

$$cpm = dpm \text{ x efficiency of detection} \qquad (7\text{-}3)$$

Only if the efficiency is 100% (which is unlikely in almost all cases) will cpm = dpm. The classic unit of radioactive decay, the *curie* (Ci), was originally defined as the rate at which 1 gram of ^{226}Ra decays; the modern definition is the quantity of radioactive substance in which the decay rate is 3.700×10^{10} disintegrations per second or 2.2×10^{12} dpm. A curie is a great deal of radioactive material; more frequently used quantities are the *millicurie* (mCi) and *microcurie* (μCi). *Specific activity* denotes the amount of radioactivity per amount of compound; it indicates the fraction of the total molecules that contain a radioisotope. Specific activity is expressed in terms of curies per gram, millicuries per millimole, cpm per microgram, and so on.

There are three common methods for monitoring the decay of biochemically important isotopes:

1. *Autoradiography* involves the exposure of X-ray film to radioactive compounds that are embedded in a solid matrix such as an electroporesis gel. The film is placed over the gel for a predetermined time and then developed. Disintegrations are indicated by the presence of silver grains formed in the emulsion due to the passage of high-

energy particles. Although the grain intensity is proportional to the radioactivity level, the former is not easily quantitated and this method is most often used qualitatively.

2. *Geiger-Müller counting* monitors the electrons ejected by normally stable atoms (in a counting gas such as butane) upon collision with β-particles emitted by an unstable atom. The ejected electrons are accelerated toward a positively charged electrode (anode) in a Geiger-Müller tube. The accelerated electrons collide with additional atoms and generate a shower of electrons which is detected as a "count" or discharge when they strike the anode.

The residual positive cations are similarly drawn to the cathode. This method can give quantitative results for high-energy β-emitters such as ^{32}P but not for low-energy ones such as ^{3}H or ^{14}C. Tritium cannot be detected by Geiger-Müller counting. The Geiger counter is the standard tool of the uranium prospector.

3. *Liquid scintillation counting* is the most sensitive and versatile method for monitoring the disintegration of both beta and gamma emitters. Levels as low as 1 billion "hot" atoms can be detected and samples can either be liquids or finely divided solids suspended in a liquid scintillation mixture. In either case, the sample to be counted is suspended in a liquid scintillation "cocktail" composed of a solvent and one or more solutes known as "scintillators" or "fluors." The function of the scintillator is the conversion of the kinetic energy of a nuclear particle to light energy which can be detected by the photocells of the counter. Some of the kinetic energy of a β–particle emitted by an unstable atom is converted to solvent excitation energy, which is transferred in turn to a solute molecule that reemits the energy as fluorescence.

Scintillation Events[1,2,3]

The kinetic energy of the β-particle is transferred by collisions with solvent molecules. The more energetic β-particle can transfer more energy and consequently excite more solvent molecules, as shown in Figure 7-2.

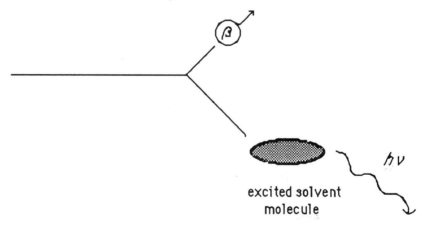

Figure 7-2 Collisional excitation of a solvent molecule

The most common solvents are aromatic liquids such as benzene, toluene, and p-xylene which can be converted to excited states that are stabilized by intermolecular associations; for example, excited dimers (excimers) can hold on to this excitation energy long enough for it to be transferred between solvent molecules.

The successive formation and dissociation of excimers plays an important role in the transfer of energy between solvent molecules.

Exciplex with random solvent molecules   **New exciplex and dissociated solvent pair**

The excitation can ultimately be transferred to the solute by a radiationless energy transfer process (Förster transfer). This process requires overlap of the absorption spectrum of the primary solute (energy acceptor) with the emission spectrum of the solvent (energy donor). The mechanism is effective over several molecular dimensions at a rate much faster than the relaxation rate of the solvent excited state. If more than one solute is used, the one with the highest fluorescence energy level is called the primary solute and the others secondary solutes. The primary solute is more concentrated (10 mM) and the secondary solutes are more dilute (0.1 mM). The purpose of the secondary solutes is to shift the emission spectrum of the system to longer wavelengths to ensure maximum spectral response of the photodetector tubes. Such compounds usually have emission spectra in the region of 415 to 430 nm. Primary solutes are usually fluorescent aromatic compounds such as naphthalene, or terphenyl or substituted polyaromatic oxazoles such as PPO, while secondary solutes are more conjugated analogs such as POPOP. PPO absorbs at 365 and 380 nm and POPOP at 420 and 441 nm.[2]

PPO

POPOP

The energy of the excited singlet state is lost by fluorescence, thermal relaxation or by dropping down in energy to the triplet state, which ultimately phosphoresces. Only about 5% of the β-particle kinetic energy is converted to light, the bulk being dissipated as molecular vibrational energy. Excited solvent and solute molecules spend varying amounts of time before transferring their energy or relaxing back to their ground states. The Förster energy transfer from solvent to solute involves a transfer time of about 0.7 x 10^{-9} s which is much shorter than solvent singlet (fluorescence) relaxation time (3.3 x 10^{-8} s).[3] Some solvent molecules, initially raised to the singlet state, may undergo intersystem crossing to the triplet state which has a much longer (phosphorescence) relaxation time (10^{-5} to 10^{-3} s). As a consequence, the light produced by a β- particle racing through the scintillation cocktail is a sustained pulse rather than an instantaneous flash; the β-particle has lost its excess kinetic energy long before the light dims below the detectable level.

Photodetection [1,2,4]

The light resulting from an excitation event is measured by a pair of photomultiplier (PM) tubes, and when both tubes respond to a simultaneous (±10 ns) light signal, the summed output signal from the PM tubes is analyzed by a pulse height analyzer, designed to convert the area of the unsymmetrical PM tube output curve to a symmetrical pulse with a height proportional to the number of excitation events. Additional discriminator circuitry permits the counting of the number of such pulses occurring between preset peak height maxima and minima. Most spectrometers have three discriminator channels, so counts can be segregated into low-, medium- and high-energy signals.

A photomultiplier (PM) tube is a marvelous device for detecting ultralow light levels. When light strikes the surface of the PM tube photocathode, a number of electrons are released from the cathode surface. The electrons are accelerated through a series of negatively charged focusing rings (dynodes) toward the anode. At each dynode the number of released electrons is increased three- to fivefold so that a single electron released from the photocathode may result in 10^6 electrons ultimately striking the anode. To avoid false counts, two PM tubes are used to measure simultaneously any light pulse arising from scintillation events. Simultaneous false counts are unlikely, so that only simultaneous counts from both tubes are selected by the spectrometer circuitry. The output from both tubes is checked by a *coincidence counting* gate, which passes through the summed signal from both tubes only when both are registering an output within ± 10ns of each other. Nonsimultaneous signals due to random thermal "noise" are discarded.

Different scintillation events have different intensity vs. time curves; particles from nuclear disintegrations, with high energy, produce a greater number of photons than particles with lower energy and the area under the intensity vs. time curves will be larger. The greater number of photons detected by the photomultiplier tube will be converted to an amplified number of electrons to produce a voltage pulse, which is measured by a scaler. The *pulse height analyzer* can be set to discriminate between different voltage pulses for different scintillation events. The device is set by means of two discriminators, which set the lower and upper levels. Voltage pulses within this range are counted in that channel, those below or above the discriminator settings are rejected. Most spectrometers have three fully independent and simultaneous channels of pulse height analysis; the channel "windows" may be positioned in overlapping, adjacent or separated segments by adjusting the discriminators. Three windows implies six discriminator limits: A to B, C to D, and E to F; these limits can be set either to optimize the counting efficiency for single-labeled tritium or carbon-14 or to permit the counting of double-labeled samples containing both nuclides.

The difference in β-particle energy is the key to double-label counting. For example, the average β-particle energy value E_{ave} for ^3H is 0.0057 MeV with an E_{max} of 0.0186 MeV, while the corresponding values for ^{14}C are 0.045 and 0.156 MeV, respectively. Double-label counting with ^3H and ^{32}P is even easier since E_{ave} for the phosphorous isotope is 0.70 MeV. The ratio of ^3H to ^{14}C activity in the same sample can be determined if the first-channel discriminators, A and B, are set at 0.005 MeV and 0.02 MeV respectively, and the second channel discriminators, C and D, are at 0.02 MeV (as is B) and 0.16 MeV. Hence all counts in the second channel are due to ^{14}C, but the counts in the first channel are due to both isotopes. A ^{14}C standard is used to obtain the ratio R = channel 1 cpm/channel 2 cpm in the absence of ^{3}H counts in channel 1. This ratio is used to estimate the ^{14}C counts in channel 1 for the dual-labeled sample:

$$^{14}C \text{ cpm in channel 1} = R \times \text{cpm in channel 2} \qquad (7\text{-}4)$$

The total counts for ^{14}C is the sum of ^{14}C cpm in both channels and the total for ^{3}H is the total counts in channel 1 minus the ^{14}C cpm for channel 1.

Consider the following numerical example:

For 400 cpm in channel 1, 900 cpm in channel 2, and an independently measured R = 0.400 the ratio of ^{3}H to ^{14}C is calculated as follows.

$$^{14}C \text{ cpm} = 900 + 0.400 \times 400 = 1060 \text{ cpm}$$
$$^{3}H \text{ cpm} = 400 - 160 = 240 \text{ cpm} \qquad (7\text{-}5)$$
$$\text{ratio } ^{3}H \text{ to } ^{14}C = 240/1060 = 0.226$$

The calculation above is based on the assumption that no tritium counts appear in channel 2. If the discriminators B and C were set much lower than E_{max} for ^{3}H, this assumption is not valid. However, one can still count ^{3}H and ^{14}C separately if the second-channel discriminator is set to allow only about 10% of the ^{3}H disintegrations within the energy range covered by channel 2; each channel has a gain setting that affords selective sensitivity.

Because ^{3}H has a lower average β-particle energy, it can be measured by a narrow window (A to B) which does not include a very substantial portion of the ^{14}C energy distribution curve. The gain control is an amplifier circuit that determines the number of times the incoming signal is amplified. Each discriminator channel has its own gain control. The gain adjustment for each of the channels can be adjusted so that the number of counts for the nuclei of interest is maximized. For this reason, the discriminator setting for the second channel does not have to be set above the E_{max} value for the isotope observed in the first channel. Figure 7-3 illustrates the β-particle energy distributions (β-spectra) of tritium and carbon-14 as well as reasonable discriminator settings (windows) for accurate double-label counting.

The gain is set high for the lower-energy nuclide ^{3}H, and the window begins at the threshold level of the spectrometer at that gain setting. The gain for the higher-energy nuclide (^{14}C or ^{32}P) is set low and the window begins above the level where efficient counting for a sample of ^{3}H *at that gain* is possible, so that almost all counts in that channel can be attributed to the higher energy nuclide.

The counting efficiency for a particular gain and discriminator setting can be determined by measurement of a standard sample:[2]

$$\frac{\text{observed counts per second}}{\text{known activity } (\mu Ci)} \times 3.7 \times 10^{4} = \% \text{ efficiency} \qquad (7\text{-}6)$$

The discriminator limit A for the lower-energy nuclide ^{3}H is set at the coincidence threshold and the gain is set quite high (50 to 100%). The lower discriminator C for the higher-energy nuclide is set to give minimal counting efficiency for a sample of just ^{3}H with a gain of less than 10%.

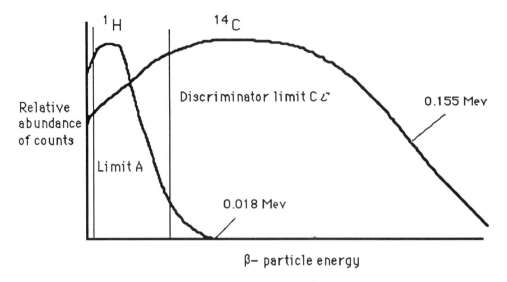

Figure 7-3 Discriminator limits for simultaneous ^{3}H and ^{14}C counting

β-Spectra

As the previous discussion implies, the energy distribution of the β-particles derived from ^{3}H, ^{14}C, and ^{32}P are all different. Each isotope has its own β-particle spectrum. The β-spectra of ^{3}H and ^{14}C can be determined by measuring the counts observed as a function of the upper and lower discriminator levels with a standardized sample of the isotope of interest. The counts per minute can be plotted as a function of the average of the upper and lower discriminator levels, which are systematically changed by a constant value. This is equivalent to sliding a window of constant width across the entire discriminator range and observing the counts at each new window position.

Quenching

Although liquid scintillation spectrometers are extremely sensitive and very reliable, like all measuring instruments they are never 100% efficient; the counts observed per minute is always some fraction of the total number of disintegrations per minute.

$$\text{Efficiency} = \text{cpm/dpm} \qquad (7\text{-}7)$$

Light from a β-decay event may be too weak to be detected by the PM tube or, even if detected, may yield a PM tube output signal below the discriminator threshold level so that it is thrown away as electronic "noise." The sequence of energy transfer processes in liquid scintillation can be interfered with in a number of places so that the maximum photon yield is not obtained.[2,3,5] Such interference, called *quenching,* can result from inefficient energy transfer, energy transfer to nonscintillating solutes, and nontransparency of the liquid scintillator solution. Transfer of energy from excited solvent molecules to scintillator (fluor) molecules is reduced by competitive collisions with other solute molecules, which relax thermally, robbing the excited states of energy without emitting a photon. This is called chemical or impurity quenching and always occurs to some extent unless very special precautions are taken; molecular oxygen and polyhalogenated hydrocarbons are common chemical quenchers. Absorption of light within the scintillator solution is called color quenching; any material absorbing light in the UV or lower wavelength visible will cause color quenching. Bubbles, suspended solids, and the like also physically obscure light from the PM tubes.

The quenching process results in a change in the β-spectrum of a radionuclide. The β-spectrum shifts to lower energy levels and the area under the curve (total light emitted) decreases (see Figure 7-3). This fact is important to both the common methods for determining the counting efficiency: the *channels ratio method* and the *external standard channels ratio method*. If a sample is counted simultaneously in two overlapping windows (A-B and C-D) where the wider window monitors the entire β-spectrum and the narrower only the upper end of the spectrum, the ratio of counts in the two channels (C-D/A-B) should decrease as the degree of quenching is increased. A set of quenched standards, all containing the same known amount of radionuclide (same dpm) but with different known amounts of quencher (different cpm) is used to construct a channels ratio quench correction curve. The samples are of defined percent efficiency, so a plot of efficiency vs. channel ratio (C-D/A-B) can be obtained for any convenient set of discriminator levels. The efficiency for an unstandardized sample can then be obtained from this plot by measuring the channel ratio for the same discriminator settings.

A closely related method employs an external standard of gamma radiation (^{137}Cs) which can be popped into position directly under the scintillation vial in the counting well. The gamma rays easily penetrate the vial and cause scintillation counts identical to those obtained by adding a large amount of radioactivity to the vial. However, external standardization is a nonevasive process and the sample is not chemically altered. In the external standardization method the sample in the vial is counted in two channels twice; a normal count resulting only from internal sources of radiation is first obtained with the external standard in the "out" position, then a count due to both the internal and external sources of radiation is measured with the external standard "in." The difference between these two counts for any one channel is the cpm arising only from the external source; for channel A-B:

$$cpm_{ext,A} = cpm_{in,A} - cpm_{out,A} \qquad (7-8)$$

The ratio of these differences for two channels gives an external channel ratio. When a set of standard quenched samples is used, a plot of the percentage efficiency vs. external standards ratio can be made. This plot can then be used to obtain quench corrections for nonstandard samples. Because the gamma rays excite many high-energy transitions that are registered in the upper channel, the external channels ratio method can be used for samples containing low levels of radioactivity or for highly quenched samples. The conventional channel ratios method is limited to high levels of radioactivity and low levels of quenching because the upper counting channel sees only transitions due to internal radioactivity.

Counting Statistics

Radiactive decay processes are random events; one can never predict when a specific nucleus will disintegrate spontaneously. The equations discussed above for the rate of nuclear decay are statistically valid only for large or high-activity samples of radioactive material measured over long times. A number of short-time activity measurements of a low-activity sample or one with a long half-life will give variable cpm values. The randomness of nuclear disintegrations is the source of this variability. Repeated measurements for a long-lived isotope, such that the total measurement time is a small fraction of one half-life, will give a distribution of cpm values centered around an average cpm value.

A plot of the number of times a given count rate is observed versus the count rate gives a normal or Gaussian distribution curve (see Figure 1-1). This normal curve has the following properties:[6]

1. The standard deviation for N measurements $\sigma = \sqrt{\Sigma(X_i - \bar{X})/N}$ can be approximated by $\sigma = \sqrt{x_{ave}}$.

2. For a large collection of measurements fitting a normal distribution, 68.3% of all measurements fall within ±1 s-value of the mean; 95.5% are within ±2 s, and 99.7% within ±3 s of the mean.

Hence for 4000 measured counts, $s = \sqrt{4000} = 63.2$ and we can say with 68.3% confidence that the true count rate is 4000 ± 63, and with 95.5% confidence that the true confidence is 4000 ± 126 counts. The standard deviation can be reduced either by increasing the number of measurements or by increasing the length of time required for each measurement. If we counted the sample for 1 minute and obtained 1000 ± 32 counts per minute, counting for 10 minutes should afford 10,000 ± 100 counts, which corresponds to 1000 ± 10 cpm, a net improvement in accuracy. Although optimizing accuracy is important, generally, practical considerations control the design of an experiment and if one is working with a large number of samples, most counts are run for 10 minutes or for a maximum count of 10,000, whichever comes first.[6] The accuracy of a count is not the same as the efficiency of a count. Neither increasing the number of measurements nor the measurement time will compensate for quenching.

Isotopic Labeling of Proteins

A wide variety of reagents are commercially available for the *isotopic labeling* of proteins.[7] These involve chemical modifications of amino acid side chains. For example, the thiol groups of cysteine residues can be labeled with "hot" iodoacetamide in the presence of quanidinium salts.[8]

$$-CH_2-SH \; + \; I-*CH_2-\overset{\overset{\text{O}}{\|}}{C}-NH_2 \longrightarrow -CH_2-S-*CH_2-\overset{\overset{\text{O}}{\|}}{C}-NH_2$$

Bolton and Hunters reagent can be used to label free amino groups with ^{125}I under mild conditions.[8]

Tyrosine residues can be labeled with Na ^{125}I under chemical or enzymatic oxidizing conditions.[9]

$$-CH_2-\bigcirc-OH \xrightarrow[\text{oxidizing cond.}]{Na*I} -CH_2-\overset{*I}{\bigcirc}-OH$$

[^{3}H]formaldehyde can be used in reductive methylations of free amino groups in the presence of sodium cyanoborohydride.[10] High concentrations of protein must be avoided since formaldehyde can lead to protein cross-linking through thiol and amino groups.

$$-(CH_2)_4-\overset{+}{N}H_3 \qquad CH_2(OH)_2$$

$$\Updownarrow \qquad\qquad \Updownarrow$$

$$-(CH_2)_4-NH_2 \;+\; CH_2O \xrightarrow{NaBCNH_3} -(CH_2)_4-N(CH_3)_2$$

This latter procedure is used to label cytochrome c in one of the experiments described below.

Sample Preparations

If the radioactive specimen can be solubilized in the scintillation cocktail to give a homogeneous mixture, then the sample can be prepared merely by diluting the specimen with enough cocktail to fill a scintillation vial. Commercially available cocktails are most convenient if only a few hundred samples are to be run. If large numbers of samples need be run or if liquid scintillation counting is a routine laboratory event, then homemade cocktails are more economical, and since there are a large number of formulations,[2,3] one can prepare a cocktail tailored to specific laboratory needs.

Two common requirements are that cocktails (1) be miscible with aqueous specimens, and (2) permit the solubilization of solid specimens such as polyacrylamide gels or biological samples such as blood, tissue, RNA, DNA, etc. The first problem can be solved by incorporating nonionic surfactants (Igepal CA 720, Triton X-100) into the cocktail. These surfactants have both hydrophilic and hydrophobic structural moieties and can solubilize bioorganic solutes in mixed aqueous organic solvents (ethoxyethanol, dioxane, toluene). Aquasol-2 is a commercially available cocktail (New England Nuclear) that is miscible with water over a wide range of proportions. The radioactive sample (0.1 to 5.0 mL) is pipetted into 15 mL of Aquasol-2 and shaken vigorously. The appearance of the sample will change from a transparent liquid in 12% or less water to a stiff gel with water concentrations in the range 20 to 50%. A nonhomogeneous two-phase system will occur at intermediate water percentages (12 to 20%). The two-phase system can be avoided by adding more cocktail or a small amount (0.2 mL) of additional water to such samples. Counting efficencies for ^{3}H and ^{14}C are 30 to 50% and 80 to 95% respectively at water concentrations below 50%.

Semi-solid biological specimens can give homogeneous samples after hydrolysis with NaOH, KOH, or quaternary ammonium bases.

$$CH_3\text{-}(CH_2)_n\text{-}N(CH_3)_3{}^+, \quad RO^-$$

The latter are especially powerful solubilizing agents and are used in a number of commercially available cocktails (NCS, Soluene, Digestin).[3] Some samples are rendered soluble only after oxidation with nitric acid or nitric and perchloric acid. Complete combustion to yield labeled CO_2 is sometimes necessary.

Equilibrium Binding Constant Measurements

Radioisotopes permit a very sensitive and nondestructive method for measuring the concentration of molecular species. This is often employed to measure the strength of binding between a radioactive ligand and a specific protein receptor. If the bound ligand has some property that permits its measurement independently of unbound ligand, the equilibrium binding constant can be measured. For example, if the complex is less soluble than the free ligand, one can capture the complex by filtration and measure its radioactivity level. The observed counts per minute (cpm) are proportional to the concentration of the complex. Radiolabeled double-stranded DNA is not retained by nitrocellulose filters unless it is associated with proteins. If a mixture of labeled DNA and a protein is passed through a nitrocellulose filter, the counts per minute recovered from the filter will be proportional to the amount of labeled DNA complexed with protein. This is the idea behind filter-binding assays, widely used to study protein-DNA interactions.[11,12]

Alternatively, a radiolabeled low molecular-weight ligand can pass through a dialysis membrane that would retain ligand bound to a high-molecular-weight protein. This is the basis of the measurement of the binding constant of *lac* repressor with the labeled inducer molecule IPTG (isopropyl β-D thiogalactopyranoside) described in this project.[13] The equilibrium binding constant K for the complex between protein P and ligand L is given as:

$$P + L = PL \qquad \text{and} \qquad K = \frac{[PL]}{[P][L]} \qquad (7\text{-}9)$$

The fraction of protein sites occupied by ligand is designated as ν and is given by the expression.

$$\nu = \frac{[PL]}{[P] + [PL]} = \frac{K[P][L]}{[P] + K[P][L]} = \frac{K[L]}{1 + K[L]} \qquad (7\text{-}10)$$

This equation can be rearranged to give a linear expression relating ν and the total ligand concentration [L] to the binding constant. A *Scatchard plot* of $\nu/[L]$ vs. ν should give a straight line with a slope - K.

$$\frac{\nu}{[L]} = K - \nu K \qquad (7\text{-}11)$$

A more sophisticated analysis[14] than the Scatchard formulation above is required in cases exhibiting positive coopertivity but the following experiment nicely illustrates the use of radioisotopes and equilibrium dialysis to measure binding constants. In this experiment, equal-sized aliquots of the protein sample are allowed to stand overnight in semipermeable dialysis bags suspended in equal volumes of buffer, containing a known concentration of radiolabeled ligand with various concentrations of excess unlabeled ligand. A sample from each dialysis bag is removed and counted before adding the protein to determine the initial counts per minute for ligand, $[cpm]_0$. The free ligand will easily pass in and out of the bag, but once bound to protein, it will be retained within. After equilibrium is reached (2 to 12 hours), a sample of buffer from outside of the bag is removed and counted to give the concentration of free (unbound) ligand, $[cpm]_f$. The concentration of bound ligand $[L]_B$ is calculated from the excess counts inside the bag; this is converted to n by dividing by the protein concentration in mg/mL evaluated by a Bradford protein assay. The following calculation illustrates the method:[13]

Initial total ligand concentration, $[L]_0 = 2 \times 10^{-7}$ M

Initial counts per minute, $[cpm]_0 = 1000$

Equilibrium counts per minute outside of bag, $[cpm]_f = 800$

Equilibrium ligand concentration, $[L]_f = (800/1000) \times 2 \times 10^{-7}$ M $= 1.6 \times 10^{-7}$ M

Equilibrium counts per minute inside of bag, $[cpm]_B = 4000$

Ligand concentration inside of bag, $[L]_i = (4000/800) \times 1.6 \times 10^{-7}$ M $= 8 \times 10^{-7}$ M

Concentration excess due to bound ligand, $[L]_B = (8 - 1.6) \times 10^{-7}$ M $= 6.4 \times 10^{-7}$M

Protein concentration (Bradford), $[P] = 10$ mg/mL

Concentration bound ligand per mg/mL, $[L]_B/[P] = 6.4 \times 10^{-8}$ M mL/mg

$v/[L] = (6.4 \times 10^{-8}$ M mL/mg $)/ 1.6 \times 10^{-7}$ M $= 0.4$ mL/mg

If the initial ligand concentration is allowed to range from 2×10^{-7} to 4×10^{-5} M, then a plot of $v/[L]$ vs. v will give a straight line with a slope (-K) equal to 1.3×10^{-6} M for IPTG with lac repressor.[13]

In Vitro **Translation Assay**

Protein synthesis involves the incorporation of free amino acids into protein that can be selectively precipitated with strong acid. The progress of protein biosynthesis can be determined by following the conversion of acid-soluble ^3H amino acids into acid-insoluble protein.[15] Trichloroacetic acid (TCA) is the reagent of choice. The assay mixture is usually a crude cellular "soup" containing ribosomes, tRNA, and the enzymes necessary for protein synthesis. This is supplemented with an energy mix of nucleotide triphosphates and an inhibitor of RNAase. The energy mix contains ATP, GTP, phosphoenol pyruvate (PEP), and pyruvate kinase, which pays the energy bill for the incorporation of amino acids into proteins. The formation of aminoacyl-tRNA's, the ribosome initiation reaction, and each chain elongation reaction require the cleavage of "high-energy" phosphate bonds. The protocol described in the experimental section is based on the use of a commercial translation system (NEN Research Products) which utilizes a lysate prepared from rabbit reticulocytes. The lysate has been treated with micrococcal nuclease to remove endogenous mRNA. Hence it is a protein synthesis system largely dependent on added mRNA, so that the system can be used to demonstrate the specific requirements of individual mRNA sequences.

There are two common complications to "amino acid incorporation" assays. First one must be assured that acid-insoluble counts are due to protein and not to insoluble aminoacyl-tRNA. The latter compounds are hydrolyzed by heating briefly in dilute trichloroacetic acid, whereas proteins are not. The protein synthesis is terminated by adding enough TCA to bring the solution to about 5%, then any aminoacyl-tRNA is decomposed by heating at 90° for 10 minutes. The cooled solution is then filtered and washed to remove all soluble counts. After washing, the filter, with adhering protein, is transferred to a scintillation vial, combined with cocktail, and counted. A second complication involves the presence of endogenous ribonucleases (RNases) which will hydrolyze any mRNA added to the translation system. The commercial system was developed to minimize this problem but precautions must be taken to keep out extraneous RNases. Fingerprints and nonsterile water are common sources of this problem. Gloves should be worn during the assay procedure and sterile plasticware, including pipette tips, must be used. Commercial reticulocyte lysate (NEN Dupont, Sigma, BRL) is very unstable at room temperature and should be subdivided into small portions that must be kept frozen until the last possible moment, with any remaining lysate frozen immediately after use.

Safe Handling of Radioisotopes

Low-activity β-emitters offer no significant health hazards unless ingested or inhaled, and sensible teaching laboratory procedures will keep the radiation exposure well below the "maximum permissible dose." However, all radiation is potentially harmful and human exposure should be minimized. The harmful consequences of exposure to radiation often follows years after the careless act that led to it, and the observation "that nothing happened" after having been inadvertently exposed to radiation from spills, improper storage or dangerous lab practices must never lull one into a false feeling of security. All such lab irregularities must be carefully logged and reported in writing to the university radiation safety officer, and if that officer deems it necessary, to the appropriate state or federal officials. It is important to remember that the use and disposal of radioisotopes is regulated by law.

A detailed list of safety precautions is given below. *The author recommends that all students be given an exam based on these safety rules which they must pass before beginning any of the projects described here.*

THE FOLLOWING RADIATION SAFETY GUIDELINES MUST BE FOLLOWED IN ALL EXPERIMENTS.

1. **Work only in a specially designated area, preferably under a hood. Absorbant blotter paper or lab bench covering must be spread over the work area to facilitate cleaning and monitoring of spilled material. The area must be clearly marked with the appropriate safety signs.**

2. Wear disposable plastic gloves and a lab coat and always wash your hands thoroughly before leaving the laboratory. The lab coat never leaves the lab.

3. Never, never pipet by mouth!

4. Label all glassware and equipment with yellow "Radiation Hazard" tape and store and wash such "hot" equipment separately from "cold" stuff. Do not discard glassware or bottles marked with the tape in the trash.

5. Do not allow "hot" material, glassware, or equipment to accumulate in the laboratory. Decontaminate, wash, or discard as soon as possible.

6. Liquid waste must be put into a designated storage vessel ultimately to be repackaged for discard by the teaching personnel. Do not pour liquid waste down the drain!

7. All solid waste (paper, broken glass, resins, and chemicals) must be wrapped in appropriately labeled plastic bags and stored in a clearly marked and tightly sealed trash container. This waste will also be repackaged for disposal.

8. All radioactive compounds must be stored in sealed containers in a locked, clearly labeled refrigerator or cabinet. Each sample must be clearly labeled with isotope, compound, specific activity, date of assay, and name of last user. Store high-decay-energy isotopes such as ^{32}P in foil-wrapped heavy glass bottles.

9. The person responsible for the lab must keep an accurate logbook recording every use of radioactivity in the laboratory, the number of microcuries removed from storage, the levels of radiation observed when lab benches are swabbed, and any unusual events accompanying the use of radioactive compounds.

10. Small spills must be cleaned up with paper towels or sponges which can be rinsed in a "hot" sink or cleaning bucket. An aerosol foam spray cleaner for degreasing "hot" spills is commercially available. To be sure the spill is cleaned up, the teaching personnel must monitor the area by taking swabs and counting them by liquid scintillation (for Tritium or low levels of carbon-14) or with a Geiger counter for other isotopes. This precaution is not only for safety- it also minimizes the background count.

11. Large spills and accidental ingestions must be reported immediately to the instructor and to the university radiation safety officer; the amount and type of radioisotope must be specified. This report must be typed, signed, and dated.

EXPERIMENTAL SECTION

Liquid Scintillation Counting

Different model liquid scintillation counters (LSCs) have different control and data acquisition technology. LSCs are very robust instruments and many older models are often found in teaching laboratories. The following procedure is for an older vintage model, with manual discriminator settings and a teletype for data output. More modern counters are controlled by built-in microprocessors, which display data differently.

Scintillation vials fit into the cylindrical loops on an endless conveyor belt or into trays that move along a track. A group of samples should be placed together with an "on" control tower preceding them and an "off" tower after the last one. A single sample can be placed directly over the sample elevator position (Figure 7-4). The vials should move smoothly along the conveyor and should gently drop down into the counting well and return within the time set by the "preset time" control. Any obvious irregularities in sample changer function should be reported immediately to the instructor. **Do not attempt to force the elevator door open or closed.**

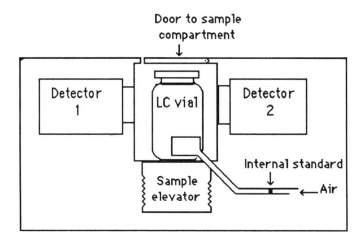

Figure 7-4 Schematic of a liquid scintillation counter

Radioactive sample spills in the spectrometer will necessitate an expensive cleaning to remove this source of background counts. Moreover, spilled samples can gum up the sample changer or elevator assembly. All vials must be clean, free from cracks, and tightly stoppered. **Never pour or pipet liquids into sample vials in the sample compartment.**

Each of the following experiments can be done by one or more students. If several experiments are planned, it is necessary for the instructor to go over all the protocols so that the counting requirements for each part can be efficiently met. Overnight counting is required for projects involving extensive counting with minimal adjustment between counts.

Determination of a ^{14}C β-Spectrum

An unquenched standardized ^{14}C reference sample should be positioned over the sample elevator. The first channel upper discriminator level should be set at its maximum setting and the lower discriminator to 95% of its maximum setting. The preset time should be 1 minute and the automatic standardization should be "out." The course gain control is set at 1% and the counting begins when the spectrometer is turned from the "stop" to the "repeat" mode. The counts observed in duplicate runs at progressively higher course gain settings are recorded; the optimum gain level is reached when the count just exceeds background levels. This adjustment assures that the β-spectrum just fills the spectrometer channel window; a gain level of slightly less than 10% should be optimal. After the gain is adjusted, the lower discriminator level is dropped to its minimum value so that the cpm corresponding to a fully opened channel window is measured. The upper discriminator level is brought down to 2% of its maximum range and the cpm is measured. Both discriminator values should then be systematically stepped across the range in unison so that the upper and lower levels differ by only 2% of the maximum range at any one window position; the cpm is recorded as a function of the average discriminator value at each new window position. A plot of observed counting rate (cpm) versus average discriminator setting should resemble the β-spectrum depicted in the background section.

Automatic Standardization

The external radioactivity source is in a metal capsule that moves from the "out" position behind 12 inches of lead to the "in" position at a precisely fixed point in the optical chamber of the detector. The capsule is literally blown into place by pneumatic forces, like a pea in a soda straw. With the automatic standardization control in the "auto" position, the sample will be counted twice, once with the external source "out" and once with it "in." The cpm measured in the first two channels is recorded both times by a slave printer or teletype. The first channel is set with a gain of about 50% and the second of about 10%. The first channel window should range from just above the threshold to its maximum value; the second should start above the cpm maximum of the ^{14}C β-spectrum and extend to its maximum value. A correlation curve between the count rate of the external γ–ray source and the counting efficiency is obtained using a set of quenched ^{14}C standards (known dpm). These calibration plots are valid only for the specific radionuclide used and the channel settings indicated.

Lac Repressor-IPTG Binding Constant[13]

The following experiment illustrates the use of equilibrium dialysis to measure the binding constant of the inducer IPTG to a crude repressor protein preparation. Ten 125-mL erlenmeyer flasks containing 20 mL of the following buffer are cooled to 4°C.

0.2 M KCl

0.01 M TRIS-HCl (pH 7.6)

0.01 M Mg(OAc)

0.1 mM EDTA

0.1 mM Dithiothreitol

To each is added 10 μL of ^{14}C-labeled IPTG (2×10^{-4} M) to give a final concentration of 10^{-7} M. Then aliquots of unlabeled IPTG from a 1.0 mM stock solution are added to each flask. The total IPTG concentration should span the range from 2×10^{-7} M to 4×10^{-5} M over the 10 flasks. The starting concentration of each flask is determined by removing 100 μL for counting. Into each flask is placed a small (20 x 18 mm) dialysis bag containing 200 μL of crude *E. coli* lysate, containing *lac* repressor. The lysate is prepared by thawing packed frozen cells in a two fold volume of the above buffer *minus the EDTA* but with 5% glycerol and 5 μg/mL bovine pancreas DNase I. After 2 hours or after overnight, 100-μL aliquots from outside and inside each bag are diluted with Aquasol-2 and counted. The protein concentration in each bag is determined separately by the Bradford method (see Chapter 4).

The experimental setup and a Scatchard plot of sample data are illustrated in Figure 7-5.

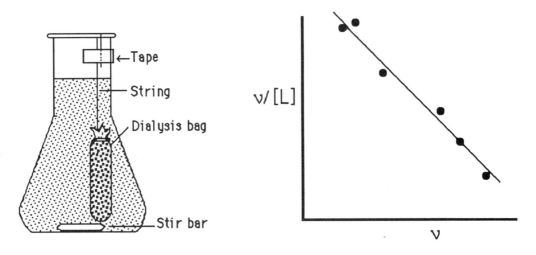

Figure 7-5 Equilibrium constant measurement setup and sample data

In vitro Protein Synthesis[15]

The following is a challenging experiment illustrating the use of radioisotopes to follow the progress of the biosynthesis of proteins. All the necessary materials and supplies (including a small-volume adjustable automatic pipet) must be placed ready before the lysate aliquot is removed from the freezer (-70°).

Material and total amount used	Amount used per assay
Reticulocyte lysate (100 μL)	10 μL
Control mRNA in distilled water	0.5 to 2 μL
50 mM Magnesium Acetate (0.5 mL)	0.5 μL
1.0 M Potassium acetate (0.5 mL)	2 μL
Translation cocktail containing:[16]	5.5 μL
Spermidine	
Creatine phosphate	

Dithiothreitol

Guanosine triphosphate

HEPES buffer

Radiolabeled leucine or methionine   5μL

Enough salts, labeled amino acid, and cocktail for five assays are combined in a "premix" (65 μL total) and stored in an ice bath. Variable amounts (0, 0.5, 1.0, 1.5, and 2.0 μL) of mRNA are added to five *sterile* 1.5-mL microfuge tubes and diluted with enough *sterile* water to bring the total volume to 2.0 μL. The premix (13 μL) is added to each tube, followed by 10 μL of freshly thawed lysate. The tube is capped and a vortex mixer is used to insure rapid complete mixing. One microliter of each assay mixture is removed and spotted onto a 1-cm^2 square of Whatman 3mm filter paper. The paper should not touch the desktop or any suface, but should be mounted on a pin above a clean sheet of Styrofoam. The protein in the aliquots transferred to paper is precipitated by adding 50 μL of 10% Trichloroacetic acid (TCA) to the paper square and allowing it to dry. The tubes are transferred to a 37°C water bath and incubated for 1 hour.

At 10-minute intervals 1-μL aliquots are removed and spotted onto paper squares as before. The radiactive TCA-fixed paper squares are air dried in the hood and placed in scintillation vials to be counted with 10 mL of Aquasol-2. Do not remove the paper from the scintillation vial before counting. The data are plotted as counts vs. μL added mRNA (a mRNA titration curve as in Figure 7-6) and as counts vs. minutes of incubation (a kinetics curve as shown in Figure 7-6).

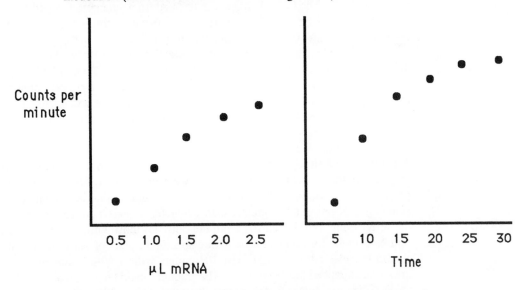

Figure 7-6 *In vitro* protein synthesis sample titration and kinetics data

[³H] reductive methylation[10]

The class should work in groups of five or six on this portion of the experiment. One "hot" column can be used if it is adequately rinsed to a constant count between runs. Some isotopic label will hang up in the column during each run. The sodium cyanoborohydride should be recrystallized at least two days before the protein methylation is begun; the recommended procedure follows.

THIS PORTION OF THE EXPERIMENT MUST BE DONE IN A GOOD HOOD.

Reagent grade $NaB(CN)H_3$ (11 g) is dissolved in 25 mL of acetonitrile and the solution clarified by low-speed (benchtop) centrifugation. The $NaB(CN)H_3$ in the supernatant is then precipitated by adding 150 mL of dichloromethane and storing the suspension overnight in the cold room (4°C). The white precipitate should be recovered by filtration and dried under reduced pressure at room temperature. Aqueous solutions of the reducing agent must be prepared on the day of use.

$NaB(CN)H_3$ LIBERATES HCN IN AQUEOUS ACID!

Unlabeled formaldehyde is combined with a small portion of [^{3}H]formaldehyde to give 1.0 mL of 10 mM formaldehyde with a specific activity of 2×10^8 dpm/ mmole; this corresponds to a 100-fold dilution of radioactive formaldehyde (10 mCi/mmole, New England Nuclear) into "cold" formaldehyde. The resulting diluted isotopic solution is said to be "hot."

Reductive methylation of cytochrome c is carried out in a 1-mL reaction mixture containing 3 mg of protein, 10 mM "hot" formaldehyde, 10 mM $NaBCNH_3$ and 50 mM HEPES buffer (pH 7.5), in a sealed microcentrifuge tube standing upright in a small clearly labeled beaker covered with Parafilm. The reaction mixture was incubated overnight at 4°C. The labeled protein can be freed from unreacted formaldehyde and low-molecular weight reaction byproducts by gel filtration through a 20 to 30 cm G-75 column eluted with 10 mM phosphate buffer, pH 7.0.

The protein will come off as a "hot" band eluting after the void volume; unreacted labeled formaldehyde as well as impurities in the formaldehyde and labeled reaction byproducts (formate, methanol , etc.) should all be retained by the column and elute after the protein. If the formaldehyde serves to cross-link the protein, higher-molecular-weight oligomers will come off before the monomeric protein, but they may not be well resolved from each other or from the monomer.

One milliliter fractions eluted from the column should be diluted to 15 mL with Aquasol-2, shaken vigorously and counted. To speed up the counting process, the "reject" control should be set at about twice the average background; this control permits the selection of a minimum counting rate during the first minute for continued counting at a "preset time" greater than 1 minute. All vials producing counts less than the reject limit are counted for 1 minute only; all those with counts above this limit will be counted for a longer time period specified by the preset time. The preset time should be selected to give at least 10^5 counts with a 50-μL sample of filtered but unchromatographed labeled protein. The column should be prepared, loaded, and eluted as described in Chapter 6, except that the column should be set upright within a large plastic basin to catch any spilled liquid during the chromatographic run. Before counting, the protein solution should be filtered through a Millipore membrane or a disposable ceramic filter.

The used column must be washed repeatedly with excess buffer until the eluted fractions give close to background counts. Then the column should be labeled with yellow radioactivity warning tape and stored in an appropriate cupboard in the cold room. If the column is to be stored for more than a couple of days, it must be protected with 0.1% Na azide in the storage buffer. The class may reuse the same column for several runs-through of this experiment, provided that all the protein solutions added to the column have been filtered and the column washed repeatedly with excess buffer between runs. It may be necessary to turn up the "reject" level for the later runs, as some label does adhere to the column and bleeds off slowly.

Labeled cytochrome c prepared by this method can be cross-linked by glutaraldehyde or with Traut's reagent as in Chapter 6, and the resulting family of "hot" oligomers (multiples of 13,000 MW) can be resolved by gel-filtration on Sephadex G-75 or G-100. This permits separations at much lower protein concentrations than required by spectrophotometric detection methods.

PROJECT EXTENSIONS

1. Students may find it informative to determine the β-spectra of a number of quenched standardized samples and to determine the channel ratios corresponding to various discriminator settings.

2. The effect of protein concentration, pH and formaldehyde concentration on the amount of label incorporation and on the amount of protein-protein crosslinking can be determined.

3. The ^{14}C-labeled cytochrome c can be cross-linked with glutaraldehyde or Traut's reagent as in Chapter 6 and the covalently linked oligomers separated on a gel filtration column. The fractions eluted from the column can be counted and the elution profile obtained at protein concentrations much lower than that possible using colorimetric methods.

4. The binding of purified *lac* repressor with ^{32}P-labeled *E. coli* chromosomal DNA can be studied by the nitrocellulose filter binding method.

DISCUSSION QUESTIONS

1. Given a 100-μL sample of ^{3}H-labeled amino acid with a specific activity of 10 mCi/ mmole), using unlabeled amino acid and water, how would you prepare 10 mL of 10 mM amino acid with an activity level corresponding to 20,000 dpm/mL?

2. Why does degassing scintillation samples and cocktail improve the efficiency of counting?

3. The percent relative standard deviation (% RSD) is computed automatically by modern liquid scintillation counters. This statistical term is defined as

$$\% \text{ RSD} = s/cpm \times 100 = (\sqrt{cpm}/cpm) \times 100$$

and describes the accuracy of the measurement solely in terms of the number of counts (i.e., 10,000 counts has an s value of 100 and a % RSD of 1.0). Calculate the % RSD for 1000, 5000, and 40,000 counts.

4. The MKS system of units defines a basic unit of radioactivity, called the **becquerel,** as one disintegration per second. One curie is equal to how many becquerels?

REFERENCES AND NOTES

1. J. B. Birks, *The Theory and Practice of Scintillation Counting*, New York: Pergamon, 1964.

2. T. G. Cooper, *The Tools of Biochemistry*, New York: Wiley-Interscience, 1977, pp. 65-135.

3. C. T. Peng, "Sample Preparation in Liquid Scintillation Counting," Radiochemical Centre Ltd., Amersham, Bucks, England, 1977.

4. Packard Instruments, "Automatic Tri-Carb® Liquid Scintillation Systems, Instruction Manual 2129," San Francisco, 1973.

5. D. L. Horrocks, *Nucl. Instrum Methods*, **1975**, *128*, 573; E. T. Bush, *Anal. Chem.*, **1963**, *35*, 1024.

6. B. D. Hames and D. Rickwood, *Gel Electrophoresis of Proteins: A Practical Approach*, Washington DC: IRL Press, 1981, pp. 265-277.

7. T. Inagami, *J. Biol. Chem.*, **1965**, *240*, PC 3453.

8. R. E. Bolton and W. M. Hunter, *Biochem. J.*, **1973**, *133*, 529-539.

9. L. Birnbaumer and T. Swartz in *Laboratory Methods Manual for Hormone Action and Molecular Endocrinology* (W. T. Schrader and B. W. O'Malley, eds.), Houston: Houston Biological Assoc., 1981, pp. 3-1 to 3-9.

10. N. Jentoft and D. G. Dearborn, *J. Biol. Chem.*, **1979**, *254*, 4359.

11. C. P. Woodbury and P. H. von Hippel, *Biochemistry*, **1983**, *22*, 4730.

12. J. M. Bailey, *Anal. Biochem.*, **1979**, 93, 204.

13. J. H. Miller, *Experiments in Molecular Genetics*, Cold Spring Harbor, NY: Cold Spring Harbor Laboratory, 1972, pp. 363-366.

14. I. M. Klotz, *Acc. Chem. Res.*, **1974**, *7*, 162.

15. D. Perlman and H. O. Halvorson, *Cell*, **1981**, *25*, 525.

16. The author recommends the use of commercially available reticulocyte lysate translation systems. DuPont (NEN) and BRL both offer this product.

Chapter 8

Disc Electrophoresis

In this experiment a mixture of proteins is separated by polyacrylamide gel (PAG) disc electrophoresis and the number of protein components and their approximate abundances are noted. The components are visualized using Coomassie blue dye.

KEY TERMS

Coomasie blue staining   *Electrophoresis*
Electrophoretic mobility   *Frictional drag*
Furguson plots   *Gel density*
Moving boundary electrophoresis   *Separating gel*
Stacking gel   *Trailing ions*
Zone electrophoresis

BACKGROUND

Proteins and nucleic acids are electrically charged and therefore will move in an electric field. They will be attracted to electrodes of opposite charge and will migrate through a solvent-filled porous matrix toward such a pole. This process is called *electrophoresis*. There are two types of electrophoresis: moving boundary and zone electrophoresis. In *moving boundary electrophoresis*, the macromolecule of interest is present throughout the matrix. This method is used primarily to characterize the physical chemistry of the electrophoretic process, not for purification of macromolecules. Although this is the original version of electrophoresis, it is now rarely used. In *zone electrophoresis*, the sample is applied as a layer or spot and electrophoretic forces are employed to separate the sample into different components with different mobilities. There are different variations to zone electrophoresis, depending on the matrix used and how the sample is applied and, most important, on the specific mechanism by which separation is achieved.

Discontinuous pH or "disc" electrophoresis is a powerful method for determining the homogeneity of a protein preparation.[1-3] Using this method, we can separate proteins according to their differences in electric charge and their molecular dimensions. The electrophoresis is carried out in a hydrated porous matrix of polyacrylamide and the mechanical rigidity of this gel eliminates convection and vibrational disturbances, permitting greater resolution than is possible in a more fluid buffer system. The resolution is greatly dependent on the buffer composition, pH, and gel porosity. The procedure outlined in this chapter is one of general utility for resolving the components of most mixtures of anionic proteins at convenient pH ranges, but additional effort will be required to optimize resolution for any particular system.

Polyacrylamide Gels

The gel used in this experiment is prepared by free-radical polymerization of acrylamide, with occasional cross-linking with BIS (N, N'-methylenebis-[acrylamide]). The density of the gel network (or pore size) depends on the degree of polymerization and the amount of cross-linking. The latter is determined by the ratio of BIS to acrylamide concentration. Gels will form at over 30% acrylamide, and such "hard" gels will discriminate between small peptides of 2000 D molecular weight. On the other hand, 2% gels (which contain 0.5% agarose for mechanical strength) will discriminate on the basis of molecular size in the range 10^5 to 10^6 D. Most gels contain between 5 and 15% acrylamide.

$$CH_2=CH-\overset{\overset{\textstyle O}{\|}}{C}-NH_2 \qquad\qquad CH_2=CH-\overset{\overset{\textstyle O}{\|}}{C}-NH-CH_2-NH-\overset{\overset{\textstyle O}{\|}}{C}-CH=CH_2$$

acrylamide   BIS

The polymerization of the gel begins with the reaction between TEMED (N,N,N',N'-tetramethylethylenediamine) and ammonium persulfate, which react rapidly to generate oxygen free radicals which initiates the free-radical polymerization of acrylamide and BIS. The amounts given in the formulation section give the optimum polymerization rate and extent of polymerization (average polymer chain length) for electrophoresis gels. Too little or too much TEMED leads to soft or irregularly polymerized gels. Ammonium persulfate is very unstable in solution.

Electrophoretic Theory

A protein will experience a force F in an applied field and will move at a velocity V determined by its net charge q, the applied field **E** and the frictional drag f. The *electrophoretic mobility* μ of a protein is the velocity per unit applied field and is a constant characteristic of the macromolecule, the pH of the medium (which determines q) and the porosity of the gel (which determines f).

$$V = \frac{Eq}{f} \tag{8-1}$$

$$\mu = \frac{F}{E} = \frac{q}{f} \tag{8-2}$$

Maximum resolution is possible when the components have widely different electrophoretic mobilities. The pH giving widely differing q values can usually be selected by studying the titration curves of the components. If such curves are not available, systematically varying the pH may be necessary to find optimum resolution conditions. Generally, the buffer pH is chosen so that all the sample components of interest are negatively charged and will migrate down the gel toward the anode. Such buffers are said to be anionic buffer systems. (Cationic buffer systems are occasionally used for protein with very high pI values because they are positively charged at convenient pH levels.) Mobility units are generally expressed as 10^{-5} cm^2 V^{-1} at 0°C. A detailed theory of electrophoresis must take into account the mechanism by which macromolecules orient themselves in an electric field and how the effect of an applied field is imposed upon an otherwise "random walk" diffusion through a porous matrix.[4,5] This diffusion is related to the sizes of the macromolecule and matrix pores. For spherical protein molecules the frictional drag can be approximated by the Stokes equation where η is viscosity and r is the radius of a charged spherical particle:

$$f = 6\pi\eta r \qquad\qquad (8\text{-}3)$$

For nonspherical proteins the mobility cannot be expressed simply in terms of r, but it is still related to molecular dimensions in much the same manner as described for gel filtration (see Chapter 6).

The pore size is inversely related to the concentration of monomeric acrylamide used to generate the polymer polyacrylamide and on the amount of cross-linking between polymer chains. Therefore, the mobility of proteins is sensitive to the concentration of acrylamide in the gel (this is expressed as *gel density* in units of w/v %) and large proteins are more sensitive than smaller ones. It is important to point out that gel density is usually less than 15% and that the bulk of the material within a gel matrix is the solvent water. The gel density is related to an effective pore size, but for any gel there is always a statistical distribution of pore sizes. The interior of a gel may be imagined as a meshwork of random polymer chains cross-linked at intervals and suspended in solvent, through which each macromolecule diffuses, going into and out of a maze of dead-end channels until passage toward the attracting electrode is effected. As the proportion of cross-linker is increased, the pore size decreases until the percentage of bisacrylamide cross linker is about 5% of the acrylamide monomer. Higher cross-linking levels lead to bundles of polymer chains with large spaces between them so that the effective pore size increases again.

Electrophoretic resolution refers to the separation of two bands relative to their bandwidths. Resolution is increased either by increasing the distance between two bands or by making the bands sharper. There is always an optimal gel concentration for the separation of any two proteins of interest, but there is no gel concentration that will give optimal resolution of all the components in a complex protein mixture. The usual research strategy is to vary the gel concentration and other running conditions (e.g., pH) to get good resolution among the components of interest. The optimal gel concentration will depend on the charge and molecular weights of the proteins of interest. This concentration can be arrived at by varying the gel concentration systematically and noting the change in mobility of the proteins of interest. The results of a series of such experiments can be plotted as a Ferguson plot,[6] relating the logarithm of electrophilic mobility to or gel density (grams monomer per 100 mL). Figure 8-1 describes the electrophoretic behavior of two pairs of proteins and suggests the strategies required for the electrophoretic separation of each pair. In part (a), two proteins of nearly identical charge but different size can be separated by increasing the gel density; in (b) two proteins of different size and different average charge can be separated by either increasing or decreasing gel density away from the density where both lines on a Ferguson plot intersect.

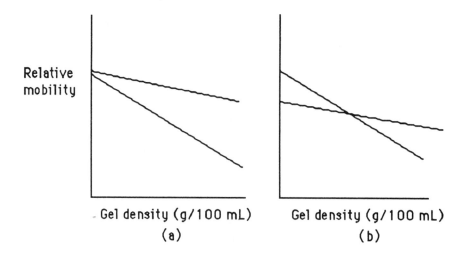

Figure 8-1   Ferguson plots of two pairs of proteins

Disc Electrophoresis

In the disc electrophoresis experiment described here, a protein mixture is subjected to an electric field along a gel rod comprised of two different sections differing in porosity and buffered at different pH values. The two sections are polymerized separately. The bottom or *separating gel* is poured first and contains a higher concentration of polyacrylamide than the *stacking gel*, which is poured after the running gel has hardened. Hence the separation gel is a "small pore gel" and the stacking gel is a "large pore gel." In a system for the separation of anionic proteins (at pH values above their pI), the gels are buffered with amine-HCl buffers with the stacking gel buffer 2 to 3 pH units lower than the pH of the running gel buffer. Different amines may be used in the two gels. It is this pH difference and the choice of buffer electrolytes that is responsible for the high resolution possible with disc electrophoresis. The upper buffer reservoir is usually buffered with the same amine as in the running gel, but the coanion is derived from a weak acid (glycine) with a pK_a slightly higher than the running gel pH. Therefore the upper reservoir is a generator of weak acid anion capable of migrating into the gel. These anions are called the *trailing ions*. If the trailing ion is glycinate, this ion will enter the stacking region when the current is applied. Since the stacking region is at a lower pH than the buffer reservoir, the glycinate anions will be protonated and become less mobile glycine zwitterions in the upper gel. Hence the mobility of anions entering the stacking gel is low.

$$H_2NCH_2CO_2^- + H_3O^+ \rightarrow H_3NCH_2CO_2^\pm + H_2O$$

On the other hand, the chloride ions moving out of the stacking gel have a charge that is independent of pH and will have a high mobility and move ahead of the anionic protein molecules. The protein molecules left behind have to move rapidly to carry as much current as the chloride ions that have preceded them. Because all regions of the gel rod are in the same electrical circuit, the current must be constant for every cross section of the rod. The cross sections are richer in chloride ions toward the bottom of the stacking gel and chloride ions carry most of the current in these sections. The farther up the stacking gel, the lower the chloride ion concentration and the higher the protein mobility must be to maintain a constant current. The stacking gel is quite porous so that protein mobility is not retarded. The electrophoresis process causes the protein mixture to separate into discrete disks in the stacking gel because of the rapid movement of proteins well behind the chloride front and the slower movement as they approach the chloride ions. In other words, as the proteins pass through the stacking gel, they are lined up or stacked as they are sandwiched between low-mobility (glycinate) and higher-mobility (chloride) buffer ions. This gives improved band sharpness. In the stacking gel the disks are too close together to be of analytical use until they migrate into the lower or separating gel, where they are separated by ordinary electrophoresis. The separating gel is less porous and discriminates according to molecular size as well as charge.

Practical   Details

Resolution is also a function of how the gels are cast and how the samples are applied. The sample protein solution is gently layered on top of the stacking gel. The sample application solution should contain sucrose or glycerol to ensure a high density so that it settles evenly onto the stacking gel. Both ends of the gel rod are suspended in buffer reservoirs and the electrodes of a DC power supply are suspended in the reservoirs. The lower buffer reservoir is usually filled with the same buffer as the upper reservoir. The electrodes are placed some distance from the rods to prevent disruption of the rods by the hydrogen evolved as current is passed through the system (see Figure 8-2).

Because most proteins are colorless, the progress of an electrophoresis experiment is monitored by observing the passage of a tracking dye. In this experiment 0.005% bromophenol blue is used. This dye is a small highly anionic molecule and its mobility is higher than that of proteins. The electrophoresis is stopped just before the tracking dye has run out the bottom of the rod. Protein electrophoretic mobility is generally expressed as the relative mobility R_m of the protein to that of the tracking dye. The distances are measured from the top of the running gel.

$$R_m = \frac{\text{distance the protein traveled}}{\text{distance the dye moved}} \tag{9-4}$$

The gel electrophoresis is usually stopped when the tracking dye reaches the bottom of the running gel.

Gel staining after the electrophoresis process is necessary to locate usually colorless proteins in the polyacrylamide gel. The stain used in this experiment is Coomasie Brilliant Blue R, a dye closely related to the dye used in the Bradford protein assay. The gel is removed from the glass tube, used to mount it in the electrophoresis apparatus, and treated with a staining formulation containing both dye and alcohol. The latter is required to precipitate any proteins within the gel matrix so they are not washed out during the staining and destaining processes; this is called fixation. The gel is also stained blue by this treatment, but the dye is only weakly bound to the gel and can be washed away by soaking the gel in circulating destaining solution. Destaining does not remove dye strongly adhering to fixed protein, so that destaining results in a colorless gel marked with sharp blue protein bands.

Gels are formed in molds. Two common types of gel molds are glass tubes, used to form gel rods, and a set of parallel glass plates used to form gel slabs. Gel rods offer the advantage that only small amounts of acrylamide are required per rod and that running parameters such as gel density are easily compared with a series of tubes during the same run. Gel slabs are more useful when large numbers of related samples are to be run under identical conditions. In either case the molds must be clean and care taken in pouring the gels free from imperfections that would distort the protein banding pattern. Reproducibility in the formation of polyacylamide gels is easily accomplished if carefully measured amounts of high-purity reagents are used and if rapid, even mixing and pouring of the gel mixture is stressed. Air bubbles and failure to remove any water or dust completely from the gel-to-gel interfaces are common problems. Disposable pipets should be used since the gel sets up quickly. Disposable gloves and a well ventilated hood are absolutely required for this experiment.

CAUTION: MANY OF THE CHEMICALS USED IN THIS
EXPERIMENT ARE TOXIC!
ACRYLAMIDE IS A NEUROTOXIN AND MUST BE USED
WITH ADEQUATE VENTILATION!

CARE MUST ALSO BE TAKEN TO AVOID
ELECTRICAL HAZARDS.
DO NOT ATTEMPT TO RUN AN ELECTROPHORESIS
EXPERIMENT UNTIL YOUR APPARATUS HAS BEEN
SAFETY CHECKED!

Acrylamide is toxic and must be handled with care. All spills must be cleaned up immediately. Polyacrylamide is not toxic, but incompletely polymerized gels will contain monomer and all gels must be handled only with disposable plastic gloves.

EXPERIMENTAL SECTION

Students should work in groups of three or four if all solutions must be prepared by students. If solutions A, B, D and the staining solutions are prepared ahead of time by the laboratory or stockroom staff, then students should work in pairs.

Formulations

The following stock solutions can be prepared in volume ahead of time. It is preferable that the acrylamide be recently recrystallized from chloroform (70 g/1 L) and stored at -20°C. Solution C must be prepared immediately before use; the remaining solutions can be made several hours ahead of time. Concentrations of gel components are traditionally expressed as weight percent.

Solution A: 3.0 M TRIS, with enough concentrated HCl to bring the pH to 8.9, also containing 0.25 mL of TEMED per 100 mL.

Solution B: 28% acrylamide, 0.73% BIS in water, can be stored at 5°C but must be filtered before use.

Solution C: 0.11% (w/v) ammonium persulfate in water. Must be freshly prepared.

Solution D: 1 M TRIS with enough HCl to give pH 6.9; also contains 0.5 mL TEMED per 100 mL.

Coomassie blue stain: 0.25% Coomassie Brilliant Blue R in methanol: acetic acid: water (5:1:5). Must be filtered before use.

Destaining Solution: 50 mL of methanol and 75mL acetic acid diluted to 1 L with water.

Preparation of the Gel

Glass (5 x 7 x 125 mm) electrophoresis tubes must be washed with cleaning solution or dilute nitric acid (20%), rinsed with distilled water and dried in a dust-free environment. The tubes should rinse free without streaking or beading. The tubes are sealed at one end with Parafilm and held upright in a rack.

The separating gel is made up first and each tube is filled to about 2.5 cm from the top with freshly prepared gel liquid containing the following mixture (for approximately 30 mL of 7.5% gel):

> 3.75 mL of solution A
> 7.50 mL of solution B
> 1.50 mL   of solution C (freshly prepared)
> 17.0 mL of water
> 15-25 µL of TEMED (added last)

The gel mixture should be made up in a small, clean sidearm flask in an ice bath and degassed by aspiration. The polymerization can be retarded during the 5 minute degassing procedure by the low temperature. Since the gel mixture sets up quickly at room temperature, the tubes should be filled in one quick, continuous transfer from a disposable Pasteur pipet, with care to avoid air bubbles or layering effects. A small amount of butanol, saturated with water, is layered on the gel to ensure a flat gel surface. After 30 minutes, a visible interface develops between the water and the gel, and the butanol can be carefully decanted or wicked from the gel with absorbant paper. The gel tubes are returned to an upright orientation and 0.2 mL of stacking gel mixture is carefully added to each tube followed by a water overlayer.

The stacking gel (20 mL, 3.75%) contains the following:

> 2.5 mL of solution B
> 5.0 mL of solution D
> 1.0 mL of solution C (freshly prepared)
> 11.5 mL of water
> 15 µL of TEMED (added last)

The tubes are filled to within 1 cm from the top of the tube. Gels that have not polymerized within 30 minutes have been improperly mixed (usually not enough or too much TEMED).

Electrophoresis Apparatus

The Parafilm bases are removed and the tubes are inserted into the rubber seals and tightly mounted into the holes of the gel electrophoresis cell (see Figure 8-2). Liquid should not be able to flow from the upper to the lower reservoir, and all unused holes should be tightly plugged with rubber stoppers. The bottom of each gel should be nearly flush with the end of the glass tube. Each tube should extend about 2 to 3 cm from the rubber seal.

The buffer solution is made by a 10-fold dilution of a TRIS-glycine stock solution containing 0.05 M TRIS (6.0 g/L) and 0.38 M glycine (28.8 g/L) in distilled water. Buffer is poured into the lower chamber until it is filled to within 2 cm from the top, then the upper chamber is put in place and filled with buffer. The electrodes and the tubes must be covered with buffer. Electrophoresis is carried out by attaching the cathode [(-) pole, black] of a dc power supply to the top electrode and the anode [(+) pole, red] to the bottom electrode. The electrophoresis apparatus is cooled by a water jacket, but cooling is efficient only if the buffer reservoirs are nearly full.

Sample Application

For most pure proteins a concentration of 1 mg/mL will give a darkly stained band. The protein is usually dissolved in buffer containing 0.005% bromophenol blue with enough glycerol or sucrose added to give an appreciably thick solution (25-40% w/v).

The sample is applied with an automatic pipet to the top of the stacking gel with a minimum of agitation. If the sample is dense enough, it will flow down the open portion of the tube to form a thin purple-blue layer on top of the stacking gel. Sample volumes of 20 to 50 μL are optimal.

Electrophoresis

A current of 2.5 mA per tube is recommended. The length of electrophoresis varies with the type of sample but 2 to 3 hours is most common. The progress of the electrophoresis can be monitored with the tracking dye. The electrophoresis is complete when the dye nears the bottom of the tube.

DO NOT TOUCH THE ELECTRICAL CONNECTIONS
WHEN THE POWER IS ON--
A FATAL SHOCK MAY RESULT!

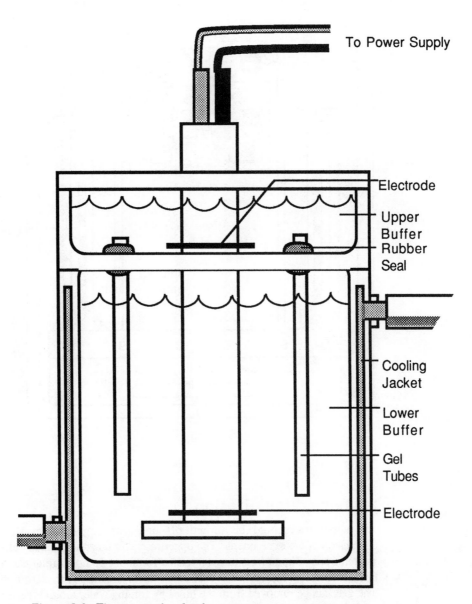

Figure 8-2 Thermostated gel tube apparatus

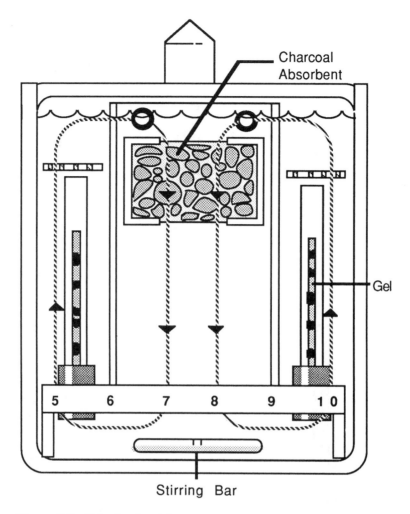

Figure 8-3 Gel tube destaining apparatus

Protein   Visualization

The apparatus should be turned off and **unplugged from the wall**   at the conclusion of the experiment. The upper reservoir must be drained before the gel tubes are removed. The buffers may be reused for about 10 hours total electrophoresis time but they must be stored in the cold room under conditions free from microorganisms. The gels can be removed from the tubes by forcing water around the sides of both ends of the gel. A circular motion with a syringe needle between gel and tube wall while water is flowing into the tube will prevent breaking and minimize handing of the gel. (Fingerprints stain just like protein bands.)

The gel should be slid onto a clean glass plate and any gel below the tracking dye cut off with a razor blade. Then the gel rods should be transferred to a clean destaining tube, with the original bottom end of the gel going in first. (After staining and destaining it is impossible to tell the original orientation of a gel rod unless all have been consistently oriented top-end up--after destaining the tracking dye is gone and the stacking gel has usually fallen off.) The gels are stained in 0.25% Coomassie blue in methanol: acetic acid: water (5:1:5) overnight. Freshly stained gels are entirely dark blue after one rinse with distilled water. After rinsing, the destaining tubes are transferred to a diffusion destainer with a fresh activated charcoal filter. The destaining tubes are plugged into the sockets in the base plate of the destainer and a perforated barrier is placed over the tubes to prevent the gels from floating away (see Figure 8-3). Destaining usually requires 6 to 12 hours but may require longer with strongly stained gels. The protein bands remain blue while the remaining gel returns to its original colorless appearance. The destaining solution contains 7% isopropanol and 7% acetic acid in water. Completed gels can be stored in 7% acetic acid for several months (if kept in a plastic zip-locked bag in a cool, dark drawer).

PROJECT EXTENSIONS

The following suggestions establish connections between some of the projects described so far and will demonstrate the power of disc-electrophoresis to reveal the interesting details about the heterogeneity of protein mixtures.

1. Disc-gels of the fractions obtained in the $(NH_4)_2SO_4$ fractionation of a natural protein mixture (Chapter 4) should show the changes in protein composition of the complex mixture upon salting out certain proteins. The redissolved pellets should be examined in a series of gels. What is the minimum number of proteins remaining in solution in each fraction? Can one make any generalization regarding the electrophoretic properties of the proteins that were salted out most easily?

2. Disc electrophoresis of the cross-linked cytochrome c prepared in Chapter 6 should reveal the complex nature of this mixture. Why would cross-linked dimeric proteins of equivalent molecular weight and approximately equal charge have different electrophoretic mobilities?

3. Disc-electrophoresis gels of radiolabeled proteins can be sliced in a gel slicer and the slices dissolved in a special liquid scintillation cocktail and counted. A good example of this is the "hot" *Xenopus* oocyte nuclei protein mixture of Richmond.[7] The oocytes are labeled with [^{3}H]-L-leucine for a only few minutes before rapid enucleation; the solubilized nuclear protein mixture is analyzed by gel electrophoresis and the radioactivity of each gel slice is counted (cpm). The "hot" fractions are compared to the stained bands in an unsliced gel.

DISCUSSION QUESTIONS

1. Why is electrophoresis done in solutions having low salt concentrations?

2. Two macromolecules with the same molecular weight and charge may still show up as separate bands on an electrophoresis gel. Explain.

3. The progress of linear macromolecules through a porous electrophoresis gel matrix has been likened to a caterpillar crawling through Swiss cheese, a process called reptation.[6,7] Discuss the merits and defects of this poetic description.

4. Plot concentration vs. distance from the top of the stacking gel, for glycinate and chloride ions at the beginning of electrophoresis and after several minutes (four plots).

REFERENCES

1. H. R. Maurer, *Disc Electrophoresis and Related Techniques of Polyacrylamide Electrophoresis*, Second Ed., New York: Walter de Gruyter, 1971.

2. R. C. Allen and H. R. Maurer, *Electrophoresis and Iso-electric Focussing in Polyacrylamide Gel*, New York: Walter de Gruyter, 1974.

3. B. D. Hames and D. Rickwood, eds. *Gel Electrophoresis of Proteins: A Practical Approach,* Washington: IRL Press, 1981.

4. L. S. Lerman and H. L. Frisch, *Biopolymers,* **1982**, *21*, 995.

5. O. J. Lumpkin and B. H. Zimm, *Biopolymers,* **1982**, *21*, 2315.

6. K. A. Ferguson, *Metabolism,* **1964**, *13*, 21.

7. C. M. Feldherr and P. A. Richmond, in *Methods in Cell Biology: Chromatin and Chromosomal Protein Research* II, G. Stein, J. Stein, and L. J. Kleinsmith, eds., New York: Academic Press, **1977**, *17*, 75.

Chapter 9

SDS-PAG Electrophoresis

The molecular weight of an unknown protein is determined by SDS polyacrylamide gel electrophoresis using standard calibration proteins. A gradient Laemmli slab method and a high gel density urea rod method are compared. The protein bands in the slab gels are visualized by silver staining methods.

KEY TERMS

Anionic surfactants

Gradient gel

Micelles

Subunit dissociation

Urea high density gels

Charge/unit protein weight

Laemmli discontinuous gel

Silver and nickel staining

Swank and Munkres gels

BACKGROUND

Anionic surfactant compounds such as sodium dodecyl sulfate (SDS) are composed of a negatively charged ionic "head group' and a hydrophobic hydrocarbon "tail."

$$CH_3CH_2CH_2CH_2CH_2CH_2CH_2CH_2CH_2CH_2CH_2CH_2OSO_3^- \ Na^+$$

The high solubility in water imparted by the ionic head group and the high solubility in nonpolar solvents implied by the hydrocarbon tail result in a compromise in aqueous solution. Surfactant molecules form aggregates called *micelles* in aqueous solution. These aggregates satisfy the solubility characteristics of both the head and tail regions of the surfactant; ionic groups are exposed to water on the surface of the aggregate while the hydrophobic tails associate with each other within the interior of a roughly spherical aggregate of 60 to 100 molecules. Hydrophobic guest molecules can be taken into the interior of the micelle. This includes the hydrophobic amino acids normally confined to the interior of a native protein. The association between hydrophobic amino acid residues and the micelle interior is strong enough to denature most proteins, turning them inside-out so that they become effectively coated with anionic surfactant molecules.

Reductive cleavage of all -S-S- bonds followed by treatment with the anionic surfactant sodium dodecyl sulfate (SDS) will disrupt the native tertiary structure of most proteins, causing them to adopt rodlike structures. It has been established that this binding occurs with a constant surfactant-totprotein weight ratio and with enough anionic surfactants to totally dominate the native charge of the protein.[1] Therefore, the *charge per unit protein weight* is nearly constant and the electrophoretic mobility of SDS denatured proteins is a function of size alone. The technique of SDS-polyacrylamide gel electrophoresis is widely used to determine the molecular weight of unknown proteins by comparing their relative electrophoretic mobility to standard proteins of known molecular weight. A direct comparison of the mobilities of known and unknown proteins run under identical conditions (in the same cell at the same time) is recommended. The mobilities of known proteins can be plotted as a function of log(molecular weight) and the mobilities of known proteins used to estimate molecular weights by extrapolation.[2]

A necessary first step in the protein sample preparation for SDS electrophoresis is treatment with an excess of 2-mercaptoethanol, which reduces all disulfide (-S-S-) bonds in the protein. This permits total disruption of the protein native structure, which is usually stabilized by disulfide linkages. Some proteins (e.g., chymotrypsin) contain polypeptides linked only by disulfide bonds. Reduction will yield two or more polypeptide fragments which will migrate independently. In the same way oligomeric proteins (e.g., hemoglobin) will dissociate into monomeric subunits when solubilized by SDS. For this reason, hemoglobin migrates as a monomer of molecular weight 16,000 rather than a tetramer of 64,500 D. The *subunit dissociation* caused by SDS is one reason why oligomeric proteins must be characterized by both SDS and disc electrophoresis. The pore size of the gel network is the critical factor in SDS-PAGE separations. The average pore size can be decreased by either increasing the total concentration of monomer (both acrylamide and BIS) or by increasing the proportion of cross-linker (BIS) to the total monomer in the gel. Table 9-1 gives the approximate molecular weight ranges that can be conveniently estimated with gels of various total monomer concentrations, expressed as **percent-gel**.

$$\% \text{ gel} = \frac{\text{g acrylamide} + \text{g BIS}}{100 \text{ mL}} \times 100 \qquad\qquad (9-1)$$

Table 9-1
MOLECULAR WEIGHT RANGES FOR SDS PAGE

Total Acrylamide (% gel)	15	10	5	3
MW Range (Kilodaltons)	3-50	10-70	25-200	1000

Gradient Gels

A wider molecular weight range of proteins can be examined on a *gradient gel* than on any one uniform polyacrylamide concentration gel. With a uniform percentage gel, there is a linear relationship between the relative mobility R_m and log (MW) so that a mixture of proteins, within a sample, cannot differ by more than a factor of ten if one wishes to determine their molecular weight accurately; light peptides will migrate off the gel before large proteins have migrated significantly. With a gradient system, the lighter peptides still move ahead of heavier proteins, but since they are always moving into a higher-percentage gel region, they do not outpace the slower-moving proteins. There is a linear relationship[3] between log (MW) and log (percent gel). For example, a gel gradient ranging over 7 to 25 % gel (with 1% BIS) can resolve protein mixtures from 14,300 to 330,000 D.[4]

Gel Slab Apparatus

In this experiment gels are poured and run in a vertical slab holder. There are a number of varieties of slab cells commercially available and two are depicted in the following figures. Figure 9-1 depicts an expensive thermostated slab cell, with a water jacket (available from Hoeffer) and which must be used for very careful research applications. Less critical SDS-gels can be run on nonthermoregulated (air-cooled) gel systems (not shown). Such a cell is less expensive and also can be easily home-made.

Figure 9-2 illustrates the pouring of an air-cooled gradient gel. Any of a number of commercially available gradient mixers can be used for this purpose. Conventional 16-cm gels usually take too long to be used to monitor continuously the progress of an isolation. The time between the steps of an isolation are often very short. Conventional large gels are also too difficult to prepare both quickly and uniformly for most rapid-screening applications. An alternative approach is to use an 8 x 8 cm ("mini") gradient gel to determine the complexity of an unknown mixture and possibly unstable mixture and to estimate the molecular weight of the components by comparing their mobility to that observed with molecular weight calibration proteins. A mini gel apparatus is shown in Figure 9-3. Mini-gels have the advantage of very short running times, making them especially advantageous when used to monitor the progress in a biochemical purification. If the mini gel is analyzed by silver staining (see below) the whole process can be complete within 2.5 hours. Another advantage of the mini gel system is that a number of equivalent gradient gels can be poured simultaneously. This enormously improves the reliability of comparisons of different gel patterns. A casting stand for the advanced preparation of up to 10 identical 8 x 8 cm gradient gels also greatly speeds up the process and ensures greater reproducibility between gels run during various stages of a complex isolation procedure. The gel casting stand offered by Bio-Rad or Hoeffer consists of up to ten 1.5-mm-thick sample well combs, glass plates with alumina and paper spacers stacked vertically in the stand which has a Teflon tube connection from the bottom of the stand to a gradient mixer. A mini-gel casting system is depicted in Figure 9-4. A disadvantage of the mini-gel system is the narrower range of molecular weights that can be resolved on a smaller gel. If all the peptides are well resolved, this may be all that is required; otherwise, conventional 16-cm uniform SDS gels can be run using a percentage gel appropriate for the molecular weight range of interest.

Laemmli Buffer System

For both gradient and nongradient applications, the Laemmli buffer and gel system[5] is the popular choice. Discontinuous systems (e.g., Laemmli) give sharper bands and require less protein (2.5 µg per band) than do continuous systems, which require about 5 µg per band. As given here this discontinuous method employs a 5 to 20% separating gel (pH 8.8), suggested by Hames,[6] topped with a 3% stacking gel (pH 6.8). This can be run very conveniently in a slab system because the stacking gel can be poured around a sample well comb. The Laemmli system contains 0.1% (w/v) SDS and TRIS-HCl buffers. For low-molecular-weight peptides (≤14,000 D) two continuous gel rod systems have been developed.[7,8] In this experiment a high density gel system[7] (15%) containing 6 M urea as an added denaturant along with 0.1% SDS in 0.1 M Na phosphate buffer (pH 7.2) will be used. (Another SDS formulation, especially designed for low-molecular weight peptides by Swank and Munkres[8] is also accurate and effective.)

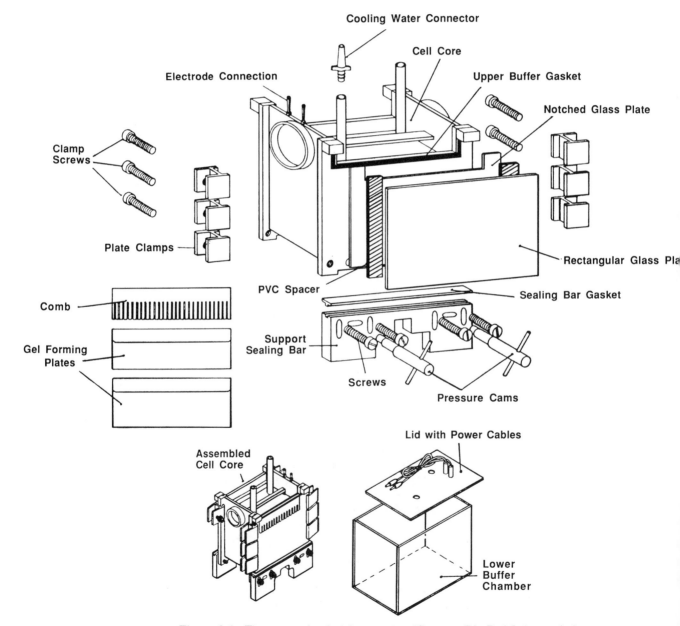

Figure 9-1 Thermostated gel slab apparatus (Courtesy Bio-Rad Laboratories)

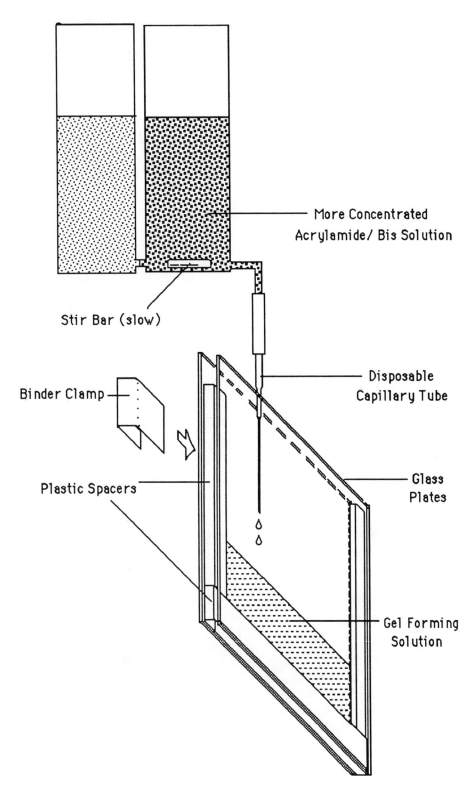

Figure 9-2 Pouring an air-cooled gradient gel

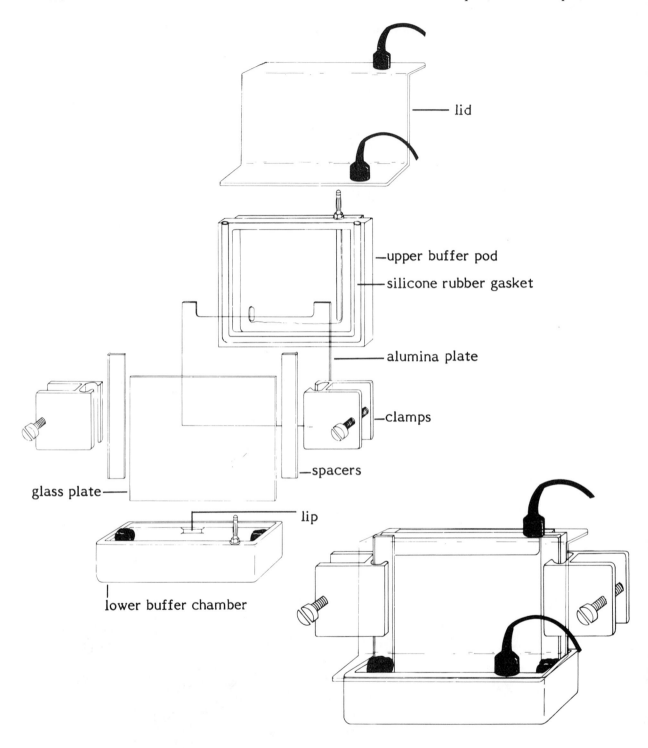

Figure 9-3 An 8 x 8 cm mini gel slab apparatus
(Courtesy of Bio Rad Corporation)

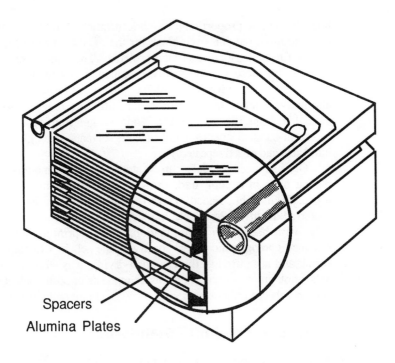

Figure 9-4   Mini gel casting chamber allows up to ten 1.5 mm gels to be cast simultaneously

Protein Sample Preparation

Protein samples are denatured by boiling in electrophoresis buffer containing 5 to 10% mercaptoethanol and 1% SDS. Any precipitated protein is removed by filtration. The sample density is increased by adding to it one volume of 40% sucrose. Bromophenol blue is also added as a tracking dye in the concentration range 0.1 to .01%.

Silver and Nickel Staining

Two new staining procedures involving the labeling of protein spots with colloidal silver or nickel, respectively, offer the advantages of speed and increased sensitivity over the Coomassie blue stain used in the preceding experiment. These methods are too sensitive to be used by students who have not yet mastered the trick of preparing slabs free from fingerprints, smears, and artifacts, but they are greatly preferred for high-quality work, especially with low levels of proteins. They cannot be used if buffer solutions containing dithiothreitol or ampholytes have been used.   The *silver staining* method of Merril[9] involves fixing the proteins with acetic acid, then treating the gel with a dilute chromate oxidizing solution followed with silver nitrate. The color is developed with sodium carbonate followed by an acetic acid stop bath. The chemistry behind the color formation is complex, but reproducible colors can be obtained if the identical procedure is used each time. Proteins are stained a dark brown or gray against a pale yellow-orange background. Occasionally, orange and rust-colored spots are seen. The color is often specific for a particular protein, adding another important dimension to the gel analysis. Unlike Coomassie blue-stained gels, the color intensity is not proportional to concentration, but the color changes with protein concentration, ranging from black for the most concentrated, to blue-brown, red, and yellow for the least concentrated proteins.[9] There have been attempts to differentiate proteins by these colors.[10] Silver-stained gels can be subsequently stained with Coomassie blue to give very revealing patterns of brown, orange, and blue stains.[11] An alternative *Nickel staining* procedure utilizing reagents sold as a Kodavue Kit is available from Eastman.[12] The Kodavue gel staining system is very sensitive to detergent-based cleaners and nonaqueous solvents, so all glass plates must be cleaned in nitric acid and rinsed repeatedly.

Once stained, gel slabs can be stored for a few months in tightly sealed zip-lock plastic bags in a cool, dark drawer. Coomassie blue stain will fade with time, especially in light. Faded or over-destained gels can be restained but not always to preferred quality. Preferable long-term data storage methods are photographing the gels with background illumination, contact printing the gel pattern on specially designed photographic paper[13] or drying the gels in a vacuum gel dryer. Because gels shrink and not infrequently tear upon drying, it is advisable that for research quality gels, photographs or contact prints be made as backup evidence. The colors of silver-stained gels can only be preserved by color photography or possibly by gel drying.

Molecular Weight Standards

A slab gel is especially useful for molecular weight determinations since the sample and molecular weight standard proteins can be run under identical conditions on a single gel. There are a number of commercially available SDS-PAG molecular weight standards which give a good spread of molecular weight lines in a gel. Most kits are also formulated to give sharp, even-intensity lines for each protein. Bio-Rad and Sigma offer standards suitable for the Laemmli slab system (see Table 9-2).

Table 9-2
LAEMMLI SLAB MOLECULAR WEIGHT MARKER PROTEINS

Protein	Molecular Weight
Lactalbumin [a]	14,200
Lysozyme [b]	14,400
Soybean trypsin inhibitor [a,b]	20,100
Trypsinogen [a]	24,000
Carbonic anhydrase [a,b]	29,000
Glyceraldehyde-3-phosphate dehydrogenase [a]	36,000
Ovalbumin [a,b]	45,000
Bovine serum albumin [a,b]	66,200
Phosphorylase B [b]	92,500

[a] From Sigma; [b] From Bio-Rad.

Bethesda Research Laboratory furnishes standards for a marker system with lower molecular weight range, suitable for the 0.1% SDS-Urea system (see Table 9-3).

Table 9-3
SDS-UREA MOLECULAR WEIGHT MARKER PROTEINS

Protein	Molecular Weight
Insulin (unresolved A and B chains)	3,000
Bovine Trypsin inhibitor	6,200
Cytochrome *c*	12,300
Lysozyme	14,300
b-Lactoglobulin	18,400
Chymotrypsinogen	25,700
Ovalbumin	43,000

The various suppliers report minor differences in molecular weights for the standards; these differences are preserved in Tables 9-2 and 9-3. The concentrations of each standard protein in the Bio-Rad sample has been adjusted to give equal staining intensity with Coomassie blue; they are in 50% glycerol and 0.5 M NaCl at the approximate concentration of 2 mg/mL. These molecular weight standards should be diluted 1:5 in the sample application buffer and applied as 25-μL samples. The Bethesda Research Laboratory standards are at a concentration of 1 mg of each protein per milliliter. The smallest protein, insulin, has an apparent molecular weight of 3000 because the individual A (2300) and B (3400) chains cannot be resolved. These standards should be heated at 100°C for 5 minutes to redissolve precipitated proteins, then diluted 1:3 in the sample application buffer and applied as a 25-μL sample.

Gel Problems-What Else Can Go Wrong?

The following are common problems with SDS gels with some first-try "solutions."

1. Lack of polymerization or too slow a polymerization. You have the recipe wrong: usually not enough or too much TEMED or exhausted ammonium persulfate. Pour it all out and start over.

2. Too fast a polymerization. The concentration of TEMED is too high or the gel solution is too warm. Cool the solutions before initiating polymerization.

3. Gel cracking during polymerization or during electrophoresis. The gel solution or running conditions are too warm. This is especially common with high-percentage gels.

4. Bubbles or detachment of the gel from the glass plates. This is probably due to unclean plates. After being rinsed with distilled water, they must drain cleanly without water spots. Photo flow rinse can be used for Coomasie-stained gels.

5. No bands after staining and destaining. Either not enough sample applied (too dilute) or destaining was too extensive. Restain to see if bands become visible.

6. Protein streaking. This is due to too concentrated a sample, insoluble protein, or not enough SDS. Dilute the sample with more SDS solution.

7. Bands on part of the slab do not move down the gel. This is usually due to air bubbles between the plates underneath the affected lanes. Bubbles must be removed.

8. Same bands in all lanes of a slab gel. Either reservoir buffer or sample buffer is contaminated. Throw them out and start over.

EXPERIMENTAL SECTION

Following are a number of experimental options illustrating the principles of SDS gel electrophoresis using the Laemmli discontinuous gel system, with or without a running gel gradient, in both conventional or mini-gel systems. An SDS-urea tube method suitable for low-molecular-weight proteins is also included. The instructor will choose one of these options. Students should work in pairs except in the preparation of gel stock solutions, which can be assigned as chores over a larger group.

Laemmli Slab Method

The slab cell must be assembled according to the manufacturer's instructions. There are a number of commercially available slab cells. Large slabs run at high currents can be connected to a circulating cooling-water bath like the Bio-Rad or Hoeffer slab apparatus illustrated in Figure 9-1. Related cell systems differing in the plate clamps, pressure cams and without beveled glass plates can be obtained from either Bio-Rad or Hoeffer Scientific. **Do not overtighten any clamp screws or cams on such cells.**

Less expensive slab systems which do not permit water cooling are available from Bethesda Research Laboratory (BRL). Many teaching laboratories use homemade slab systems much like the BRL model. Such systems employ plastic spacers that frame the poured region of the gel. These spacers must be clean, smooth, and fit tightly together. They must all be of the same thickness. Remember that the bottom spacer must be removed after the gel has been poured and allowed to set. Care must be taken to ensure that the glass plates are clean and free from fingerprints. If grease is used to prevent leakage, it must be used sparingly. The glass sandwich is mounted upright in the vertical electrophoresis cell and clamped in place with binder clamps. It is not recommended that students fill their cell with distilled water to check for leaks before actually pouring the gel; the cells are too hard to drain without dismantling. However, placing a pan or towel under the cell is recommended to catch any leaks. **Remember: Acrylamide is a Poison. Wear gloves and clean up all spills immediately.**

The **Laemmli electrode buffer** is a TRIS-glycine (pH 8.3) solution prepared from

TRIS (free base)	6.0 g
Glycine	28.8 g
10% (w/v) SDS	10.0 mL

with enough distilled water to give 1 liter of solution. Some gel systems require 1.5 to 2 L of electrode buffer.

In this experiment gradient slab gels are poured with the gradient maker indicated in Figure 9-2. The gel mixture must be degassed and TEMED added (10 µL per 30 mL of gel mixture) just prior to transferring the gel solution by small-diameter Tygon tubing, under gravity flow, from an elevated gradient maker. One has only a few minutes to complete the entire pour before the gel begins to set up. The more concentrated acrylamide solution enters the glass plate sandwich first and runs down the inside of the sandwich. The sandwich is then filled with increasingly less dense gel formulation from the gradient mixer. There is little mixing if the plates are clean and held upright. Rapid stirring in the gradient mixer and interruptions in filling the sandwich must be avoided since they lead to the bubble formation which often plagues gradients poured from the top. [If your laboratory has a mini-gel system that is poured from the bottom (see below), the gels are poured upside down, with the most concentrated acrylamide mixture still entering first.]

For a 5 to 20% acrylamide linear gradient separating gel the two solutions outlined below are added to the two chambers of a gradient mixer (see Figure 9-2). This recipe is suitable for one conventional 0.75 mm x 14 cm slab gel but must be scaled up 5 to 10-fold if a number of mini-gels are poured in a casting chamber (see Figure 9-4). Note that a very small gradient mixer must be used if single mini-gels are cast one at a time. It is more efficient to cast 10 or more mini-gels simultaneously. The sucrose gradient is included to minimize mixing caused by convection as the mixture warms upon polymerization. The high-concentration acrylamide/BIS mixture is added to the stirred reservoir, which serves as a mixing chamber as the more dilute mixture empties into it. The polymerization catalyst TEMED (10 µL) is added to both reservoirs, with gentle but complete sirring, just before pouring the gel; all connections must be tightly made and everything ready to go when TEMED is added. The flow rate of the gel mixture into the casting stand should be about 3 to 5 mL/min. If too long is taken in casting the gels, the gel mixture becomes too viscous to transfer through Tygon tubing.

Immediately after the gradient has been poured, the gradient maker must be flushed out with water to prevent acrylamide from polymerizing in the apparatus.

If a discontinuous buffer system is used, the top of the freshly poured gel is overlayered with water-saturated butanol. When the gel mixture is hard, the butanol is removed and a stacking gel is poured with a sample well comb in place.

20% Acrylamide mixture

Acrylamide : BIS (30% : 0.8%)	5.0 mL
3.0 M TRIS-HCl (pH 8.8)	3.75 mL
10% (w/v) SDS	0.3 mL
1.5% Ammonium persulfate	0.7 mL
Sucrose	4.5 g
Water	2.75 mL

5% Acrylamide mixture

Acrylamide : BIS (30% : 0.8%)	5.0 mL
3.0 M TRIS-HCl (pH 8.8)	3.75 mL
10% (w/v) SDS	0.3 mL
1.5% Ammonium persulfate	0.7 mL
Water	20.25 mL

The gradient mixer must be flushed out with water **immediately** after use. The separating gel layer should be overlayered with water-saturated butanol, which floats on the acrylamide, giving a very smooth boundary. Overlayering only with water, which diffuses into the upper portion of the gel, gives an irregular boundary. Once the separating gel is hard, the butanol is removed and a stacking gel is poured. The stacking gel formulation for either gradient or nongradient gels is

Distilled water	6.3 mL
0.5 M TRIS-HCl, pH 6.8	2.5 mL
10% (w/v) SDS	0.1 mL
Acrylamide : BIS (30% : 0.8%)	2.0 mL
10% Ammonium Persulfate	0.1 mL
TEMED (added last)	5-10 µL

The stacking gel is allowed to polymerize for 30 to 45 minutes. After polymerization, 0.5 M TRIS-HCl (pH 6.8) buffer is poured around the comb, which is gently removed. The sample wells should be washed with buffer. When the upper electrophoresis chamber is filled with electrode buffer, care must be taken to ensure that there are no bubbles in the sample wells and no leaks around the slab spacers. Bubbles must also be removed from the bottom of the gel sandwich when it is submerged in the bottom buffer. These bubbles can usually be "sucked out" with a bent syringe needle and a syringe.

Nongradient Laemmli Gel

If time or equipment limitations preclude the use of a gradient gel, a nongradient Laemmli system can be poured. A plastic syringe with a long needle or a Pasteur pipet may be used to fill the glass plate assembly from the top, without bubble formation. The separating gel is prepared by the following formulation (for two 0.75 mm x 14 cm slabs, total volume 30 mL)

Distilled water	10.1 mL
1.5 M TRIS-HCl	7.5 mL
10% (w/v) SDS	0.3 mL
Acrylamide : BIS (30%: 0.8%)	12.0 mL
10% Ammonium persulfate (fresh)	0.1 mL
TEMED (added last)	10-25 µL

This formulation must be doubled for two 1.5 mm x 14 cm slab gels.

The gel solution is prepared on ice and degassed under reduced pressure for 15 minutes. The TEMED is added only after deaeration and just prior to filling the assembly. The separating gel should polymerize overnight. After polymerization is complete, the unit must be inverted to drain and the gel surface should be washed with 1.5 M TRIS-HCl (pH 8.8) buffer. Any Parafilm or plastic barrier on the bottom should be removed at this time. Then a sample well comb should be inserted between the glass plates. The stacking gel solution (prepared as above) should be pipetted around both ends of the comb. It is not necessary to overlayer the stacking gel.

Sample Preparation for all Gels

Sample protein is dissolved in 0.1 mL of running buffer containing 10-µL of 2-mercaptoethanol with 0.1% SDS (and 6 M urea when needed). The mixture is heated in a boiling-water bath for 5 minutes. After cooling, the solution is diluted with an equal volume of 0.02% bromophenol blue in 40% sucrose. The amount of protein to be applied depends on the sample but applications in the range 1 to 5 µg of protein per protein band usually give sharp, easily viewed bands. Sample volumes of 10 to 100 µL are most common. For unknown proteins the optimum concentration has to be determined experimentally; a range of sample volumes is applied, a different one in each well, with 25-µL reference samples every fifth well.

Slab Electrophoresis Conditions

For discontinuous gels a current of 2.5 mA per gel is optimal to stack the proteins, and the electrophoresis running time can be shortened to less than 2 hours if the current is raised to 30 mA per gel (at 100 V) after the dye front enters the gel. Electrophoresis on a large gel slab can require about 16 hours if run at 10 mA (constant current) or 40 V (constant voltage). Mini-gel slabs require less than 2 hours. If research-quality resolution on large slabs is required, cooling water should be used for currents greater than 10 mA per gel, but a fan directing air toward the center of the slab gel will serve almost as well.

SDS Urea Tube Method

Gel tubes are filled and mounted according to the procedure described for the separating gel in the preceding experiment (Chapter 8). The 15% gel contains acrylamide : BIS (15% : 0.4%) in 0.1% SDS, 6 M urea, and 0.1 M sodium phosphate (pH 7.2). Ammonium persulfate (0.4 mL) and TEMED (100 µL) are added with stirring before pouring the gels. Currents of 2.5 mA per tube are optimal. The running buffer contains 0.1 % SDS, 6 M urea and 0.1 M sodium phosphate (pH 7.2).

SDS-Coomassie Staining and Destaining

After slab runs, the glass plates are gently pried apart under a direct stream of water from a syringe. The stacking gel of a Laemmli slab will probably tear off at this stage. This is not a serious problem if the remaining slab is marked to identify the gel lanes. A common method is to cut off a small portion of the right-hand corner of the separating gel slab before staining so that the original orientation of the slab can be identified after staining. Tracing numbers onto a glass plate corresponding to the gel lanes will also assist in assigning patterns of bands, especially if some of the lanes do not contain visible bands. Very careful note-taking is required to keep the identity of bands straight after a lengthy experiment.

The slab should be transferred to a "square" of clean plastic mesh (plastic window screen) slightly larger than the gel (about 18 x 20 cm). The slab can be reoriented with puffs of distilled water from a squeeze bottle. The slab is then covered with a second "square" and gently folded into a cylinder about 5 cm in diameter. The gel must be kept moist and not kinked; it will crack or tear very easily, especially if allowed to dry out for 10 to 20 minutes. The rolled cylinder is transferred to a staining tube and from there (after several hours) to a diffusion destainer.

Silver and Kodavue Staining

Merril[9] developed a procedure where proteins are fixed in a solution of 50% methanol and 12% acetic acid for at least 20 minutes followed by three rinses in 10% ethanol/5% acetic acid to remove excess SDS. The gels are then soaked in 200 mL of oxidizer

(0.0034 M $K_2Cr_2O_7$/0.0032 M nitric acid), washed repeatedly in distilled water, and placed in 200 mL of silver reagent (0.012 M $AgNO_3$), followed by rapid rinsing in three 200-mL portions of developer (0.28 M $NaCO_3$/0.5 mL of commercial formalin per liter). Gentle agitation in the third portion of developer will give a gel image of the desired density. Development is stopped by replacing the developer with 1% acetic acid.

Kodavue staining follows much the same procedure but requires the prepackaged developer, reducing agent, presensitizer, and sensitizer reagents. The gel is first fixed with an acetic acid-methanol-water system, then run through a series of sensitization and development steps. The reader is referred to the instructions that accompany this kit for full instructions. The chemical identity of the solutions is proprietory information.

Photographic Images of Gels

Kodak offers products to enable one to prepare photographic images of stained gels without using a camera; a variety of contact printing is used. Kodak Electrophoresis Duplicating Paper (EDP) produces positive photographic images of coomassie blue stained electrophoresis gels. A related product, Electrophoresis Duplicating Film (EDF), is used for Silver and Nickel-stained gels. The contact printing process is done in darkness or with a very dim red darkroom lamp, so one must carefully practice using a junk gel and ordinary paper to be sure all the steps can be done in the dark. The gel is fragile and cannot be picked up often; it should be wrapped in Saran Wrap first.

The following procedure for EDP paper must be memorized; one cannot stop to consult the instructions part way through the process.

1. The EDP paper is placed on a clean, dry surface in a darkroom.

2. An amber filter sheet (from Kodak) is placed directly on top of the EDP paper.

3. The wet gel is placed gently on the filter sheet.

4. A desk lamp with a 15-watt incadescent bulb, is placed 36 inches directly above the wet gel.

5. The light is turned on for 5-10 seconds (a darkroom timer is required).

6. The exposed EDP paper is transfered to Kodak Dektol (1:1) developer for one minute. An image should appear on the paper within 15 seconds. Prepare the Dektol solution as indicated on the container.

7. The paper is then washed in running water and fixed with Kodak fixer for 2 to 4 minutes. It should be air dried in a dust-free place.

This is a positive paper and it will develop black where it has not been exposed to light and white for exposed parts. If the image does not appear and the paper is all white, it has probably been overexposed. The exposure time should be reduced or the light moved farther from the paper.

EDF paper requires a cellulose acetate sheet, a much stronger light source (two 15-W bulbs at 8 in), and a longer exposure time (2 to 3 minutes). The development process is identical.

Optional Gel Drying

The following procedure is recommended for gel drying. The gel is first soaked in an aqueous 1% glycerol/10% acetic acid solution for 30 to 40 minutes. It is carefully removed and drained of all obvious moisture then sandwiched between a double layer of stiff absorbant paper and a plastic (Mylar) sheet on the dryer. The paper is at the bottom facing the screen and the whole affair has to be flush with the inside wall of the heating block. The gel must be centered on the paper and any trapped air bubbles must be smoothed out before drying begins. The dryer is a heating block with a vacuum connection so that the gel sandwich is warmed under reduced pressure. A silicone rubber panel is placed over the gel sandwich to give a good vacuum seal. Gels are most frequently damaged by blistering, which occurs when the gel has begun to dry before being tightly compressed, and by tearing when the gel sticks to the plastic sheet which must be removed from the finished gel. The latter problem is from incomplete drying; 1.5-mm gels require about 2 hours. A porous polyethylene overlay sheet may be necessary for high density and gradient gels.

DISCUSSION QUESTIONS

1. There are a number of micelle-forming ionic surfactants that could be used like SDS. What changes in prodedure would be required if cetyl trimethylammonium bromide (CTABr) were used instead of SDS?

2. The relative mobility R_m of glycoproteins generally falls off the linear plots of R_m vs. log (MW) established with nonglycoprotein markers. Explain.

3. Note that in disc gels the upper reservoir buffer has a different pH than the stacking gel; does this matter with SDS gels?

REFERENCES AND NOTES

1. J. A. Reynolds and C. Tanford, *Proc. Nat. Acad. Sci. USA*, **1970**, *66*, 1002.

2. A. L. Shapiro, E. Vinuela and J. V. Maizel, *Biochem. Biophys. Res. Comm.*, **1969**, *28*, 815; K. Weber and M. Osborn, *J. Biol. Chem.*, **1969**, *244*, 4406.

3. B. D. Hames and D. Rickwood, *Gel Electrophoresis of Proteins: A Practical Approach*, Washington D.C.: IRL Press, 1981, pp. 14-18.

4. J. F. Paduslo and D. Rodbard, *Anal. Biochem.*, **1980**, *101*, 394.

5. U. K. Laemmli, *Nature*, **1970**, *222*, 680.

6. Ref. 3, p. 75.

7. Bethesda Research Laboratory Product Profile, catalog number 6000SA.

8. R. T. Swank and K. D. Munkres, *Anal. Biochem.*, **1971**, *39*, 462.

9. R. C. Switzer, C. R. Merril and S. Shifrin, *Anal. Biochem.* **1979**, *98*, 231; B. R. Oakley, D. R. Kirsch and N. R. Morris, *Anal. Biochem.*, **1980**, *105*, 361; D. W. Sammons, L. D. Adams and E. E. Nishizawa, *Electrophoresis*, **1981**, *2*, 135; T. Marshall, *Anal. Biochem.*, **1984**, *136*, 340.

10. J. K. Dzandu, M. E. Deh, D. L. Barratt and G. E. Wise, *Proc. Nat. Acad. Sci. USA*, **1984**, *81*, 1333.

11. J. K. Dzandu, M. E. Deh and P. Kiener, *Biochem. Biophys. Res. Commun.*, **1985**, *127*, 878.

12. Kodak, *Lab. Chem. Bull.*, **1984**, *55*, No. 4; **1983**, *54*, No. 1.

13. For more information on the Kodavue protein visualization method or details regarding EDP and EDF products, contact
    Customer Technical Services
    Kodak Laboratory Chemicals
    Health Sciences Market Division
    Eastman Kodak Company
    Rochester, NY 14650

Chapter 10

E. Coli DNA and Agarose Electrophoresis

E. coli bacteria are grown in a nutrient medium until late exponential phase growth, then harvested by centrifugation. Warm SDS solution is used to disrupt the cells and the liberated nucleic acid is freed from protein by partitioning between aqueous 1 M sodium perchlorate and chloroform/*iso*amyl alcohol. The nucleic acid remains in the upper aqueous layer. The slow addition of two volumes of ethanol at dry ice temperatures precipitates DNA. The DNA precipitate is purified by dispersal in dilute sodium chloride/sodium citrate buffer followed by precipitation with added salt and ethanol. An optional DEAE-cellulose chromatography step can be used as a final cleanup. The length heterogeneity of the DNA is determined by stepwise precipitation with various concentrations of polyethylene glycol. The fractions in each case are characterized by electrophoresis on agarose with ethidium bromide fluorescence detection.

KEY TERMS

Agarose gel electrophoresis	*Auxotrophs*
Ethidium bromide intercalation	*Growth curves*
Growth media	*Nucleases*
PEG fractionations	

BACKGROUND

The bacterium *Escherichia coli* is widely studied; much research in molecular genetics has utilized this bacterium and it is the usual choice for a procaryotic organism in recombinant DNA technology. Although *E. coli* is a major bacterium in intestinal flora, the common strains of *E. coli* employed in molecular biology are derived from noninfectious *E. coli* K12,[1] which grows well in chemically defined medium with very little tendency to "get loose." This strain and those derived from it cannot easily infect humans and present no serious danger in a laboratory setting, even when manipulated by inexperienced experimentalists. It is far more likely that a student will cause accidental contamination of the *E. coli* culture rather than experience infection of his or her person by this microorganism.

Growth of *E. coli*

If care is taken to minimize contamination by other bacteria, *E. coli* can be grown either on a microscale (a few milliliters of medium) or on a macroscale (hundreds of liters at a time) with a relatively inexpensive medium consisting of a few inorganic compounds, trace levels of several ions, and an organic compound to serve as carbon and energy source. It grows with a maximum growth rate corresponding to cell divisions every 20 minutes or so. Glucose can be used as the sole carbon and energy source and ammonium chloride as the nitrogen source. The medium should be buffered near pH 7.0 and must contain important salts such as Mg^{2+}, SO_4^{2-} and HPO_4^{3-} Such a growth medium is called a ***minimal medium***.

This medium can be supplemented with additional compounds which the cell would otherwise have to synthesize; such compounds are most often added to cultures in the form of commercially available media components such as tryptone and yeast extract. Yeast extract contains the necessary cofactors for many enzymatic processes. Tryptone is obtained by enzymatic hydrolysis of milk and is very rich in amino acids. *E. coli* grows rapidly on the nutrient medium suggested in the procedure section.

Some common strains of *E. coli* feature mutations that prevent the synthesis of some necessary amino acid, purine, pyrimidine, cofactor, or some other biomolecule which must be added to a minimal medium to permit growth. A strain that cannot synthesize all the compounds needed for normal growth from a single common carbon source such as glucose is an ***auxotroph***. For example, a histidine auxotroph requires histidine to be present in the medium. Such a mutation functions as a genetic marker since the auxotrophic strain is easily detected by observing the absence of growth when the required substance is lacking. The standard notation for a genetic locus is a three letter italicized symbol[2]; for example, the genes for histidine synthesis are designated *his* and a deficient gene is designated *his⁻*. The phenotype of a cell is written with a capital letter and not italicized, e.g., His⁻.

E. coli can grow either in the presence or absence of molecular oxygen, O_2. The growth rate in the absence of O_2 (anaerobic growth) is relatively slow and the maximum cell density reaches only about 10^7 cells/mL. If air is bubbled through the culture, rapid growth is possible and maximum cell densities of 10^8 to 10^{10} cells/mL can be obtained. After a sterile, aerated medium is inoculated with slowly multiplying cells from an anaerobic medium, there is a lag time of up to several hours before cells begin to multiply rapidly. This time is required to synthesize the components of the aerobic electron transport system, as well as ribosomes and other cell components necessary for increased protein, RNA and DNA synthesis. Once the necessary cellular materials are on hand, the culture grows exponentially at a rapid rate until a stationary-state culture is reached. At steady state the population may hold steady for some time, increased growth being hindered by insufficient oxygen or nutrients or the accumulation of waste products. The population ultimately decreases due to such factors.

A bacterial population growth curve has four phases: the ***initial lag***, the ***exponential or log phase***, the ***stationary phase***, and the period of ***decline and death***. During the exponential phase the number N of bacteria is related to time t by the expression

$$log(N) = Kt + const$$

(10-1)

where K is related to the generation time t_g of the culture (i.e., the time required to double the number of cells) by the expression[3]

$$t_g = \frac{0.30}{K} \qquad (10\text{-}2)$$

A growth curve, such as that depicted in Figure 10-1, can be obtained either by counting bacteria removed from the medium, diluted, and spread onto an agar plate to establish new colonies, or by measuring the absorbance of a suspension of cells at 550 nm. The plating method takes time but gives unambiguous results since the number of visible colonies on the agar plate equals the number of bacteria in the volume of sample *spread on the agar plate*; from the dilution factor one can calculate the number of bacteria per milliliter in the original culture. The absorbance method can be done quickly, but care must be taken in relating absorbance directly to cell number. A cell suspension scatters light so that the light reaching a photomultiplier tube is diminished as the cell count is increased, but the absorbance is a measure of cell mass, which depends on both cell number and cell size. Cell size varies with the growth medium, bacterial strain and growth stage. A calibration curve can be prepared relating absorbance to cell number determined by the plating method, and this curve can be used with all systems that have the same average cell size as the calibration system. To save time in this experiment, cell numbers will be estimated from the calibration curve in Figure 10-2, even though the precautions discussed above have not been fully satisfied. We will assume that the absorbance reading at 550 nm is directly proportional to cell density up to absorbance readings of about 0.70; above this level absorbance is no longer directly proportional to population. If the culture is too dense, it must be diluted prior to reading the absorbance and the concentration of cells calculated by multiplying the measured cell number for the diluted sample by the appropriate dilution factor.

Contamination of bacterial cultures is always a potential source of frustration, and it is particularly troublesome in teaching laboratories where chemicals and glassware are shared. The bacteria should be grown in 1-L flat-bottom culture flasks each containing 250 mL of medium. Flasks containing media, *except glucose*, should be stoppered with sterile cotton and autoclaved for at least 30 minutes and allowed to cool to 37°C before inoculation with *E. coli* stock. Addition of glucose before autoclaving leads to the formation of char and organophosphates that inhibit bacterial growth. The sugar should be added as a sterile solution, autoclaved separately or sterilized by Millipore filtration, by the instructor, just prior to inoculation. The culture flasks can be aerated by rapid shaking in a shaker bath either at room temperature or at 37°C. Dow antifoam A (50 μL perflask) reduces foaming. Samples withdrawn from the growing culture must be removed with sterile pipets.

If the medium is inoculated in the early morning (2:00 to 4:00 A.M.), the culture should be in log phase by early afternoon. The growth curve can be obtained by pulling aliquots with a sterile pipet every hour or so starting in the early morning (8:00 A.M). to monitor the increase in turbidity measured as absorbance at 550 nm. After 12 hours or when the absorbance reaches 0.7 (for twofold diluted culture), the culture is harvested by centrifugation. One liter of *E. Coli* culture grown to early log phase contains 10^8 cells/mL or 10^{11} cells/liter. In late log phase the yield may be 10^{10} cells/mL. One gram of *E. Coli* cell paste contains about 10^{12} cells.

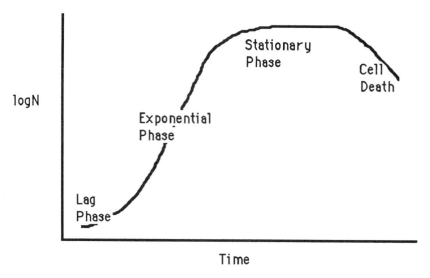

Figure 10-1 E. coli growth curve

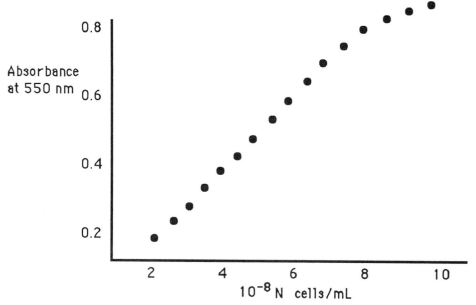

Figure 10-2 Cell count and absorbance values calibration curve

DNA Isolation and Purification

Most purifications of DNA employ the principles that high concentrations of salt or organic solvents weaken the binding between nucleic acids and proteins and that at high ionic strength the nucleic acid becomes less soluble. The double-stranded-to-single-stranded (dsDNA to ssDNA) transition also depends on salt concentration. Added cations shield the phosphates from one another, reducing interstrand electrostatic repulsion. In distilled water, DNA denatures at room temperature. Extraction with either chloroform: isopropanol or with water-saturated phenol is required to denature and remove proteins that bind to DNA *in vivo*. Sodium dodecyl sulfate and saline-citrate minimize nicks due to DNase activity by inactivating these enzymes.

Low concentrations of EDTA (1 mM) also bind, by strong chelation forces, the metal ions required by nucleases. Nicks due to mechanical shearing forces can be minimized by avoiding rapid stirring and using syringes, and so on. In this experiment *E. coli* DNA is extracted from cells and freed from protein, then fractionated according to length by polyethylene glycol precipitation.[4] It is not possible to distinguish between fragments due to nuclease cleavage and fragments due to normal replication.

DNA Length Heterogeneity

During exponential growth a substantial percentage of the cells are replicating at any one time and the DNA recovered from such cells exhibits considerable length heterogeneity even if caution is taken to minimize strand nicks during isolation. Generation times on the order of 20 minutes imply that every cell contains a bacterial chromosome in the process of replication. The intact, circular supercoiled chromosome isolated from nonreplicating *E. coli* contains 4×10^6 base pairs (bp). Degradation of this highly polymerized DNA and partial replication of the bacterial chromosome both give DNA fragments upon isolation. Polyethylene glycol (PEG) can quantitatively precipitate DNA and DNA fragments;[4] see Table 10-1. The PEG method requires no sophisticated equipment and can be scaled up to any size. The DNA is not harmed by PEG and can be redissolved in any appropriate buffer. Highly polymerized DNA and fragments larger than 1500 base pairs are precipitated nearly quantitatively by 5% PEG in 0.5 M NaCl, 10 mM TRIS-HCl buffer (pH 7.4) containing 10 mM EDTA, and 4.4 mM $MgCl_2$. The initial DNA concentration must be at least 50 µg/mL. Successively higher concentrations of PEG are required to precipitate shorter DNA fragments, and fractionation of fragments below 200 bp is not practical.

Table   10-1

PEG FRACTIONATION OF DNA BY LENGTH[a]

% PEG	Minimum Length DNA Precipitated (bp)
5	1500
6	800
7	600
8	400
9	300
10	200

[a]Source: Ref. 4

Agarose Gel Electrophoresis of DNA

DNA is usually characterized by its mobility on agarose gel electrophoresis. Polyacrylamide gel electrophoresis can also be used with DNA, and this technique easily fractionates short oligonucleotides, however the electrophoretic behavior of DNA in polyacrylamide gels is sensitive to the base composition of the DNA and to the temperature. Also, polyacrylamide cannot be used to resolve fragments larger than 1000 bp; the concentration of acrylamide and cross-linkers would be so low that the gels would be too soft. Size analysis of DNA fractions is most conveniently done with horizontal bed agarose gel electrophoresis. Agarose gels can resolve dsDNA fragments from 70 to 800,000 bp depending on the agarose percentage.[5] The reader is referred to Chapter 19 for a more detailed account of the factors that govern DNA mobility during agarose gel electrophoresis.

Agarose gels are very easy to pour and run. Horizontal slab gels are almost always used because the low agarose concentrations necessary for the separation of large DNA molecules gives a soft gel that must be supported from underneath. The powdered agarose is combined with the electrophoresis buffer and the resulting slurry is heated in a boiling-water bath until the agarose dissolves. The solution is cooled to 50°C, combined with enough ethidium bromide solution to give a 0.5-μg/mL solution and poured onto a clean, dry glass mold. The gel should be at least 3 mm thick. A sample application comb is quickly inserted into the warm liquid and the gel is allowed to set for 30 to 45 minutes at room temperature. If the gel has been poured properly, it will be free from bubbles and the sample application comb will pull away, leaving square wells with at least 1 mm of gel at the bottom. The sides of the gel mold are removed and the glass plate is transferred to a gel apparatus and covered with about 1 mm of electrophoresis buffer. This is called the "submarine" technique.

The presence of DNA on a gel can be detected by the specific binding of the fluorescent dye ethidium bromide to DNA. The DNA bands in the gel show up as orange patches under UV illumination.

EXPERIMENTAL SECTION

Since the large-scale growth of bacteria requires a shaker bath or fermentor, this part of the experiment is best done by the class as a whole. The *E. coli* are collected by centrifugation and the bacterial pellets parceled out to pairs of students who isolate DNA and characterize it by electrophoresis.

Growth of *E. coli*

E. coli can be grown in a commercially available medium with reagents supplemented with NaCl. The most common formulation is the Luria-Bertani or *LB medium*

Bacto-tryptone	10.0 g
Bacto-yeast extract	5.0 g
NaCl	10.0 g
Distilled water	1 L

The pH is adjusted to 7.5 with dilute NaOH. The solution is divided equally among four 1-L culture flasks. The flasks are stoppered with sterile cotton and sterilized in an autoclave for 30 minutes. An additional 90 to 120 minutes is required for exhaust and cooling. To the sterilized medium is added 10.0 mL of 10% glucose dissolved in distilled freshly boiled water. The glucose solution should be sterilized either by filtration through a 0.45μm Millipore filter or by heating in an autoclave separately from the rest of the medium. The medium is inoculated with one loopful from an agar stock culture. The instructor may choose one of several options.

If optimum cell density is required for an afternoon laboratory, the culture should be started at 2:00 A.M. (Faculty member or student volunteer) and the growth curve monitored by pulling 2-mL aliquots with a sterile pipet every hour starting at 8:00 A.M. The absorbance at 550 nm is measured versus a blank of sterile medium. Absorbance values above 0.7 are unreliable, and dense samples must be diluted with an equal volume of sterile medium and monitored immediately. The absorbance of the undiluted sample can be estimated from the dilution factor (x 2). An absorbance value of 1.0 corresponds to 4×10^8 cells/mL.[6]

If such heroic efforts are not practical, the culture can be established the previous day and grown overnight. Laboratories with very serious time or space constraints, (e.g., with very large size classes) may use a paste of frozen cells and go right on to the cell lysis portion of this experiment.

Isolation of DNA

Freshly grown cells should be harvested by centrifugation when the absorbance reaches 0.7 or after 12 hours (2:00 P.M.). Centrifugation at 6000 g for 15 minutes pellets most of the cells; *E. coli* cells cannot be collected by filtration since they quickly plug any filter small enough to retain them. Freshly pelleted cells (1 g) are washed once with 25 mL of 0.15 M NaCl containing 0.1 M EDTA (pH 8.0) and recovered by centrifugation for 15 minutes at 3000 x g. The cells are lysed by suspension in 10 mL of the same buffer containing 1% SDS. The mixture is heated to 60°C for 10 minutes in a hot-water bath or a heating block, then allowed to cool to room temperature. The rupture of the cells results in diminished turbidity and increased viscosity. Two milliliters of 5 M sodium perchlorate is added to give a final solution volume of 11 to 12 mL. The cooled aqueous solution is shaken with an equal volume of chloroform-*iso* amyl alcohol [24:1 (v/v)] in a stoppered centrifuge tube for 30 minutes. The emulsion is settled by centrifugation at 5000 x g for several minutes. This first deproteination leads to a three-phase mixture.[5,7] The upper, aqueous phase contains nucleic acid and is carefully removed with a wide-bore glass pipet. The middle layer contains denatured proteins and entrapped DNA, and the bottom layer is largely chloroform.

DNA in the middle layer can be removed by back-extraction with a volume of 0.15 M NaCl/0.1 M EDTA equal to one-half the upper layer volume followed by centrifugation. The supernatant from this extraction is combined with the previously recovered upper layer in a 50-mL beaker and cooled in an ice bath.

Two volumes of cold 95% ethanol are gently overlayered on the cold aqueous nucleic acid solution and the layers are mixed with a glass rod. The precipitated nucleic acid can be spooled onto the glass rod; the key to success is gentle, even rotation of the glass rod. Excess alcohol can be pressed from the spooled precipitate. This DNA is suitable for many purposes and some laboratories may prefer to go on to the gel-electrophoresis part at this point.

The recovered nucleic acid can be further purified by suspension in 0.015 M NaCl/0.0015 M trisodium citrate (pH 7.0) to dissolve the DNA and chelate divalent cations required for nuclease activity. The solution may have to be stirred gently for as long as 30 minutes to ensure complete DNA dispersion. The salt concentration is adjusted to 0.15 M NaCl and 0.015 M trisodium citrate by the slow addition of a more concentrated saline buffer (1.5 M NaCl/0.15 M citrate, pH 7.0). A second extraction with an equal volume of chloroform/*iso* amyl alcohol (as above) leads to nucleic acid reasonably free of proteins. The DNA is precipitated with 2 volumes of ethyl alcohol at low temperature and redissolved in 0.15 M NaCl/0.015 M trisodium citrate or into the TRIS-phosphate electrophoresis buffer (see page 136). Pure DNA can be obtained by repeated deproteinations until the aqueous layer gives a negative Bradford protein assay.

Alternatively, a pass through a 1-mL DEAE-cellulose column (in a Pasteur pipet) equilibrated to 0.3 M NaCl buffered with 0.01 M TRIS-HCl (pH 7.5) and elution with several column volumes of 1 M NaCl/0.01 M TRIS-HCl with 0.25-mL fractions gives

DNA fractions "clean" enough for most applications. The DNA-rich fraction can be selected by the strong UV absorbance at 260 nm and relatively weak absorbance at 280 nm. The absorbance of DNA at 260 nm can be used to calculate the final yield; 50 μg/mL DNA has an absorbance of 1.0 at 260 nm.

PEG Fractionation

In this experiment the heterogeneous DNA prepared above is fractionated by size-specific precipitation with polyethylene glycol. The initial DNA concentration must exceed 50 μg/mL and the Mg^{2+} concentration must be less than 0.5 mM. The DNA concentration can be estimated from the absorption for a 1-cm light path at 260 nm; a 50-μg/mL DNA solution has an absorbance of 1.0. Enough 5 M NaCl is added to bring the salt concentration to 0.55 M, then the sample is divided into five portions and to each is added a different amount of PEG to bring the concentration to somewhere between 5 and 10%. The samples are incubated at 37°C overnight and the precipitated DNA recovered by centrifugation at 10,000 g for 10 minutes. Because the higher molecular weight DNA precipitates quickly, it is possible to fractionate the DNA by adding progressively higher levels of PEG to the same sample and recovering the DNA precipitated after an hour or so at each increment. Longer times and colder temperatures (0 to 4°C) are needed for the very short pieces of DNA. Precipitated DNA can be redissolved in 0.15 M NaCl/0.015 M trisodium citrate, pH 7.0, and freed from trapped PEG by ethanol precipitation.

DNA preparations can be freed of RNA by incubation in 50 μg/mL *DNase-free* RNase in 0.15 M NaCl/0.015 M citrate followed by another chloroform deproteination and ethanol precipitation. DNA purified in this way can be concentrated by PEG precipitation from 0.5 M NaCl buffered with TRIS-HCl.

Agarose Electrophoresis Buffers

TRIS-acetate (TAE) or TRIS-phosphate (TPE) buffers are commonly used electrophoresis buffers, with the latter preferred because it is a more concentrated buffer with greater buffering capacity.

TAE buffer

0.04 M TRIS	4.84 g/L
0.019 M Acetic Acid	1.14 g/L
0.001 M EDTA (as sodium salt, pH 8.0)	0.416 g/L

TPE buffer

0.08 M TRIS	10.8 g/L
Phosphoric acid (as 85%, conc. solution)	1.55 mL/L
0.002 M EDTA (as sodium salt, pH 8.0)	0.832 g/L

In both cases it is preferable to make up the EDTA solution as a concentrated stock solution (0.5 M) adjusted to pH 8.0 with sodium hydroxide. Two and 4 mL of this stock solution would be used for 1-L volumes of TAE and TPE buffers, respectively. The final buffer solutions should be diluted to 1000 mL with distilled, deionized water.

Pouring Agarose Gels

Horizontal agarose gels can be made in any gel percentage from 0.1 to 3% by dissolving the agarose in the chosen buffer to give the desired final concentration; for this experiment 1% gels will work nicely. The agarose must completely dissolve when the buffer is heated in a boiling-water bath. Electrophoresis-grade agarose (Bio-Rad standard low -m_r) must be used.

A gel mold is prepared by framing a glass or Lucite sheet with autoclave tape to make a shallow "pan" with removable sides. Some commercial gel systems (e.g., Hoeffer) feature Lucite pouring molds with detachable sides. The gel is poured into the horizontal mold until the layer is 3 mm thick (see Figure 10-3). The solution must be free of bubbles as it is allowed to cool to 45 to 55°C. The thickening temperature depends on the gel strength and is usually higher for the denser gels. The sample well forming comb must penetrate the gel at least 2 mm, and the slots must be free of bubbles. It is important that the slots not penetrate all the way through the gel. Once the gel has hardened the comb is gently rocked back and forth until free. The removable walls are stripped from the mold and the gel is carefully placed in the horizontal apparatus. The gel shrinks when dried and must be *just* covered with electrophoresis buffer about 10 minutes before electrophoresis. Any air bubbles in the wells should be teased out with a clean glass capillary.

Sample   Application

The DNA samples are added to the filled well combs with an Eppendorf automatic pipet. For heterogeneous DNA (as the case is here), 2 μg of DNA in sample application buffer is applied to each well. For homogeneous DNA, 0.01 μg of DNA is sufficient for analytical results. The *sample application buffer* should contain

> 0.10 % bromophenol blue
>
> 0.10 % Xylene cyanol
>
> 0.2 M TRIS pH 7.6
>
> 0.1 M sodium acetate
>
> 0.01 M EDTA pH 7.6
>
> 15 % (w/v) Ficoll 400 in water

Ficoll is a water-soluble polysaccharide that makes the application buffer dense enough to sink in the slot.

In this experiment the different DNA length fractions obtained by PEG fractionation are placed in different sample wells so that the specificity of this fractionation process can be evaluated from the resulting pattern of "DNA streaks." The higher the final % PEG, the farther down the gel will be the center of the migrated DNA. Well-defined bands will not be seen. The DNA is much too heterogeneous to migrate as distinct bands but each PEG fraction will have a different average molecular weight and will extend down the gel to a different extent. The molecular weight range of these fractions can be estimated by reserving one or two wells for 2 to 5 μg samples of DNA-size standards. These are restriction enzyme cleavage products of phage DNA, so that the pieces are of well-defined length. The Hind III digest of lamda DNA (sold in 200-μg/mL samples by New England BioLabs) produces eight fragments:

23.72, 9.46, 6.67, 4.26, 2.25, 1.96, 0.59, and 0.10 kilobases

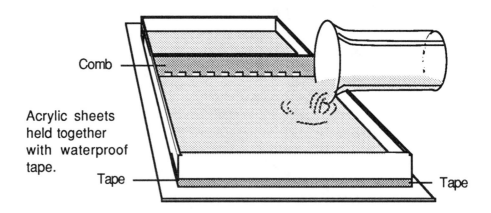

Comb

Acrylic sheets
held together
with waterproof
tape.

Tape   Tape

The form and comb are removed and the agarose
slab is submerged in buffer in the submarine method.

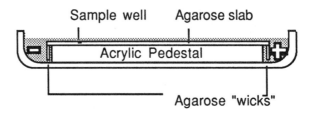

Sample well   Agarose slab

Acrylic Pedestal

Agarose "wicks"

Figure 10-3   Agarose gel is poured into a detachable form

Running the Gel and Ethidium Bromide Staining

The gel is run at 35 to 40 V (at 50 mA) overnight or at 100 V at 50 mA for 5 to 6
hours or less. The progress of the electrophoresis can be monitored by following the
migration of bromophenol blue which comigrates with 300 bp dsDNA. Alternatively,
ethidium bromide (0.5 μg/mL) is incorporated into the gel and electrophoresis buffer
and the gel can be examined directly under UV illumination. Gels can also be immersed
for about 45 minutes in buffer containing ethidium bromide after electrophoresis is
complete to visualize DNA. RNA and ssDNA are also visualized by ethidium bromide,
but the fluorescent yield is considerably less than with dsDNA.

CAUTION: ETHIDIUM BROMIDE IS A POWERFUL
MUTAGEN.
WEAR GLOVES AND AVOID DIRECT CONTACT TO THE
SKIN.
DO NOT DISPOSE OF ETHIDIUM BROMIDE
DOWN THE DRAIN.

DISCUSSION QUESTIONS

1. Why does the rupture of cells lead to diminished turbidity and increased viscosity?

2. What is the purpose of EDTA in the original extraction buffer?

3. Why is it *absolutely necessary* to wear UV-absorbing goggles when observing ethidium bromide-stained gels?

4. If 50 μg/mL of dsDNA in a 1-cm cuvette has an absorbance of 1.0 at 260 nm, what is the $E^{1\%}_{260}$?

5. Why is it important to avoid potassium salts in the extraction and lysis buffers? *Hint:* What is the solubility of potassium dodecyl sulfate in water?

REFERENCES

1. B. J. Bachmann, *Bacter. Rev.*, **1972**, *36*, 525.

2. J. H. Miller, *Experiments in Molecular Genetics*, New York: Cold Spring Harbor Laboratory, 1976.

3. M. Demerec, E. A. Adelberg, A. J. Clark, and P. E. Hartman, *Genetics*, **1966**, *54*, 61.

4. J. T. Lis and R. Schlief, *Nucleic Acids Res.*, **1975**, *2*, 383; J. T. Lis in *Advances in Enzymology*, **1980**, *65*, 347.

5. T. Maniatis, E. F. Fritsch and J. Sambrook, *Molecular Cloning, a Laboratory Manual*, New York: Cold Spring Harbor Laboratory, 1982, p. 150.

6. P. G. Sealey and E. M. Southern, in *Gel Electrophoresis of Nucleic Acids, a Practical Approach*, D. Rickwood and B. D. Hames, eds.,Washington D. C: IRL Press, 1982, pp. 39-76.

7. R. F. Schleif and P. C. Wensink,"Practical Methods in Molecular Biology,"Springer-Verlag, New York, 1981, p.204.

Chapter 11

Human Hemoglobin and Hemoglobin Chain Separation

Red blood cells are separated from plasma by low-speed centrifugation, washed repeatedly with isotonic 0.9% NaCl solution, and disrupted in a tissue homogenizer in cold hypotonic solution to afford a hemolysate sample. Reasonably pure hemoglobin can be recovered by $(NH_4)_2SO_4$ precipitation of impurities, and gel filtration through Sephadex G-25. All the above can be done in one 3-hour laboratory period. Long-term storage of hemoglobin is possible if it is converted to the carbonmonoxy form, then dialyzed overnight. In two additional laboratory sessions, the alpha and beta chains of carbon monoxyhemoglobin can be separated by first reacting carbon monoxyhemoglobin with p-hydroxymercuribenzoate followed by CM-Sephadex ion-exchange chromatography with a pH-gradient elution between pH 6.8 and 7.5.

KEY TERMS

Centriprep concentration   CM-Sephadex ion-exchange
Diafiltration   Dialysis
Hydroxymercuribenzoylation   Osmotic lysis
Plasma

BACKGROUND

Human blood is a complex substance containing a variety of cells suspended in an amber liquid known as blood *plasma*. The plasma is a solution of salts, gases, carbohydrates and proteins, including the proteins that confer immunological and clotting properties to blood.

The most abundant cell is the red blood cell or erythrocyte, but at least a dozen other types of cells as well as cellular debris can be found in blood. The principal protein in erythrocytes is hemoglobin; the hemoglobin content of packed erythrocytes is usually between 14 and 18 g per 100 mL, which makes it an exceptionally abundant protein. Human erythrocytes do not have nuclei or mitochondria, but they do carry on many of the metabolic activities of most living cells and they contain a multitude of other proteins as well as hemoglobin. Many of these proteins are membrane bound. For example, SDS PAG electrophoresis of erythrocyte cell membrane proteins yields 1 to 6 prominent bands and at least 35 weaker ones, ranging in molecular weight from 10,000 to 360,000 (see Chapter 29). Hemoglobin can be separated from membrane-bound proteins by removing the membrane from lysed cells; the water-soluble hemoglobin will stay in solution when the membranes are pelleted by centrifugation.

Because of its abundance, hemoglobin can be easily separated from most of the other constituents of blood.[1,2] The plasma can be removed by low-speed centrifugation which packs the erythrocytes as a pellet. If a citrate-phosphate-dextrose (CPD) buffer is used to inhibit the blood -clotting reaction and preserve the cells, little irreversible aggregation will occur and the cells can be resuspended in an isotonic 0.9% NaCl solution. The washing procedure can be repeated several times with little loss due to aggregation. Erythrocytes lyse in distilled water, which is hypotonic; excess water enters the cells, swelling them to the bursting point. Neutral 20% saturated $(NH_4)_2SO_4$ precipitates most stromal impurities as a gelatinous mass, which can be removed after precipitation is complete. Dialysis against distilled water removes the $(NH_4)_2SO_4$ as well as most low-molecular weight impurities and gel filtration through a Sephadex G-25 column, which retains solutes of 5000 D or less, completes the purification of hemoglobin for most practical purposes. Additional purification by electrophoretic or chromatographic procedures is possible and is required for special purposes, including purification of abnormal human hemoglobins or glycosylated derivatives such as hemoglobin A_{Ic}, which makes up about 5% of the hemoglobin in normal blood cells. In diabetes mellitus, Hb A_{Ic} can occur to the extent of about 7 to 10%.[3]

The iron in native hemoglobin is Fe(II) even when bound to oxygen but upon prolonged standing it can be converted irreversibly to Fe(III) to form methemoglobin. Methemoglobin is the hydrated Fe(III) species derived from hemoglobin; it cannot reversibly bind oxygen (see Chapter 13). Since native hemoglobin is readily oxidized in air, these preparations must be stored in the cold at neutral pH. Hemoglobin solutions should be used within a week, as methemoglobin builds up. Prolonged storage is possible if hemoglobin is converted to the very stable carbon monoxyhemoglobin by gently bubbling CO through a bright red oxyhemoglobin solution.

Separation of α and β Chains

The separation and characterization of hemoglobin alpha and beta chains is a convenient project that utilizes the techniques developed so far in this laboratory. Although the alpha and beta chains differ in pI, this difference is not large enough to permit separation of these chains by some large scale method such as ion-exchange chromatography. This difference in pI is magnified by a reversible covalent modification. Tetrameric carbon monoxyhemoglobin can be dissociated into the alpha and beta chains by reaction with either p-chloro- or p-hydroxymercuribenzoate (POMB) at pH 5.5.[4] The chloro- or hydroxyl- group ligands to mercury are readily displaced by the sulfur of cysteine. The alpha chain contains one free cysteine sulfhydryl and the beta chain two; these groups are converted to anionic groups by reaction with POMB, as in Figure 11-1.

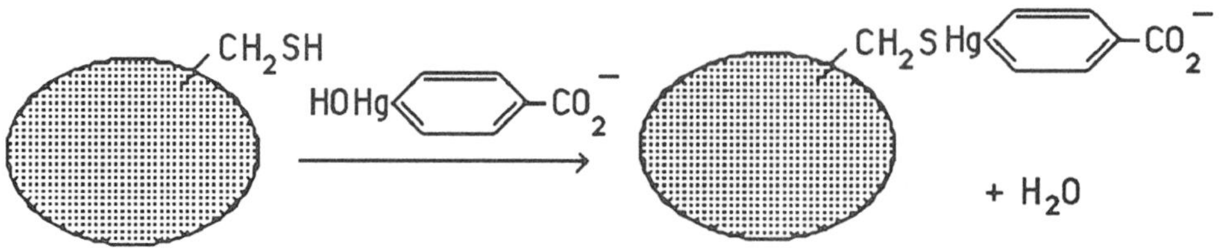

Figure 11-1 Reaction of POMB with sulfhydryl groups

These chains will then differ in net charge and can be separated by ion-exchange chromatography on a carboxymethyl (CM) cellulose[4] or CM-Sephadex[5] cation exchange column. The beta chain comes off first at pH 6.8, the alpha at pH 7.3. Carboxymethyl (CM) cellulose and Sephadex are weak acid cation exchange materials, which binds cations at pH values above the pK_a of the carboxyl residues of the exchanger. The use of a pH gradient to elute bound cations from such a resin is an especially good way of amplifying the small differences a single charge makes. At lower pH the chromatographic column packing is less negatively charged and bind cationic peptides less tightly so that slight differences in pI can translate into respectable differences in affinity constants between protein and ion exchanger.

The separated chains may be reconverted to the sulfydryl form by incubation with a 50-fold excess of cysteine in 0.05 M phosphate buffer, pH 7.3. The cysteine displaces the mercury from the protein sulfhydryl groups.

Dialysis and Centriprep Concentration

The purification of hemoglobin, the formation of the peptide POMB derivatives and the regeneration of the isolated chains all require the separation of high- and low-molecular-weight solution components. The latter two cases are markedly speeded up either by the use of *continuous dialysis* in a pressure cell or centrifugation in a *Centriprep concentrator*, which amounts to filtration through a molecular weight discriminating membrane in a special low-speed centrifuge tube. Dialysis separates the low-molecular weight mercury reagents and their cysteine derivatives from the high-molecular-weight peptide chain, which cannot pass through the dialysis membrane. Dialysis and Centriprep concentration both exploit the size difference between proteins and molecules of lower molecular weight.

In dialysis the hemoglobin salt mixture, obtained by $(NH_4)_2SO_4$ precipitation, is enclosed in a semipermeable membrane bag with "pores" large enough to permit the diffusion (in either direction) of small-diameter molecules but too small to permit passage of the protein. The bag is suspended in a dialysis tank filled with a large volume V_T of the appropriate buffer, as shown in Figure 11-2. Given enough time, the smaller molecules in the bag of volume V_B are diluted into the total solvent volume. The final concentration of the low-molecular-weight species within the bag will be reduced by the factor.

$$V_B / (V_T + V_B) < 1 \qquad\qquad (11\text{-}1)$$

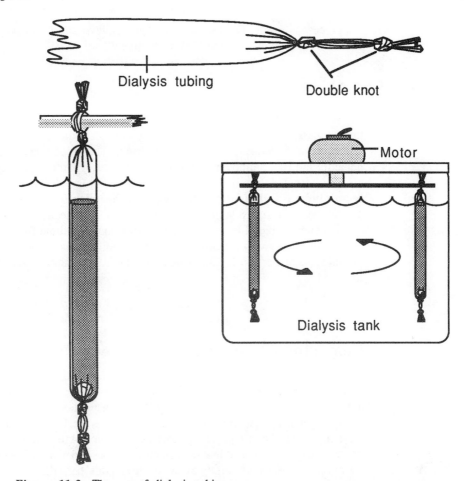

Figure 11-2 The use of dialysis tubing

The concentration of protein will remain unchanged, save for the loss of some material due to precipitation. Dialysis is often used to change buffers for a protein solution; the alternative of precipitation and redissolving almost always leads to some denaturation. Dialysis is a very mild method by comparison.

A faster way of doing dialysis, which is especially useful for large volumes is *continuous dialysis* , in which the protein solution is forced by gas pressure through a molecular-weight-selective membrane filter, which permits passage of solvent and all low-molecular-weight species but not the protein. This process is much faster than conventional dialysis because the flow through the membrane is driven by a positive nitrogen pressure rather than simple diffusion. In continuous dialysis fresh buffer from a reservoir is added to replace the buffer that has passed out the bottom of the filter, so that the buffer in the protein solution can be exchanged for a new one. The same apparatus can also be used to concentrate the proteins in solution in a process called *diafiltration* by driving the solution through a membrane which retains the larger protein molecules, their concentration increases in the solution that remains above the filter. Centrifugal force can also drive a solution through a membrane. There are several ranges of molecular weight discrimination available with these concentrators. When working with hemoglobin or its POMB derivative, it is recommended that the concentrator with a 30,000 D cutoff be used rather than the 10,000 D cutoff because the process is faster. The use of continuous dialysis and Centriprep concentration is illustrated in Figures 11-3 and 11-4, respectively.

EXPERIMENTAL SECTION

Blood for this project can be obtained from volunteers if the services of a licensed phlebotomist are available. A more likely (and less painful) approach is to acquire a unit of expired blood from a local blood bank. The blood must be certified free from hepatitis and HIV antibodies (almost surely the case) and can usually be obtained as a courtesy from a blood bank, **especially if replaced by fresh contributions from student and faculty volunteers.** In handling blood, one should use gloves and do the initial stages in the sink. All spills must be cleaned up immediately.

One unit of blood is enough for 6 to 10 students and the first part of the isolation is best done as a group experiment. Students should work in groups of two or three for all subsequent portions. Convenient overnight stopping points in this project are indicated.

First Laboratory Period

Hemoglobin Isolation

One unit (450 g) of human blood in citrate buffer is diluted to 1000 mL with 0.9% NaCl solution and centrifuged at low speed (2000 x g) at10°C for 10 minutes. The amber plasma supernatant is removed quickly by aspiration and discarded. The red cells are resuspended in isotonic NaCl (0.9%) to give a total volume. of 1000 mL. This washing procedure is repeated at least twice again. After the last centrifugation the erythrocytes are suspended in three volumes of ice-cold deionized water and kept cold for 20 minutes. Many erythrocytes burst during this period. Complete disruption of erythrocytes is assured by brief homogenization of the cold hemolysate sample. The erythrocyte suspension is placed in a *square* plastic bottle and the probe of a Brinkman Polytron homogenizer is submerged nearly to the bottom of the bottle. The probe must be submerged at least 3 inches to ensure flow of cold solvent through the probe head.

High-efficiency homogenization can be obtained at speeds of 5 to 7, especially if the probe is slightly off from the center of the bottle. The homogenization speed should slowly be turned up to speed 7 and held 30 to 60 seconds then stepped down. **The Polytron should never be run for more than 60 continuous seconds.** Total homogenization times of 2 to 3 minutes are all that is necessary. Short bursts may be necessary with very foamy samples. **The homogenizer probe must be cleaned by "homogenizing" several samples of deionized water. A clean probe will not cause any foaming when run in water.**

After homogenization, 20 mL of cold (4°C) neutral saturated $(NH_4)_2SO_4$ is added slowly to every 80 mL of hemoglobin solution and after 15 minutes the stromal impurities are removed by centrifugation at 10,000 x g (4°) for 30 minutes.

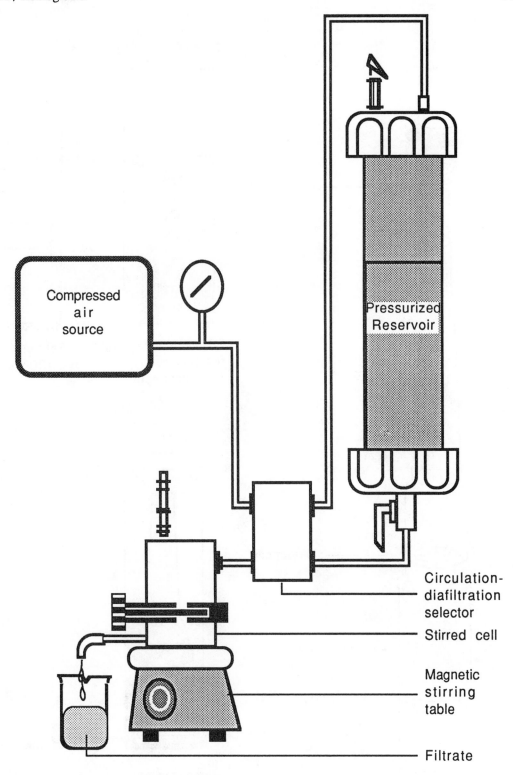

Figure 11-3 Apparatus for diafiltration/continuous dialysis

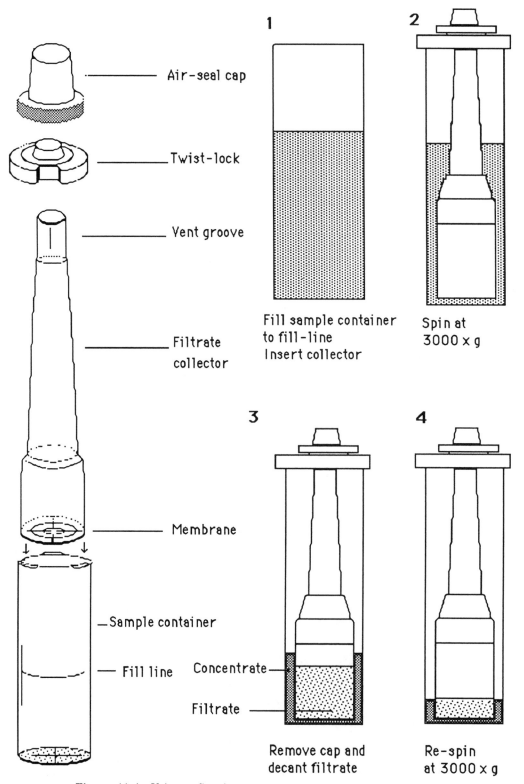

Figure 11-4   Using a Centriprep concentrator

The red jellylike precipitate is discarded. The bright red supernatant contains primarily oxyhemoglobin [Hb(II)O$_2$], which can be separated from low-molecular-weight impurities by dialysis against distilled water overnight or by gel filtration through Sephadex G-25. Each group of students is given a portion of the hemoglobin mixture to be dialyzed overnight. Some of this solution is used for the chain separation procedure and some can be used in projects described in Chapters 12 and 13.   The oxyhemoglobin can be stored for only a few days in the cold but must be converted into the carbon monoxy form for prolonged storage.

Hemoglobin Carbonylation

Carbon monoxyhemoglobin [Hb(II)CO] is prepared by bubbling CO gas through a hemoglobin solution as shown in Figure 11-6. If a fritted gas dispersion tube is used, the conversion is complete even at low flow rates in a minute or so; the color change to a darker red is very pronounced. The [Hb(II)CO] solution is diluted to 2 to 4% in 0.02 M KH$_2$PO$_4$, and enough freshly dissolved POMB is added to give a POMB-to-protein mole ratio of 10:1. POMB dissolves only slowly at neutrality, but a stock solution can be prepared by first dissolving it in dilute NaOH solution followed by just enough acetic acid to give a slight precipitate. The ionic strength is brought to 0.1 M with NaCl and the pH adjusted to 6. The reaction mixture is allowed to set overnight in the coldroom to ensure complete reaction. Then excess POMB and (NH$_4$)$_2$SO$_4$ are removed by one of three methods, the choice depending on the time available between the first and second laboratory periods, the volume of sample, and the availability of equipment.

CAUTION: CARBON MONOXIDE IS A POISONOUS GAS.

The protein concentration of the carbon monoxy form stock solution can be calculated from the absorbance observed at 540 nm. A 1% solution has an absorbance of 8.4 (i.e., $E^{1\%}_{1cm} = 8.4$). Since visible spectrometers measure absorbance accurately only over the range 0.1 to 1.5, the hemoglobin solution must be diluted by a known factor and the $E^{1\%}$ value calculated from the dilution factor.

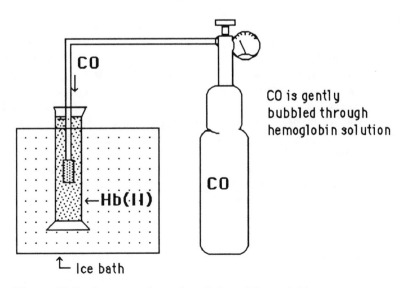

Figure 11-5 Apparatus for carbonylation of hemoglobin

Second Laboratory Period

Conventional overnight dialysis against 0.01 M phosphate buffer at pH 6.7 is the least expensive method and can be scaled to any volume, but it requires at least overnight dialysis with a large volume of solvent or several buffer changes to ensure complete removal of unreacted POMB. A more rapid method, which is especially applicable if large volumes are to be processed in a large-scale purification, is the use of continuous dialysis with a 10-kD molecular weight discriminating membrane and a solvent reservoir containing 0.1 M NaCl in 0.01 M phosphate buffer. The sample can then be concentrated by diafiltration before application to a column. A quicker method suitable for a small sample volume is the use of Centriprep concentration.

Dialysis (Option 1)

Dialysis tubing comes on rolls like tape. In this form the tubing is stiff and a little brittle. The tubing is softened by boiling for 30 minutes in 1 mM EDTA/10 mM sodium bicarbonate followed by 10 minutes in boiling distilled deionized water. The EDTA is necessary to remove cupric ion adhering to the tubing, since that ion reacts with sulfhydryl groups. The tubing is simply cut to the desired length and put into a beaker of boiling EDTA solution. A glass rod is used to hold down the tubing, which will tend to float.

BOILING WATER IN AN OPEN BEAKER CAN LEAD TO
SERIOUS BURNS IF THE SOLUTION BUMPS
A BOILING CHIP MUST BE USED.
CAUTION: WEAR SAFETY GLASSES!

After this treatment the tubing should be soft and pliable when wet. The tubing is removed and allowed to drain but not dry out. A double knot in one end converts the tubing into an open bag. The tubing tears very easily, so undue stretching should be avoided when tying the knots. The bag is held upright with one hand while a funnel is inserted into the open end. The protein solution is poured through the funnel into the bag. The funnel is removed, and if the bag is to hang downward, a glass marble is dropped into the bag. The bag should not be filled completely because some slack is needed in the upper end to tie another knot and an air space should also be left so that the tubing bag floats in the dialysis bath. Dialysis tubing can be mounted within the rotating wheel of a dialysis tank (see Figure 11-2) or simply allowed to float in a tall beaker or graduated glass cylinder (3000 to 5000 mL).

An air bubble rather than a weight within the bag is necessary to allow it to float. A magnetic stir bar can be used to stir the contents of the cylinder to ensure even buffer mixing. Dialysis tubing cannot be allowed to dry out once boiled because it becomes stiff and very brittle.

Continuous Dialysis (Option 2)

Your instructor will demonstrate the assembly of a stirred continuous dialysis cell and how to fill it with Hb(II)CO solution. There are a number of different models, but the one illustrated in Figure 11-2 is representative. Air or nitrogen pressure above the solution is used to force the liquid through the filter membrane so that large volumes can be processed much more quickly than by the bidirectional process of simple dialysis. Continuous dialysis can be used to change a buffer solution without altering protein concentration. The protein solution is placed in the cell and the new buffer in the pressurized reservoir (see Figure 11-3). Alternatively, a large volume of Hb(II)CO solution in the reservoir can be concentrated, before application to a column (see below). The volume of protein solution in the reservoir is reduced as buffer is squeezed through the filter. Because the protein is retained behind the filter, its concentration goes up. This process is called diafiltration. Diafiltration membranes are easily clogged with precipitated protein. All solutions must be clarified by centrifugation or Millipore filtration prior to dialysis, and the cell must always be stirred to keep the membrane surface clean. A protein solution should not be reduced to dryness. Occasionally membranes leak; the filtrate should never be discarded until it has been shown not to contain valuable protein. Continuous filtration requires low pressure compressed air or nitrogen (10 to 20 psi).

Pressures higher than 30 psi will damage the cell or selector (a plastic switch regulating flow through the reservoir). The compressed gas cylinder must be firmly mounted to the wall and all tubing tightly connected.

WEAR SAFETY GLASSES WHEN WORKING WITH COMPRESSED GASES.

The membrane filter must be equilibrated to water of buffer by soaking overnight, then placed within the cell, shiny side up. A pair of forceps are required. Membrane filters are quite expensive (unlike filter paper) and can be used 4-6 times unless clogged with protein. When diafiltration is operating properly, filtrate flow rates of 0.5 mL/min at 15 psi should be realized. A clogged filter or a leaky cell will lead to lower flow rates.

Centriprep Concentration (Option 3)

The use of centriprep concentrators, which contain a molecular-weigh- descriminating filter held in a low-speed centrifuge tube, permits the rapid concentration of a small volume of protein solution. Diafiltration is more convenient for larger volumes, but the Centriprep method is best if one wishes to reduce a 50-mL volume by a factor of 10 or so. The sample container is filled to the line, then the plastic filtrate collector is inserted and the twist-lock and air-seal caps are put in place and tightened. Twenty minutes centrifugation at 4000 x g forces buffer but not protein into the filtrate collector, from which it can be discarded by inverting the concentrator and pouring the pink-to-nearly colorless liquid from the top.

Since the filtrate collector can be removed and additional sample added to the sample container, the process can be repeated several more times to give a concentrated sample of Hb(II)CO, considerably reduced in volume. The sample container is then filled several times with 0.1 M NaCl at pH 6 to exchange the buffer without loss of protein. One should use cold buffer and protein solutions and keep the concentrator on ice whenever possible.

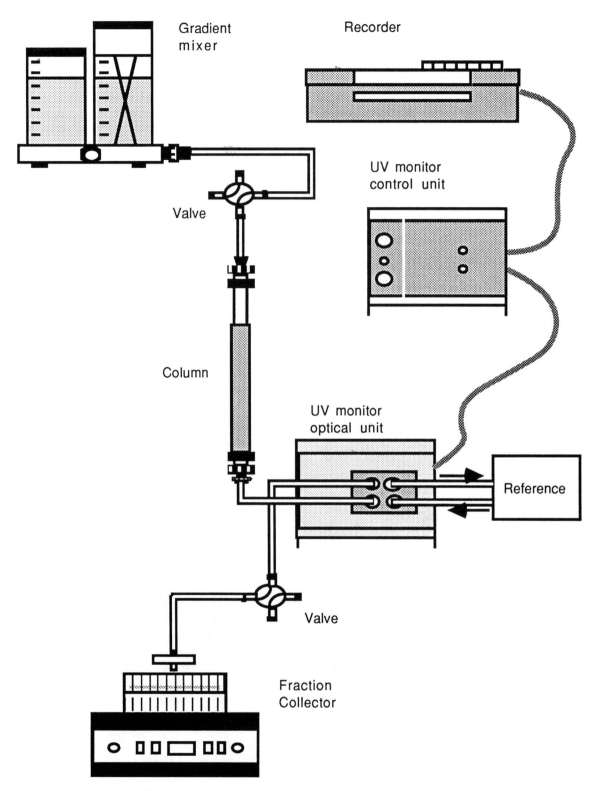

Figure 11-6 Gradient elution with a fraction collector

Third Laboratory Period

Separation of Hemoglobin Chains

The chains can be isolated by ion-exchange chromatography on a 7-cm CM-Sephadex column[5] equilibrated with 0.01 M phosphate (pH 6.0) and eluted with an increasing pH gradient. The gradient can be obtained by mixing 0.01 M KH_2PO_4 phosphate buffer (pH 6.7) with 0.02 M K_2HPO_4 (pH 8.5) in a gradient mixer, as shown in Figure 11-6. The lower pH solution is in the forward stirred chamber of the gradient mixer. The flow rate is 0.5 mL/min and 2-mL fractions are taken. An elution profile obtained by measuring the absorbance for each fraction at 310 nm and plotting this value versus the elution volume should give three peaks. In order of increasing volume they correspond to the beta subunit, undissociated tetramer, and alpha subunit, respectively.

The protein solutions for each fraction are rather dilute but can be concentrated by either diafiltration or the Centriprep method. The isolated subunits occasionally precipitate when concentrated. Solutions on the order of 1 mg/mL can be analyzed by disc electrophoresis either before or after reaction with excess cysteine in a dialysis bag. The cysteine removes the POMB moiety.

PROJECT EXTENSIONS

Human blood and hemoglobin can also be used in a number of other projects:

1. Isolation of Hemin from either hemoglobin or myoglobin; see Chapter 12 for details.

2. Visible spectroscopy of hemoglobin in its various ligation states; see Chapter 13 for details.

3. Characterization of blood plasma proteins by disc or SDS electrophoresis as in Chapters 8 or 9.

4. Two-dimensional PAGE of erythrocyte membrane proteins as described in Chapter 29.

DISCUSSION QUESTION

How would you experimentally determine the chemical identity of the three heme protein peaks that come off the CM sephadex column?

REFERENCES

1. A. Rossi-Fanelli, E. Antonini and A. Caputo, *J. Biol. Chem.*, **1961**, *236*, 391.

2. E. Antonini and M. Brunori, *Hemoglobin and Myoglobin in Their Reactions with Ligands*, North-Holland, 1971, Chap 1.

3. E. Bucci and C. Fronticelli, *J. Biol. Chem,* **1965**, *240*, PC 551.

4. L. J. Kaplan, *J. Chem. Educ.,* **1976**, *53*, 64.

5. K. H. Winterhalter and A. Colisimo, *Biochemistry*, **1971**, *10*, 621.

Chapter 12

Isolation of Hemin and Apohemoglobin

Hemin is extracted from hemoglobin or myoglobin with 2-butanone under mildly acidic conditions. Apohemoglobin can be prepared by this method and reconstituted with unique hemin derivatives.

KEY TERMS

Apohemeproteins   *Heme prosthetic groups*
Reconstitution of hemeproteins

BACKGROUND

The heme [Fe(II)] prosthetic group of hemoglobin or myoglobin is bound to protein by noncovalent interactions involving salt links to the porphyrin propionate anions, van der Waals interactions with the porphyrin vinyl groups and histidine ligation to the iron.[1] All these interactions are weak enough to permit the reversible extraction of hemin [Fe(III) protoporphyrin IX] from hemoglobin or myoglobin in acidic solutions.[2,3] The iron is oxidized to the Fe(III) state when freed from the protein. Apomyoglobin is reasonably stable and can be preserved at low temperatures for several days. Apohemoglobin is less stable but can be prepared and used in reconstitution experiments employing unusual hemin derivatives or even chlorophyll derivatives to give "green globulins." [4] Other heme proteins, such as the cytochromes, have their heme prosthetic groups locked into place by covalent bonding through the vinyl group. The prosthetic group is removed only under rather drastic conditions and the proteins cannot be reconstituted.

Apohemoglobin purification is a good test of the experimenter's skill in handling unstable proteins and has been used as an end-of-first-semester test in the author's laboratory. The visible spectrum of free hemin is characteristically different from that of Fe(III) methemoglobin and lacks the several peaks between 450 and 650 nm that are observed with the protein. Hemin can be further purified by esterification, silica gel TLC, and base-catalyzed hydrolysis.[5]

EXPERIMENTAL SECTION

The following procedure can be run with either hemoglobin or myoglobin but not with cytochromes. An ice-cold, salt-free hemoglobin solution (0.4 g per 10 mL) is clarified by centrifugation (5000 x g, 10 minutes, 2°C) and acidified by the slow addition of an equal volume of 0.03 M HCl. The final pH should be in the range 3.4 to 3.6. The ice-cold acidified hemoglobin solution should be slowly layered with an equivalent volume of ice-cold 2-butanone in a graduated cylinder standing upright in an ice bucket. The cylinder is capped with Parafilm and quickly inverted three times. The upper butanol layer will take on a purple color as the hemin is extracted into it. The butanone layer is carefully decanted away and the extraction procedure repeated twice again to yield a dark butanone solution containing hemin which can be recovered or concentrated for other uses. The last extraction should yield a pale pink upper layer and a nearly colorless lower layer; if not, an additional extraction is required.

The lower apohemoglobin layer is freed from 2-butanone by dialysis against at least 2 liters of distilled ice-cold water overnight followed by dialysis against 0.1 M phosphate buffer (pH 7.01) at 2°C. The phosphate buffer contains 22.8 g of $K_2HPO_4 \cdot 3H_2O$ and 13.6 g KH_2PO_4 per 2 L of solution.

The UV absorbance of apohemoglobin is characterized by strong absorbance at 280 nm. Apohemoglobin gives a clear, very pale straw colored solution in water and precipitates under all extremes of pH or temperature or mechanical shock. Precipitation during the dialysis step usually means that the protein was roughly treated during purification.

The following procedure for reconstituting apohemoglobin with hemin is recommended. Hemin is dissolved in a minimum volume of 0.1 M NaOH and this solution diluted about 10-fold, then added dropwise to a stirred apohemoglobin solution in an ice bath. At least 15 minutes is required for the hemin addition. The reconstituted hemoglobin solution is allowed to set in the ice bath for 1 hour, then the pH is raised to 10 with 0.1 M NaOH. Dialysis overnight against ice-cold distilled water and centrifugation (5000 x g, 10 mim., 2°C) affords a hemoglobin solution with the spectral properties of native hemoglobin. The reconstitution experiment is very challenging and works only if the apoprotein remains stable. Any vigorous stirring or temperatures above 2°C will precipitate apoprotein.

PROJECT EXTENSION

A number of porphyrin derivatives differing in ring substitution patterns and metal are commercially available and many have been reconstituted into heme proteins, mainly for NMR investigations of the interaction between the porphyrin ring and the protein.[6]

DISCUSSION QUESTIONS

1. Why is the extraction of hemin more easily accomplished at low pH; that is, what interaction becomes less important below pH 4?

2. Why must the pH be raised to pH 10 during the reconstitution of hemin into apoprotein?

3. What is the solubility of 2-butanone in cold water? Does this explain the need for several changes of the dialysis buffer?

4. Another possible experimental extension would be the preparation of apo-alpha or apo-beta hemoglobin chains from the subunits purified in the preceding project. Why are these apo-subunits very unstable?

REFERENCES AND NOTES

1. M. F. Perutz, *Ann. Rev. Biochem.*, **1979**, *48*, 327-86.

2. F. W. J. Teale, *Biochem. Biophys Acta*, **1959**, *35*, 543.

3. L. L. Shen and J. Hermans, *Biochemistry*, **1972**, *11*, 1845.

4. S. G. Boxer and K. A. Wright, *J. Am. Chem. Soc.*, **1979**, *101*, 6791.

5. K. Smith, *Porphyrins and Metalloporphyrins*, New York: Elsevier, 1975, pp. 835-837.

6. Consider the important work of G. LaMar and K. Smith on heme proteins reconstituted with isotopically labeled hemin.

Chapter 13

Visible Spectroscopy of Hemoglobin Derivatives

The visible spectra of heme proteins depends on the oxidation state and the sixth ligand. In this experiment the visible spectra of hemoglobin with various ligands attached to the heme iron, in both the Fe(II) and Fe(III) states, are compared.

KEY TERMS

Charge transfer bands   *Heme ligation states*
Low spin and high spin   *Soret bands*
Metaquo-, methydroxy-, and metcyanohemoglobin

BACKGROUND

Oxygen is only sparingly soluble in water and all large multicellular animals require oxygen-carrying protein molecules to convey oxygen to tissues efficiently. Heme proteins serve this function in "higher" animals. Hemoglobin is the oxygen carrier in red blood cells which transports oxygen from the lungs, where the oxygen tension is high, to tissues with high oxygen requirements such as muscles. Hemoglobin releases oxygen when the oxygen tension is low. In muscles, released oxygen is taken up by myoglobin, which facilitates oxygen movement in muscle while maintaining a reserve concentration of oxygen at a much higher level than that possible if oxygen were merely dissolved in water.

Both myoglobin and hemoglobin have an iron-containing *porphyrin ring* linked to the protein by histidine-iron coordination and by noncovalent forces. This porphyrin ring is what gives these proteins their distinctive color. The iron is ligated to four nitrogen atoms at four positions in the large planar aromatic organometallic molecule.

The iron can form up to two additional bonds, one above and one below the heme plane. In hemoglobin and myoglobin, one of these additional linkages is to the imidazole side chain of one of the protein histidines. The sixth coordination site can be occupied by a variety of ligands including molecular oxygen and carbon monoxide. The sixth site even can be unoccupied. The iron atom in the porphyrin can be in one of two possible oxidation states: the Fe(II) [ferrous or +2 state] or the Fe(III) [ferric or +3 oxidation state]. In either case the positive charge is largely delocalized into the protoporphyrin ring, which has a 2- charge distributed among the four nitrogens.

Not all combinations of ligands and iron oxidation states give rise to stable compounds. For example, only the Fe(II) state can bind molecular oxygen. The interaction between the sixth ligand and the remaining electrons on the iron accounts for the properties of each complex (see Table 13-1).

Table   13-1
PROPERTIES OF HEMOGLOBIN COMPLEXES

Oxidation State	Sixth Ligand	Spin State	Soret λ_{max}[a]
Fe(II)	O_2	Diamagnetic	415
Fe(II)	CO	Diamagnetic	419
Fe(II)	None (deoxy)	S = 2	430
Fe(III)	CN^- (metcyano)	S = 1/2	420
Fe(III)	OH^- (methydroxy)	S = 1/2	405 (pH 8.5)
Fe(III)	H_2O (metaquo)	S = 5/2	405 (pH 6.5)

[a] Data from reference 1.

The spin state, color, and stability of each complex differ. The visible spectra of heme proteins depends on the oxidation state and the sixth ligand, and changes in the visible spectrum are useful in studying the structure and function of hemeproteins.[1-3] Although the visible spectra of these complexes differ in significant ways, the general features of a porphyrin electronic spectrum are preserved in almost all complexes and the position and intensities of certain characteristic bands give information regarding the geometry and coordination scheme of the protein. Porphyrins and metalloporphyrins, including heme complexes, have a visible spectrum characterized by one or two very intense bands near 400 nm called *Soret bands* and several weaker bands at longer wavelengths. The Soret band is characteristic of the delocalized macrocycle and is due to a $\pi \rightarrow \pi^*$ transition. Additional bands (usually four) occur in descending order of intensity between 450 and 650 nm and are also associated with the porphyrin ring. These are called alpha, beta, gamma and so forth for the first, second, and third bands appearing at longer wavelengths from the Soret band. In some cases these less intense transitions are accompanied by charge transfer bands in which the electronic transition can be viewed as an electron transfer between the metal and ligand. Each of the five d-orbitals of a transition metal can contain up to two electrons. Fe(III) can therefore have five unpaired d-electrons or one unpaired d-electron. These two forms are both paramagnetic and are called *high-spin* S = 5/2 and *low-spin* S = 1/2 complexes respectively. Fe(II) can exist as a low spin S = 2 paramagnetic complex or as a diamagnetic (no unpaired electrons) complex. Orbitally forbidden metal d→d* transitions are especially weak for Fe(III) because they are also spin forbidden.

The oxidation state of iron can be increased with an oxidizing agent such as potassium ferricyanide or lowered by a reducing agent such as sodium dithionite. The metaquo form can be converted to the methydroxy form by raising the pH above 7.5, and the metcyanoform can be generated by adding KCN to either the metaquo or methydroxy forms.

In this experiment we will prepare several of the important complexes of hemoglobin and determine the wavelengths of the Soret and minor bands with a recording spectrometer. In principle all the reactions can be done in a series starting with the least stable deoxyhemoglobin and ending with the very stable Fe(III)CN and Fe(II)CO compounds (Figure 13-1). However, in this laboratory each derivative will be prepared from a freshly dialyzed sample of oxyhemoglobin, prepared according to the procedure of Chapter 11, and the excess reducing or oxidizing agents and all low-molecular-weight reaction byproducts will be removed by gel-filtration on a Sephadex G-25 column. The deoxy form must be prepared in an inert atmosphere (N_2) in a glove bag with solvent and gel material that has been very extensively deoxygenated. The solution must be transferred to a glass cuvette with an airtight stopper. The momentary introduction of air into the cuvette will permit conversion of the deoxy to the oxygenated form of hemoglobin. Repeated scans from 360 to 600 nm during this process will show the buildup of the latter at the expense of the former. Working with the deoxy form is especially challenging.

EXPERIMENTAL SECTION

Because recording spectra and running columns both take time and require special equipment, these experiments should be done in groups of four to six students. Enough purified oxyhemoglobin (Chapter 11) is diluted to 100 mL with 0.01 M phosphate buffer (pH 7.0) to give a solution with maximum Soret band absorbance of about 2.0. The oxyhemoglobin form should be bright red, while the metaquo form is reddish brown. If the sample is in the oxy form, the extinction coefficient of the Soret band at 415 nm is 1.25×10^5 M^{-1} cm^{-1}. Any precipitated protein should be removed by decanting or centrifugation. The spectrum of the oxy form should be recorded from 350 to 600 nm vs. a water blank. The instructor will explain the operation of the spectrometer.

Deoxyhemoglobin (Optional)

This protein is very unstable in air and reacts upon exposure to oxygen to form the oxy form and ultimately the metaquo form. All steps, including the preparation of a 5-cm G-25 column, must be performed in a glove bag continuously purged with nitrogen. Attempts to deoxygenate a column with large volumes of deoxygenated solvent after it has been poured are generally not successful. The solvent must be briefly boiled and allowed to cool down under a stream of nitrogen. The dry gel material must be made up in deoxygenated water while it is still hot. Note that the nitrogen flow must be sufficient to remove the excess water vapor that would otherwise condense on the glove bag, obscuring visibility into the bag. Do not pass nitrogen into the gel mixture and start over if you get excessive bubble formation in the gel.

FLOW CHART

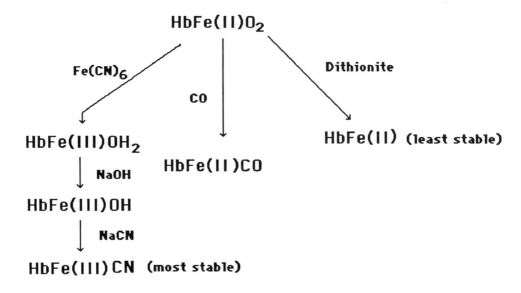

Figure 13-1 Flow chart of heme iron oxidation and ligation states

Some practice is required to be able to pour a column in a glove bag. Deoxyhemoglobin is prepared by adding 1 mg of $Na_2S_2O_4$ to 10 mL of undiluted hemoglobin solution. The dithionite reduction byproducts can be removed with a short (5 cm) G-25 column before the material is transferred to a tightly stoppered cuvette. If a small portion of the reduced hemoglobin is transferred to a sealed cuvette without removing excess dithionite and these byproducts, its spectrum may have absorbances due to nonprotein impurities left by the reducing agent. The course of oxygenation of the hemoglobin sample can be monitored by repeatedly scanning spectra after the stopper has been removed. The presence of excess dithionite and its byproducts, not removed by gel filtration, will influence this process. The time course of this reaction is less reproducible without the gel filtration step.

Carbon Monoxyhemoglobin

This protein is more stable than the oxyhemoglobin and can be prepared from it by flushing a stream of CO from a lecture bottle into an open container of the oxy-form as described in Chapter 11; a fritted glass tube is useful, but the reaction is complete in a minute or so even with a gentle flow through a Pasteur pipet on the end of a rubber hose.

CAUTION: CARBON MONOXIDE IS A POISON.

USE A HOOD!

This cherry red compound is stable enough for the spectrum to be run in an open cuvette but it does undergo a photochemical dissociation in bright light after several hours, to yield the deoxy form if the CO is removed in a stream of nitrogen. This alternative method of generating the deoxyform is best for large-scale preparations.

Metaquohemoglobin

This compound can be prepared by oxidizing oxyhemoglobin with potassium ferricyanide. The ferrocyanide resulting from the oxidation of the heme iron binds strongly to the protein and must be removed by passage through Sephadex G-25. It is not necessary to work in a glove bag for this step. The spectrum of the *metaquo* can be recorded as a function of pH so that the conversion to the *methydroxy* form can be accomplished in the cuvette. Similarly, the *metcyano* form can be prepared by adding NaCN or KCN to methydroxyhemoglobin below pH 8.5.

PROJECT EXTENSIONS

1. The preparation of a wide variety of other ligation derivatives of hemoglobin is possible. This protein binds to all sorts of nitrogen, sulfur and oxygen ligands.

2. A kinetic study of the conversion of the deoxy to the oxy-form, upon adding a small volume of oxygenated solvent through the septum with a syringe, is possible. Since oxygen is a reactant, meaningful kinetic results can be obtained only if one monitors the oxygen concentration remaining in solution with an oxygen electrode (See Chapter 31).

REFERENCES

1. M. R. Waterman, *Methods in Enzymology,* **1978**, *52*, 460.

2. R. W. Hanson, *J. Chem. Educ.*, **1981**, *58*, 75.

3. E. Antonini and M. Brunori, *Hemoglobin and Myoglobin in Their Reactions with Ligands,* Amsterdam: North-Holland, 1971.

Chapter 14

Enzyme Isolation, Purification, and Characterization

Practical guidelines for the purification of proteins, especially enzymes, are developed in terms of the information needed and the choices one must make in designing a purification scheme. An introduction to enzyme assays, enzyme kinetics and the proper format for reporting a protein purification are given.

KEY TERMS

Activity	*Affinity chromatography*
Assays	*Autolysis*
Cofactors	*Homogeneous sample*
Hydrophobic interaction chromatography	*Isoelectric focusing*
Lineweaver-Burk plot	*Michaelis-Menten law*
Specific activity	*Total activity*

GENERAL COMMENTS

Purification of biomolecules is the *forté* of biochemistry. Purification schemes currently in the literature reflect a century of practical experience. The isolation and purification of a desired compound from a natural source involves selectively concentrating it, while discarding all other compounds, including some that are closely related in structure and properties. All too frequently the purification scheme is developed by trial and error, without an overall strategy for optimizing yield and purity. It is often more art than science. One must know a great deal about the desired molecule, as well as the other chemical components of its natural source, in order to devise a theoretically based optimization strategy, but paradoxically, this information is usually not known when the purification is undertaken. The purification process affords information about the physical and chemical properties of the desired molecule. Hence most schemes are optimized only after the purification has been effected by a less efficient process. However, even with a limited amount of information about the desired compound, there are general schemes of purification that afford relatively pure material and provide important information that can be exploited in subsequent steps.

Protein Purification Schemes

Following are practical guidelines for the purification of proteins, especially enzymes.

1. **Define a purpose for the purified compound.** Purification is a means of obtaining sufficient amounts of a relatively pure compound for some practical use, usually further study. The first consideration one must have when devising a purification is to define the purpose to which the desired compound will be put and how pure it must be for this end. If 90% purity with a modest amount of heterogeneity will suffice, further effort is wasted. It is pointless to purify a compound to homogeneity if that is not required by its practical application; it is a waste of time, supplies and probably an unneccesary source of exasperation. The ideal of homogeneity is not always achieved, even after a number of fractionations. Since each step necessarily involves loss of material, including enzyme, it may be wise to settle for a partially purified fraction, especially if the enzyme is unstable.

2. **Choose a reliable assay.** A mixture of proteins from any natural source will have a wide variety of components, differing only in very subtle ways. If the physical properties of the desired protein are not known ahead of time, then the researcher must use the only handle he or she has on the enzyme-its *activity*. After each step of a purification scheme, one must determine the activity of the desired protein and be able to compare this to the total protein present in the sample. It allows one to measure the efficiency of specific purification steps and to pinpoint any failure to recover the desired molecule. The *total activity* of a fraction is the micromoles (μmoles) of product formed per minute (under defined conditions) by the entire fraction. *Specific activity* is the total activity divided by the mg of protein in the fraction and is a measure of the purity of that enzyme. A partially purified enzyme fraction may have a greater total activity than a homogenouse enzyme isolated after several more steps, because so much enzyme is lost in the purification process. However specific activity is always highest for a homogenous enzyme. The protein concentration is determined by any one of a number of convenient methods (Lowry, Bradford, etc.; see Chapter 3). The activity measurement employs a chemical reaction specific to the desired molecule. The assay reaction must be used under conditions (pH, substrate concentrations, etc.) that give optimum rates. A reliable assay requires at least a partial understanding of the kinetics of the assay reaction even before the separation begins (see below).

3. **Gather knowledge of the enzyme under investigation.** How stable is the protein? Is it heat or acid labile or sensitive to chelating agents? What can one learn about its molecular weight, quaternary structure and pI with a few preliminary experiments? Is it a lipoprotein stable only in the presence of added surfactants? Are there other required stabilizing compounds, including *cofactors* and substrates? Knowledge of such details often can be obtained by simple experiments on unpurified protein, yet such information is of great value in selecting separation steps that preserve the activity of the protein. It is also worthwhile to look into the nature of possible contaminating compounds present in the source since the purification must strip these away. Are any of them inhibitors or compounds necessary for optimum activity?

4. **Select a convenient, high-yield source.** One must always select that source which affords the highest return on the time and material invested in a purification. It must be a reliable source, easily obtained, easy to extract, and one that gives a reproducible yield of stable product. Knowledge of the location of the desired

compound within the source is of practical importance (e.g., is it in the leaves or in the roots?). This can include information about its intracellular location, whether it is membrane bound, associated with chromatin, mitochondria, or a soluble enzyme in the cytoplasm. Your choice of extraction and fractionation methods is dictated by such information.

5. Choose a method of extracting the protein from the crude source. The choice of the extraction method is dictated by the source and the stability of the enzyme. Intracellular enzymes may be released by any one of the following methods, the choice depending on the enzyme and its source: (1) extraction of the minced or pulverized tissue with buffer; (2) homogenization by ultrasonication or high-speed blending followed by low-speed centrifugation to remove insoluble material; (3) repeated freezing and thawing, (4) chemical lysing of cells with detergents such as Triton X-100; (5) enzymatic lysing with such agents as lysozyme or cytohelicase which cleave bacterial and yeast cell walls, respectively. The method that gives total maximum activity is usually selected. The enzyme source (tissue or microorganism) is first mechanically broken down by some method which ruptures cell membranes, releasing the intracellular components. This must be done quickly and in an ice bath. Cell rupture leads to the exposure of the desired compound to proteolytic agents, oxidizing agents and inhibitors that are normally compartmentalized in the intact cell. *Autolysis* of the desired product during extraction can be reduced by the addition of a general cocktail of protease inhibitors and buffers that stabilize the protein against rapid changes in pH. Reducing agents such as β-mercaptoethanol are added to counteract oxidation of cysteine sulfhydryl residues. EDTA is sometimes added for the same purpose unless the desired protein is a metalloenzyme. Surfactants are added to stabilize some proteins, especially hydrophobic ones.

6. Choose one or more preliminary fractionation methods to remove gross impurities. Methods that exploit the difference in solubility of the desired protein from most of the extract components are used at the start of the separation protocol. They are high-capacity methods that usually knock out 90% or more of the proteins within the extract. Salt fractionation with ammonium sulfate is the most common technique (see Chapter 4).

Precipitation with alcohol, acetone or polyethylene glycol (PEG) are also used. Heat treatment is perfect for the isolation of heat-stable proteins which are not precipitated by brief heating. If more than one of these preliminary fractionation methods is used, inactive fractions are put aside and active fractions are fractionated by a second method, which discriminates among proteins on a basis different from the first. Again only the active fractions are retained. By doing so, the number of components in the active fractions is reduced and the enzyme has been partially purified.

7. Select a series of high resolution separation methods that give the desired degree of purity. Only rarely will a single high-resolution technique allow a one step purification of a protein. A series of low-capacity but high-resolution chromatographic steps are usually required. Each method should select on the basis of a different molecular or chemical property, such as molecular size, shape charge, isoelectric point (pI), relative hydrophobicity, or affinity to certain ligands. (see Table 14-1) If we were able to measure three independent properties for all the different proteins in a protein extract, we would find that the components differ to some extent in one or more of these properties. These differences are illustrated in Figure 14-1.

A series of high-resolution methods must be able to select out a small subpopulation of molecules from a larger mixture of molecules that copurifies during the earlier low resolution steps. The selectivity can be applied positively, to pick out the desired compound, or negatively, to pick out molecules other than the desired ones.

A number of high-resolution chromatography methods are usually used because these methods combine resolution and high recovery. The recovery for each step must be high because the product fraction of one step is the starting material for the next so that recovery is cumulative. For example a recovery of 80% for each of three steps would correspond to (0.80)(0.80)(0.80) = 0.51 or 51% yield. A number of high-resolution protein purification methods are illustrated in Table 14-1. A protocol usually involves two or more of these methods. The order of high resolution steps depends on the properties of the desired compound and the most likely impurities. Generally, high-capacity methods such as ion-exchange or *hydrophobic interaction chromatography* (HIC) are employed earlier in the protocol than *affinity chromatography*, which requires a low capacity custom designed resin. Ideally, a series of fractionations will ultimately yield a single component--the homogeneous enzyme.

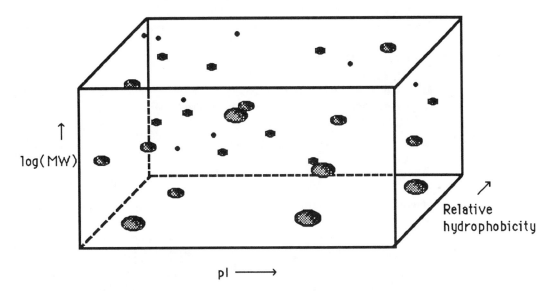

Figure 14-1 Proteins exhibit a wide variety of molecular or chemical properties that can be exploited in a purification (Source: Pharmacia)

Purification to maximum specific activity will not ensure enzyme homogeneity. The enzyme may associate very tenaciously with other proteins under conditions required for the assay and assay independent means, under dissociating conditions, must be taken to ensure that the enzyme is homogeneous. Electrophoretic methods are usually used, although screening for activities other than the desired one is also used in cases where a contaminate of known catalytic behavior is suspected. A purified monomeric enzyme will usually give single bands in SDS-PAG electrophoresis and in *isoelectric focusing*. The sample must also be free from extraneous activities. A single spot on a two-dimensional (O'Farrell) gel implies homogeneity. Enzymes with several subunits will show up as several spots. The activity should be restored upon recombining the separated subunits; a subunit not required for activity may be an impurity.

Table 14-1
HIGH RESOLUTION CHROMATOGRAPHIC PURIFICATION METHODS

Technique	Separation Parameter	Reference
Gel filtration	Molecular size or shape	(see Chapter 6)
Ion exchange	Molecular charge	
Bio-Rex 70	Weakly acidic cation exchange	1-3
AG MP-1	Strongly basic anion exchange	4
DEAE-Cellulose	Weakly basic anion exchange	5,6
CM Bio-Gel A	Weakly acidic cation exchanger	7
Chromatofocusing	pI	8,9
Hydrophobic interaction chromatography (HIC)		10
Affinity chromatography	Highly specific interaction for:	
Immobilized Avidin	Biotin-containing enzymes	11
Immobilized Protein A	Immunoglobulins	12
DNA-cellulose	DNA-binding proteins	13
Cibacron blue	NAD^+-binding proteins	14
Lysine-Sepharose	Plasminogen	15

ACTIVITY MEASUREMENT

The simplest enzyme-catalyzed process involves a single substrate S binding to an enzyme E, which catalyzes the formation of product P:

$$E + S \underset{k_{-1}}{\overset{k_1}{\rightleftharpoons}} ES \xrightarrow{k_2} E + P \qquad (14\text{-}1)$$

This scheme corresponds to the kinetic expression for the initial velocity V given by

$$\frac{V}{k_2 [E]_o} = \frac{[S]_o}{\dfrac{k_{-1} + k_2}{k_1} + [S]_o} \qquad (14\text{-}2)$$

where $[E]o$ and $[S]o$ represent the stoichiometric concentrations of enzyme and substrate. This expression is usually written as the Michaelis-Menten equation

$$\frac{V}{V_{max}} = \frac{[S]_o}{K_m + [S]_o} \qquad (14\text{-}3)$$

K_m is known as the Michaelis-Menten constant and has units of concentration. At any substrate concentration the initial velocity is given by

$$V_i = \frac{[S]\, V_{max}}{K_m + [S]} = \frac{k_2 [E]_o}{1 + \dfrac{K_m}{[S]}} \qquad (14\text{-}4)$$

This equation describes the curve obtained when initial velocity is plotted against substrate concentration. This figure is a rectangular hyperbola with a vertical limit of $V = V_{max} = k_2 [E]_0$ when $[S]_0 > \Delta K_m$.

If *in vitro* assay conditions are selected so that the substrate is in excess (10^{-2} to 10^{-6} M) with enzyme in limiting amounts (10^{-7} to 10^{-12} M), the initial velocity is directly proportional to $[E]$. For an assay period short enough to ensure that only a small fraction of substrate is consumed (less than 5%), the amount of product formed increases linearly with time and the initial velocity is well defined and V_i can be used to quantitate the enzyme present.

$$V_i = \frac{\Delta[P]}{\Delta t} \tag{14-5}$$

In developing an *in vitro* assay, one must find conditions where the initial velocity V_i is independent of $[S]$ (zero order in substrate concentration) and where the $[P]$ vs. time plot is linear. A portion of the enzyme fraction may have to be diluted to give $[E]_0$ values in the limiting range. The concentration in the original fraction can be calculated from the dilution factor.

Enzyme velocities are also dependent on pH, temperature, ionic strength, and the concentration of other species, including coreactants. Considerable care must be taken to ensure that the activity is measured under optimum conditions.

In multiple-substrate reactions, one of the substrates is generally held high and constant. If each substrate combines only with its binding site on the enzyme and the binding of one substrate does not affect the binding of the other (nonallosteric enzymes), then the equation for initial velocity vs. the second substrate concentration will formally correspond to equation (4), but K_m will be for the binding of substrate 2 under the assay conditions.

Any of a number of methods can be used to measure $[P]$ as a function of time. Spectroscopic methods are most applicable if the substrate and product differ in spectra (e.g., a colored product formed from a colorless substrate). This is the case with a number of enzymes described in this book. In some cases where the product can be conveniently separated from the substrate, assays using radioactive substrates can be used. For a variety of enzymes the catalyzed process of interest may be coupled to a second reaction which very rapidly converts the product to a readily quantitated form. This second reaction employs a second enzyme or may either be accomplished by a rapid nonenzyme reaction.

Estimation of K_m and V_{max}

The kinetic characterization of more fully purified enzymes necessarily requires rate determinations over a wide range of substrate concentrations. In order to measure K_m and to evaluate any allosteric or inhibitory features of an enzyme, the substrate must range above and below K_m. When $[S]$ is in the neighborhood of K_m, plots of $[P]$ versus time are curved. The amount of product formed during any given time interval is given by the integrated Michaelis-Menten equation:

$$\frac{V_{max}\, t}{K_m} = 2.303 \log\left\{ \frac{[S]_o}{[S]_o - [P]} \right\} + [P] \qquad (14\text{-}6)$$

This expression applies to all $[S]_0$ values but it can be simplified for initial substrate concentrations well below K_m ($[S]_0 < K_m$):

$$\frac{V_{max}\, t}{K_m} = 2.303 \log\left\{ \frac{[S]_o}{[S]_o - [P]} \right\} \qquad (14\text{-}7)$$

If $[P]$ can be evaluated directly then an apparent first-order rate constant k_{obs} can be taken from the slope of a plot of $\log\{[S]_0 - [P]\}$ vs. t. Spectrophotometric measurements of rates are especially convenient because the log term can be replaced by $\log(A_\infty - A_t)$ where A_∞ and A_t are the absorbances measured at infinite time (reaction complete) and at time t, respectively.

$$k_{obs} = \frac{k_1 k_2 [E]_o}{k_{-1} + k_2} \qquad (14\text{-}8)$$

For substrate concentrations above or near K_m the integrated equation can be used in a rearranged form, which is the equation for a straight line.

$$\frac{2.303}{t} \log\left\{ \frac{[S]_o}{[S]_o - [P]} \right\} = \frac{-1}{K_m} \frac{[P]}{t} + \frac{V_{max}}{K_m} \qquad (14\text{-}9)$$

Thus K_m and V_{max} can be determined by measuring the concentration of product produced at several times during the reaction and plotting:[15]

$$\frac{2.303}{t} \log\left\{ \frac{[S]_o}{[S]_o - [P]} \right\} \quad vs. \quad \frac{[P]}{t} \qquad (14\text{-}10)$$

The slope is $-1/K_m$ and the intercept is V_{max}/K_m.

The most widely used procedure for evaluating K_m and V_{max} is to use the *Lineweaver-Burk* double reciprocal form of the *Michaelis-Menten* rate law:

$$\frac{1}{v} = \frac{K_m}{V_{max}} \frac{1}{[S]_o} + \frac{1}{V_{max}} \qquad (14\text{-}11)$$

A plot of the reciprocal of the initial velocity $1/v$ vs. $1/[S]_0$ gives a straight line (for non-allosteric enzymes) with a slope K_m/V_{max} and y-intercept $1/V_{max}$. The concentration of substrate must range above and below the neighborhood of K_m. Care must be taken to ensure that $\Delta[P]/\Delta t$ is linear with time for all substrate concentrations used in a Lineweaver-Burk plot. This condition is generally satisfied by following the formation of P for only a few percent of the reaction time. The

Lineweaver-Burk equation and related expressions will be discussed in detail in chapter 17.

Reporting Enzyme Purifications

Table 14-2 illustrates the variety of techniques that can be used to fractionate a protein mixture and the information that should be included in any report of a purification scheme. Note that the table includes information on yield (calculated from total activity), purification (from specific activity), and recovery. The activity and heterogeneity of the crude extract should be determined by electrophoresis. The number of separate bands in a disc or SDS electrophoresis should decrease as the purification progresses. In the purification of primase by Lee and Kornberg,[10] described in Table 14-2, the overall purification was approximately 27,000-fold with an overall recovery of only 6%. Even after this lengthy treatment the product was only "nearly homogeneous." The assay method used for this enzyme nicely illustrates the round-about approach that sometimes must be used to measure activity. In this case, activity was determined by a complementation assay; that is, from the ability of various fractions in the purification scheme to restore the *in vitro* DNA synthesis ability of an extract from a temperature-sensitive mutant strain of *E. coli*, with limited primase activity. The level of *in vitro* DNA synthesis was monitored by following [^{3}H]TTP incorporation. Valyl-Sepharose chromatography discriminates through hydrophobic interaction.

Table 14-2
PRIMASE PURIFICATION SCHEME[a]

Fraction	Protein (mg)	Total Activity (units x 10^{-3})	Specific Activity (units/mg x 10^{-3})	Yield %	Purification (x fold)
Extract	283,000	11,500	0.041	(100)	-
$(NH_4)_2SO_4$	46,900	11,500	0.25	100	6
Bio-Rex-70	3,960	8,800	2.2	77	54
DNA-cellulose	124	6,600	53	57	1,290
$(NH_4)_2SO_4$ (again)	60	6,200	103	54	2,500
Valyl-Sepharose	8.7	3,700	430	32	10,500
DEAE-cellulose	0.64	700	1,090	6	26,600

[a] Source: Ref. 10.

REFERENCES AND NOTES

1. L. Wen and G. R. Reeck, *J. Chromatog.*, **1984**, *314*, 436.

2. J. A. D'Anna, G. F. Strniste, and L. R. Gurley, *Biochemistry*, **1979**, *18*, 943.

3. B. Blazy and A. Ullmann, *J. Biol. Chem.*, **1986**, *261*, 11798.

4. M. P. Menon, S. Miller, and B. S. Taylor, *J. Chromatog.*, **1986**, *378*, 450.

5. M. F. Lukacovic, M. B. Feinstein, R. I. Sha'afi, and S. Perrie, *Biochemistry*, **1981**, *20*, 3145.

6. R. D'Alisa and B. F. Erlanger, *J. Immunol.*, **1976**, *116*, 1629.

7. A. H. J. Ullah and V. P. Cirillo, *J. Bacteriology*, **1976**, *127*, 1298.

8. L. A. AE Sluyterman and E. Elgersma, *J. Chromatog.* **1978**, *150*, 17.

9. L. A. AE Sluyterman and J. Wijdenes, *J. Chromatog.* **1978**, *150*, 31.

10. L. Rowen and A. Kornberg, *J. Biol. Chem.*, **1978**, *253*, 758.

11. K. P. Henrikson, S. H. G. Allen, and W. L. Maloy, *Anal. Biochem.*, **1979**, 94, 366.

12. H. Hjelm et al., *FEBS Lett.*, **1972**, *73* .

13. B. Alberts and G. Herrick, *Methods Enzymology*, **1971**, *21*, 198.

14. R. B. Dunlap, ed. *Immobilized Biochemicals and Affinity Chromatography*, New York: Plenum Press, 1972, pp. 123-134.

15. D. G. Deutsch and E. T. Mertz, *Science*, **1970**, *170*, 1095.

16. I. W. Segal, *Biochemical Calculations*, 2nd ed., New York: Wiley, 1976, Chap. 4.

17. H. U. Bergmeyer, *Principles of Enzymatic Analysis*, New York: Verlag Chemie, 1978.

18. Figure 14-1 was adapted from one appearing in "Separation News 13·6" from Pharmacia Laboratory Separation Division.

Chapter 15

Wheat Germ Acid Phosphatase Isolation

Acid phosphatase activity is extracted from freshly ground, unprocessed whole wheat germ with acetate buffer, pH 4, and the enzyme is partially purified by a series of precipitation reactions with acetone and $(NH_4)_2SO_4$. The product of each purification step is characterized both kinetically and electrophoretically (SDS-PAGE). The enzyme is further purified and its molecular weight estimated by gel filtration through a Sephadex G-75 column. The purified enzyme must be stabilized by 0.1% Triton X-100 in the chromatography step and with 0.1% bovine serum albumin in the assay procedures. Acid phosphatase activity is measured by the hydrolysis of p-nitrophenylphosphate at 37°C ultimately to form the yellow p-nitrophenoxide ion, which is monitored spectrophoto- metrically.

KEY TERMS

Acetone precipitation   Activity staining
Isoenzymes

BACKGROUND

Acid phosphatases are relatively nonspecific enzymes that catalyze the hydrolysis of phosphate esters of primary and secondary alcohols and phenols at neutral or acid pH. Acid phosphatase levels can be detected in semen, prostate fluid, and in blood, where changes in these levels have both clinical and forensic significance. Fresh plants, especially potatoes and germinating seeds, also contain acid phosphatase. The acid phosphatase activity increases markedly in seeds during germination indicating a possible role in phosphate mobilization.[1] Despite the wide distribution of this type of enzyme in nature, the enzyme has seldom been reported purified to homogeneity, and little is known with certainty about the properties of the purified enzyme. Metalloenzymes, glycoproteins and purely protein forms of acid phosphatase have been found. Acid phosphatases occur in very low levels and frequently appear as isoenzymes which differ from source to source even within the same species.[2] *Isoenzymes* are closely related forms of the enzyme which usually differ in reaction rate or some aspect of regulation.

The isolation of wheat germ acid phosphatase is a traditional experiment in an introductory biochemistry laboratory and is included in almost all current laboratory texts.[3-8] The traditional procedure usually employed in teaching laboratories is adapted from the very old methods of Singer[9] and Joyce and Grisola[10] and affords only about a ten-fold enhancement of activity. The procedure is a popular one because the activity is easily assayed and the preparation does not require expensive chemicals and can be done in one or two laboratory periods; however, it falls short as a demonstration of modern enzyme isolations and the purification factors they permit. The procedure suggested here is based on the work of Waymack[2] and Van Etten[11] and leads to a more highly purified enzyme as well as a more complete characterization of its heterogeneity. The heterogeneity of the product of each purification step can be characterized by electrophoresis and the protein concentration by the Bradford method, so that the purification can be seen both in terms of increasing specific activity and decreasing heterogeneity. The gel-filtration step also leads to an estimate of molecular weight.

There are reports that germinating wheat seeds contain from seven[12] to more than nine[13] acid phosphatase isoenzymes. There are even conflicting reports regarding the presence of metals in these enzymes. Highly purified acid phosphatase may give a single band in SDS-PAG electrophoresis but still be a mixture of related proteins. SDS electrophoresis will not discriminate between true isoenzymes, which differ little in molecular weight. However disc electrophoresis and DEAE-cellulose chromatography both show multiple bands of phosphatase activity in all commercial samples of wheat germ acid phosphatase and in phosphatase extracted and purified as described here. The activity of protein bands in a gel can be estimated by a phosphatase activity staining procedure.[2,14] It is not clear what the total number of wheat germ isoenzymes is or what significance can be attached to the large differences in the levels of isoenzymes that is observed between different varieties of wheat. For the purposes of a teaching experiment this variability is not a serious source of variation in the results from student to student since all students in the same class will be using wheat germ from the same source.

However, it may lead to significant variations from year to year if the source of wheat germ is not kept constant. The past history of the wheat germ also determines the total enzyme activity that can be recovered by an extraction. Wheat germ that has been roasted has no activity and wheat germ that has been stored refrigerated is more likely to be a good source than that which has been stored for months in a "health food" store. The author has found that large supermarkets often carry *fresh* raw wheat germ that is a superior source of enzyme than that from some smaller stores that specialize in such products but do not have the product turnover.

One wheat germ isoenzyme has been purified 7000-fold to homogeneity to yield a monomeric purely protein enzyme of 58,000 D molecular weight,[2] and that procedure is the model for the procedure described here. The protein spontaneously precipitates when highly purified unless stabilized with detergent or added protein.[2] The protein concentration must be kept high through all the precipitation steps, and small volumes must be used. Failure to do this will result in significant loss of recovery. The purification scheme involves the precipitation of the desired enzyme and many other proteins from the acidic aqueous extract by the careful addition of very cold acetone to give 55% (v/v) acetone in water. The high capacity of this method is required because of the large volume of crude extract that must be used to obtain enough enzyme to warrant further purification. The enzyme is present in very low levels; the initial extract will have a specific activity of 0.05 to 0.09 unit/mg. Acetone denatures many proteins in the initial extract, which precipitate but will not redissolve, but not acid phosphatase precipitates as a stable protein that can be redissolved. This organic solvent does a much better job of precipitating acid phosphatase activity than the $MgCl_2$ or $(NH_4)_2SO_4$ methods used by other workers,[2,5] and gives a larger purification factor and a higher percent recovery of activity than the traditional methods. The acetone precipitation produces a stable product that can be allowed to settle out overnight so that the supernatant can be discarded, thus avoiding centrifugation of large volumes of initial extract. The precipitated proteins, remaining in a small volume of acetone buffer, must be centrifuged to a firm pellet that excludes most of the acetone then resuspended with very gentle stirring in a *small volume* of cold water overnight. The resulting suspension must be cloudy and yellow and it must have activity at least 10 times greater than the original extract. Failure to redissolve precipitated protein is the most frequent problem students have with this preparation.

The suspension is clarified by centrifugation and solid ammonium sulfate is added to the yellow supernatant to give 65% saturation. This salt "salts out" both the desired protein, which precipitates, and the acetone which forms a layer on top after centrifugation. The pellet contains enzyme that can be redissolved in acetate buffer, then freed from excess salt by dialysis. The dialysate is diluted to give a protein titer of 8 mg/mL with a specific activity of at least 1 unit/mg. It is clarified by centrifugation and combined with enough solid ammonium sulfate to give 35% saturation $(NH_4)_2SO_4$. The protein that precipitates is discarded and the supernatant is brought to 50% saturated $(NH_4)_2SO_4$ in order to precipitate the desired enzyme. The centrifugation pellet is dissolved in a small volume of cold water containing 0.1% Triton X-100 and chromatographed through a Sephadex G-75 gel filtration column. The nonionic surfactant Triton X-100 must be present to stabilize the partially purified enzyme, during the final purification steps. The enzyme is stabilized for the assay reaction with bovine serum albumin.

Each fraction in the purification scheme, including the fractions from the G-75 column, must be assayed for enzyme activity (at saturating levels of substrate) so that the total activity can be reported for each step along the purification pathway. The protein concentration of each fraction can be determined either from the UV absorbance at 280 nm or by the Bradford dye binding assay described in Chapter 4. The enzyme activity assay employs the substrate p-nitrophenylphosphate which is hydrolyzed to p-nitrophenol. The assay reaction is initiated by adding a small aliquot of stabilized enzyme solution to a 5 mM p-nitrophenylphosphate in sodium acetate buffer (pH 5.7, 37°C) and allowing the p-nitrophenol product to build up over exactly 5 minutes. The reaction is then quenched by the addition of 1 M KOH, which also converts the reaction product to the yellow p-nitrophenoxide ion, which can be quantitatively determined from the absorbance at 405 nm. The anion has an extinction coefficient of 18,300 m^{-1} cm^{-1} at 400 nm. Purified acid phosphatase very readily precipitates and significant activity is lost after only a few hours unless the protein is stabilized by 0.1% Triton X-100 or bovine serum albumin. Albumin must not be added to enzyme solutions before protein concentrations are determined, but it must be present in the activity assay.

Preliminary experiments must also be done to verify that 5 minutes corresponds to the first 5% of the reaction (see discussion in Chapter 14). The enzyme is inhibited by inorganic phosphate and fluoride ions, so these must be avoided in the purification and assay steps.

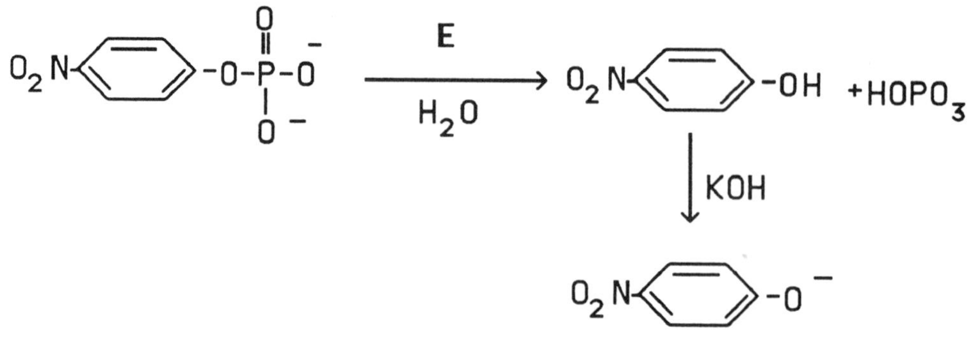

Figure 15-1 Assay reaction catalyzed by acid phosphatase

The enzyme activity can be localized on a disk electrophoresis gel by the phosphatase activity stain of Barka[14] as used by Waymack.[2] Gels are extruded from their tubes and incubated for 5 to 10 minutes in a solution containing 1 mg/mL naphthyl phosphate and 1 mg/mL Fast Garnet GBC salt in 0.1 M acetate buffer, pH 5.2. The dye is a stable diazonium salt which couples with phenolic moieties. The hydrolysis product α-naphthol, liberated in the vicinity of the gel-bound enzyme, couples with the dye diazonium salt to give a thin brown band. Weakly active bands must be stained for longer periods but strong bands will broaden considerably, if stained for more than 10 minutes, and the separation of closely spaced bands will be obscured. Diazo coupling of the dye to tyrosyl residues will stain tyrosine rich proteins in the absence of enzyme activity; consequently controls not employing naphthyl phosphate should also be run. The diazonium salt dye level may have to be adjusted downward to give only very faint banding in the absence of α-naphthyl phosphate.

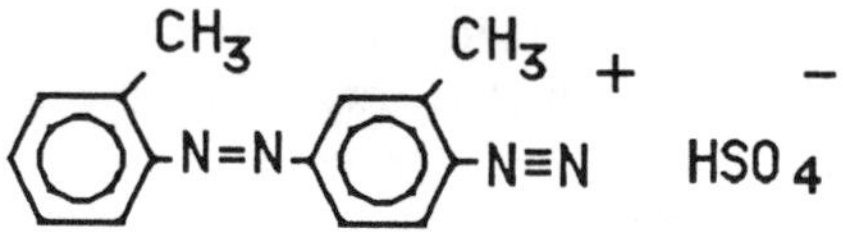

Fast Garnet GBC Salt

EXPERIMENTAL SECTION

Because of the very large volumes required for the initial extraction and acetone precipitation procedures, these steps should be done in one laboratory period as a group effort; the following procedures are scaled for groups of four students. This procedure makes stringent demands on the planning, speed, and foresight of student workers. All the solutions must be kept as cold as possible. The protein concentration must be kept high, before the chromatography step, to avoid spontaneous loss of enzyme activity. The characterization of the products of each step should be done the same day *as that step*, therefore students should reserve time for setting up the electrophoresis apparatus and running protein and activity assays. Mini SDS gradient gels are recomended because of the short running time involved.

First Laboratory Period

Crude Extract

Thirty grams of fresh, uncooked wheat germ is finely ground in a *cold* Waring blender; care is taken to avoid heating the sample. When fully pulverized, the powdered wheat germ is combined with 300 mL of ice cold 0.3 M acetate buffer, pH 4.0, and the mixture is stirred to a uniform slurry, placed in an ice bath, and shaken at 10-minute intervals for an hour. After centrifugation at 10,000 x g for 15 minutes (4°C), the supernatant is quickly freed of floating material by filtration through several layers of cheese cloth. A portion of the clarified extract is removed for characterization by the Bradford protein assay (see Chapter 4), disc electrophoresis and/or SDS electrophoresis (Chapters 8 and 9) as well as a determination of the enzyme activity (see below). The crude extract *must* have appreciable activity to warrant continuing the experiment.

Acetone Precipitation

Cold acetone (-20°C) is added slowly with gentle stirring to the crude extract, suspended in an ice bath, until the final acetone concentration is 55% (v/v).

WORK IN AN ICE BATH UNDER THE HOOD.
ACETONE IS A FLAMMABLE, TOXIC SOLVENT.

The acetone should be placed in an explosion-proof freezer well before the beginning of the laboratory period to ensure that it is cold enough. The precipitate that partially floats is resuspended by gentle swirling of the flask and allowed to precipitate for at least 30 minutes. The bulk of the supernatant can be gently decanted to leave about 80 to 120 mL of cloudy residual liquid and precipitate. This material is pelleted by centrifugation, at low speed (<3000 x g) to exclude as much acetone as possible, then stirred in 30 to 50 mL of cold water (4°C) until the next laboratory period, preferably just overnight. The resulting suspension must be yellow to straw colored.

Second Laboratory Period

The dissolved enzyme can be freed from the largely insoluble pellet by centrifugation at 10,000 x g for 15 minutes to yield a straw-colored solution. A 1-mL portion is removed for electrophoretic characterization, protein concentration and activity measurements. The specific activity must be about 1 unit/mg at this stage.

65% $(NH_4)_2SO_4$ precipitation

The dissolved enzyme is further concentrated and the acetone removed by precipitation in 65% saturated $(NH_4)_2SO_4$. Solid salt (0.432 g/mL) is added to the amber solution and the mixture is allowed to stand for a half hour at 4°C then pelleted at 5000 x g for

10 minutes. Acetone that is "salted out" forms a layer on top; this material is carefully removed with a Pasteur pipet. The pellet is rinsed with a small amount of distilled water to remove occluded ammonium sulfate solution then dissolved in 15 to 25 mL ice-cold 0.3 M acetate buffer, pH 4, and transferred to a presoftened dialysis tube (see Chapter 11) and dialysed against a 20-fold volume of 0.3 M acetate buffer overnight. If possible, the buffer should be changed *at least* twice. A 1-mL portion of the dialysate is reserved for electrophoretic, Bradford and kinetic characterization. The protein and activity assays must be done before beginning the next step. The protein concentration must be at least 8 mg/mL, with a specific activity of about 2 units/mg.

Third Laboratory Period

35 to 50% $(NH_4)_2SO_4$ fractionation

The dialysate above is clarified by centrifugation at 15,000 x g for 10 minutes and diluted to 8 to 10 mg/mL with 0.3 M acetate buffer, pH 4. The solution is brought to 35% saturation $(NH_4)_2SO_4$ by adding 0.210 g/mL salt. The solution is again allowed to stand for a half-hour, then centrifuged at 15,000 x g for 10 minutes.

The pellet is discarded and more solid $(NH_4)_2SO_4$ is added to bring the solution to 50% saturation (0.091 g/mL additional). After another 30 minutes, the solution is centrifuged as above and the pellet containing the enzyme is retained. The pellet is washed with a small amount of ice-cold water to remove occluded $(NH_4)_2SO_4$ and then shaken gently in enough ice water containing 0.1% Triton X-100 to give at least a 4 to 6 mg/mL protein solution with a specific activity of about 6 units/mg.

Gel Filtration

The solution above is applied to a 2.5 x 50 cm Sephadex G-75 column equilibrated with 0.01 M acetate, pH 5.2, with 0.1 M NaCl and 0.1% Triton X-100. If necessary the protein solution can be clarified by centrifugation or concentrated by Centriprep centrifugation (see Chapter 11) before application to the column. The column should be set up with a fraction collector and solvent reservoirs to run unattended overnight. The protein eluting from the column is collected as 1-mL aliquots and monitored by the absorbance at 280 nm and by activity assays. Elution rates of 30 to 40 mL per hour are reasonable; care should be taken to insure that enough buffer is in the solvent reservoir to ensure that the column does not run dry when unattended.

Fourth Laboratory Period

A fourth laboratory period is necessary to assay each fraction and to dismantle the column. An elution profile of both protein concentration (absorbance at 280 nm) and enzyme activity vs. eluent volume is the best way to present the final results. A data table analogous to Table 14-2 and a well-resolved SDS or disc electrophoresis gel of each fraction are also expected. A disc electrophoresis gel stained for acid phosphatase activity would also be valuable.

Enzyme Assay

A stock solution (100 mL) containing 5 mM p-nitrophenyl phosphate in 0.1 acetate buffer pH 5.2 and 0.1% Triton X-100 or 0.1 % bovine serum albumin is prepared on the first day of the experiment. (It does not seem necessary to add the $MgCl_2$, recommended by several laboratory text books.) Enzyme assays are initiated by adding 1 to 100 μL of enzyme solution to 3 mL of the substrate stock solution in a small test tube which is thermally equilibrated at 25°C in a water bath or at 37°C with constant temperature block. The tubes can be briefly removed from the block to ensure complete mixing with a vortex shaker or by "finger flicking." The reaction is allowed to continue for exactly 5 minutes; then it is rapidly quenched with 250 μL of 1.0 M KOH. Vortex mixing is necessary to resuspend any precipitated material. The absorbance at 405 nm is directly proportional to the concentration of p-nitrophenoxide ion. Blank reactions initiated with 100 μL of distilled water rather than enzyme solution must be run to rule out spontaneous hydrolysis of the substrate or contamination with p-nitrophenol. For especially active enzyme samples, reaction times of just 1 or 2 minutes must also be examined to ensure that the data taken after 5 minutes are still in the "linear" portion of a plot of product concentration vs. reaction time (see Chapter 14).

Gel Activity Staining (Optional)

Disc gels are incubated in a solution containing 1 mg/mL naphthyl phosphate and 1 mg/mL Fast Garnet GBC salt in 0.1 M acetate buffer, pH 5.2. The solution must be filtered just prior to use. The solution is not stable and must be prepared in the small amounts needed on the day of use. The staining can be followed while the gels are in the bath, especially if a glass dish is used and there is a source of background illumination such as a light box. The staining reaction is quenched by the addition of enough acetic acid to bring the concentration up to 7%. Gel bands broaden upon storage, so photographic methods are required to preserve this information.

PROJECT EXTENSIONS

1. The partially purified enzyme obtained in this project can be further purified by chromatofocusing (see Chapter 16) or used in kinetic studies (Chapter 17).

2. An additional methanol precipitation step after the ammonium sulfate fractionation is reported to give an additional 10-fold purification.[2] At a methanol concentration of 22% (v/v), the enzyme activity precipitates with very little other protein. This activity can be redissolved in ice-water, stabilized with Triton X-100, and purified by gel filtration as above.

3. Acid phosphatase may be present in other germinating seeds. We have found that mung bean sprouts have a low level of activity that cannot be recovered after the addition of acetone. An alternative purification scheme must be devised.

4. Polyethylene glycol (PEG) precipitation may be an attractive method of selectively precipitating acid phosphatase activity without the problems of toxicity and flammability associated with acetone.

DISCUSSION QUESTION

Enzyme concentrations above 1 mg/mL are stable for a month, but those below 0.1 mg/mL show rapid loss of activity with a half-life of less than 4 days at pH 5.2 (4°C). This inactivation is inhibited by adding serum albumin or Triton X-100. What does this suggest about the protein structure?

REFERENCES

1. V. Macko, G. Honold and M. Stahman, *Phytochemistry*, **1967**, *6*, 465.

2. P. P Waymack, "Isolation, Properties and Mechanism of an Isoenzyme of Wheat Germ Acid Phosphatase," Ph.D. thesis, Purdue University, 1978.

3. J. M. Clark and R. L. Switzer, *Experimental Biochemistry*, (2nd ed.) San Francisco: W. H. Freeman, 1981, pp.105-109.

4. G. D. Crandall, *Selected Exercises for the Biochemistry Laboratory*, Oxford: Oxford University Press, 1983, pp. 24-43.

5. J. Stenesh, *Experimental Biochemistry*, Boston: Allyn and Bacon, 1984, pp. 181-194.

6. T. G. Cooper, *The Tools of Biochemistry*, New York: Wiley-Interscience, 1977, pp. 391-404.

7. D. C. Wharton and R. E. McCarty, *Experiments and Methods in Biochemistry*, New York: Macmillan, 1972.

8. R. R. Alexander, J. M. Griffiths, M. L. Wilkinson, *Basic Biochemical Methods*, New York: John Wiley & Sons, 1985.

9. T. P. Singer, *J. Biol. Chem.*, **1948**, *174*, 11.

10. B. K. Joyce and S. Grisola, *J. Biol. Chem.*, **1960**, *235*, 2278.

11. M. E. Hickey, P. P. Waymack and R. L. Van Etten, *Arch. Biochem. Biophys.*, **1976**, *172*, 439.

12. D. M. Foster and D. J. Weber, *Plant Sci. Lett.*, **1973**, *1*, 169.

13. T. Akiyama, *Kagaku to Seibutsu* (Japan), **1982**, *20*, 703.

14. T. Barka, *Nature*, **1960**, *187*, 248.

Chapter 16

Acid Phosphatase Chromatofocusing

A partially purified sample of wheat germ acid phosphatase is fractionated by chromatofocusing on a polybuffer PBE 94 (Pharmacia) column over the pH range 9.4 to 6.0. The pH, protein concentration, and acid phosphatase activity of each fraction is determined. The protein concentration is estimated from optical density measurements at 280 nm and enzyme activity from the rate of hydrolysis of p-nitrophenyl phosphate in acetate buffer at pH 5.2. The protein heterogeneity of each fraction can be evaluated by SDS-PAG electrophoresis using the Laemmli gradient slab system.

KEY TERMS

Ampholytes
pH gradient
Polybuffer gel

Chromatofocusing
Starting buffer

BACKGROUND

Chromatofocusing is a column chromatographic method for separating proteins according to their isoelectric points (pI values) by taking advantage of the ion-exchange and buffering action of a special anion-exchanger gel. A *pH gradient* can be introduced within a column by passing a buffer of a certain initial pH through a column of the ion exchanger preadjusted to a higher initial pH. Even though the ion exchanger contains only one kind of ionizing group (usually an amine) with one intrinsic (thermodynamic) pK in the absence of salt, there will be a range of apparent pK values owing to polyelectrolyte effects. As the buffer and gel exchanger with different pH values are mixed in the top portion of the column, a pH gradient is formed, the top being more acidic than the lower portions.

The pH of buffer aliquots running through successive sections of a column changes in a manner that can be predicted theoretically.[1,2] Conceptually, the column of exchanger can be divided into equal cylindrical sections or disks and the elution buffer into aliquots equal to the void volume of any one column section. The pH of each aliquot passing through successive sections is readjusted in each section by acid-base reactions and the retention of the more negatively charged components of the mobile phase buffer by the anion exchanger.

The pH in each section is given by the expression

$$pH = \frac{A_m\, pH_m + A_s\, pH_s}{A_m + A_s} \tag{16-1}$$

where A_m and A_s are the buffering capacities per unit column length of the mobile and stationary phases, respectively. The pH of the mobile phase entering the section is pH_m and the pH of the stationary phase (before the aliquot arrives) is pH_s.

For initial pH values of the running buffer and ion exchanger column of 6.0 and 9.4, respectively, the change in pH down the column for three different aliquots is shown in Figure 16-1. The three aliquots represent buffer applied to the column at the beginning, during the middle and toward the end of the experiment. The pH of the aliquot of buffer when it emerges from the column is equal to the pH in the last section. Since the pH gradient decreases as the experiment proceeds, the pH of the liquid emerging from the column changes with elution volume in a sigmoidal fashion.

A protein moves with the buffer down the column when both protein and exchanger are positively charged (at pH values below the protein pI), but the protein is retained by the cationic exchanger resin when the protein is anionic. A protein with a pI of 9.0 will be carried along with its original buffer aliquot until reaching a column section with a pH slightly above 9.0. As more eluting buffer is added to the column, the pH of each section decreases and the protein moves down the column eluting according to its pI as shown in Figure 16-1.

This behavior is the explanation of the extremely high resolving power of chromatofocusing. Proteins are focused into narrow bands or disks because protein molecules above the pH = pI region are repelled from the exchanger resin and move rapidly down the column, catching up with the remainder of protein moving with the pH = pI region. Band broadening due to sample volume or the nonuniform layering of sample are unimportant. Broadening due to back diffusion and nonuniformity of flow can be minimized by keeping a constant, relatively rapid flow through an evenly packed column.

In this experiment a commercial enzyme preparation of wheat germ acid phosphatase (Sigma P3627) is further purified by chromatofocusing. This preparation contains a number of proteins, including one conferring lipase activity,[3] which can be separated from a broad band of acid phosphatase activity emerging from the column between 7.0 and 6.5 (see Figure 16-2). The method does not resolve the known acid phosphatase isoenzymes, but further refinements, including a longer column, may permit this. When applied to the Sigma sample, the method does separate the bulk of proteins which do not have acid phosphatase activity from those that do; the fractions with the highest absorbance at 280 nm emerge from the column between pH 8.5 and 7.5. The fractions containing phosphatase contain much less protein and absorb less at 280 nm. Students may also take the option of using chromatofocusing to further purify acid phosphatase, already partially purified by gel filtration or methanol precipitation (Chapter 15).

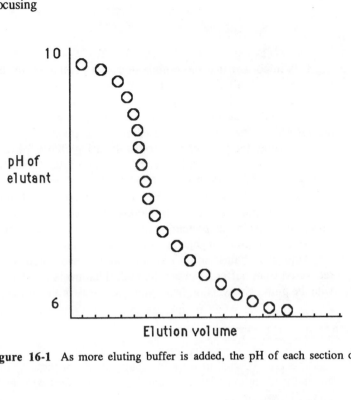

Figure 16-1 As more eluting buffer is added, the pH of each section decreases

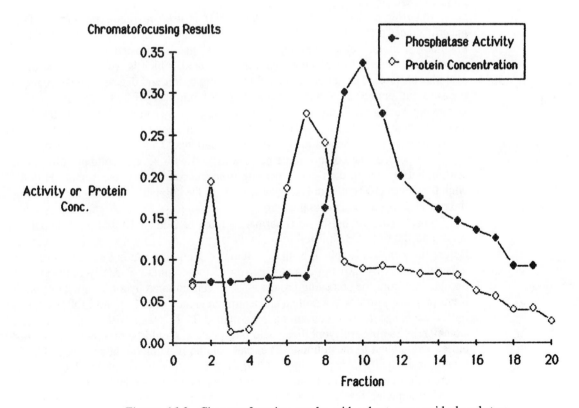

Figure 16-2 Chromatofocusing results with wheat germ acid phosphatase

It is important that the sample does not contain more than 0.05 M salt. High levels of salt limit the buffering range of the exchanger. Gel filtration through Sephadex G-25 or dialysis may be necessary to remove excess salt if the sample is from any other source than commercial acid phosphatase or the pooled G-75 gel filtration highest-activity fractions of the preceding preparation.

In general, the proteins in chromatofocusing eluent fractions are accompanied by buffer components but can be freed from these salts either by precipitation in 60% saturated $(NH_4)_2SO_4$ or by desalting with added unswollen Sephadex G-25 resin.[4] The G-25 resin slowly swells in cold buffer to incorporate salt within the gel matrix but excluding the protein. Care must be taken to avoid adding too much resin, which will irreversibly precipitate protein by absorbing all the water. Short columns of AG 501-X8 can also be used for separating peptides from *ampholytes*.[5]

Acid phosphatase is an unstable protein and some activity is lost when it is recovered from buffer by either method. Pharmacia Polybuffer 96 does not interfere with the p-nitrophenyl phosphate hydrolysis assay or the Bradford protein assay.

EXPERIMENTAL SECTION

Students are expected to work in groups of two or three on this project. Because the column packing material is expensive, students must be cautioned not to discard the polybuffer exchanger.

Polybuffer exchanger PBE 94 (Pharmacia) is used in this experiment. This material comes preswollen, but it must be equilibrated with start buffer and packed into a suitable column before running. The PBE 94 gel is adjusted to pH 9.4 with a *starting buffer* of 0.025 M ethanolamine/acetic acid, which is well above the pI of acid phosphatase. The gel is suspended in starting buffer and degassed under reduced pressure and poured as a slurry into a short column partly filled with start buffer. Chromatofocusing media have very high buffering capacity and it is difficult to overload a column with excess sample. A 5-cm column with a total volume of only 15 to 20 mL is adequate for the separation of samples containing 20 mg or more of protein.

Once poured, the column must be equilibrated with several column volumes of starting buffer or until the liquid emerging from the column is at a constant pH (9.4). High flow rates (100 cm/h) are acceptable because the polybuffer exchangers are cross-linked and do not deform under pressure.

Prepackaged Polybuffer 96 (Pharmacia) must be diluted 10-fold with deionized water and the final pH adjusted to 6.0 with acetic acid before use as *eluting* buffer. Before the sample is applied, a small volume (< 1 mL) of eluting buffer Polybuffer 96/acetic acid (pH 6.0) is run onto the top of the column. Partially purified acid phosphatase from the preceding project must be concentrated by diafiltration or with a Centriprep concentrator to a protein concentration greater than 1 mg/mL, followed by a buffer exchange to one containing less than 0.05 M NaCl with 0.05 % Triton. Commercial samples of lyophilized acid phosphatase should be dissolved in *starting* buffer and applied to the column. The total volume is unimportant as long as all the sample is applied before protein is eluted from the column. Some insoluble brown material clings to the top of the resin bed if the commercial material is used.

After applying the sample and allowing it to drain into the column, one must fill the solvent reservoir with about 200 mL of eluting buffer and set the column running.

Fractions ranging in volume from 1 to 3 mL are collected in an ice bath every 5 to 10 minutes. It is recommended that the fractions be no larger than the minimum volume necessary to measure absorbance at 280 nm (i. e., the minimum volume required to fill a cuvette above the light path level). Small aliquots of each fraction 100 μL should be assayed for enzyme activity (see Chapter 15).

The pH of each fraction should also be measured within minutes to ensure an accurate profile of the pH gradient; the buffer solutions will absorb CO_2 if allowed to set. The pH meter must be standardized for cold solutions since the samples will be stored in an ice bath.

Polybuffer exchanger may be reused -- which is comforting since it is expensive. All protein is flushed from the column with 1.0 M NaCl (10 column volumes), the column is reequilibrated with 3 to 5 column volumes of start buffer. Long-term storage of Polybuffer exchanger requires about 40% (v/v) ethanol as a preservative. Azide ion cannot be used for this purpose. Stored samples of exchanger can be kept in the cold room for months if tightly stoppered.

The fractions emerging from the chromatofocusing column cannot be analyzed by SDS electrophoresis unless freed from the buffer components. The following trick for desalting small volumes of protein solutions works. Fractions eluting from the column may be separated from buffer components by adding *just enough*, dry G-25 to give a thick but free-flowing suspension (200 μL total sample volume) in a small plastic microfuge tube, which is then pierced with a needle--one must make the smallest hole possible. The small plastic centrifuge tube is then squeezed into a larger plastic centrifuge tube, balanced with a similarly contrived pair of tubes, and run on at high speed for 10 minutes in an Eppendorf or Beckman tabletop microcentrifuge. The liquid portion containing protein is forced away from the gel-bound salt solution and passes through the hole into the larger tube. A small but usable volume of protein can be obtained and applied to an electrophoresis gel.

PROJECT EXTENSIONS

1. It may be possible to separate the isoenzymes of wheat germ acid phosphatase by using a chromatofocusing ion-exchanger gel that covers a narrower pH range (e.g., pH 7.0 to 6.0). A longer column and smaller fraction volumes would also improve resolution too.

2. A series of desalted samples corresponding to different fraction pH values gives a convenient two dimensional analysis of pI vs. protein molecular weight for a complex protein mixture when applied to an SDS gradient gel.

REFERENCES

1. L. A. Æ. Sluyterman and O. Elgersma, *J. Chromatog.*, **1978**, *150*, 17.

2. L. A. Æ. Sluyterman and J. Wijdenes, *J. Chromatog.*, **1978**, *150*, 31.

3. T. P. Singer, *J. Biol. Chem.*, **1948**, *174*, 11.

4. Pharmacia Fine Chemicals, "Chromatofocusing," a technical bulletin, 1978.

5. J. A. Bakker, J. L. Van den Brande and C. M. Haggerbrugge, *J. Chromatog.*, **1981**, *209*, 273.

Chapter 17

Acid Phosphatase Kinetics and Inhibitors

The initial velocity for the wheat germ acid phosphatase catalyzed hydrolysis of p-nitrophenyl phosphate is determined as a function of substrate concentration and in the presence of the inhibitors phosphate and fluoride ion. The rate data are analyzed by the Lineweaver-Burk and Dixon plot methods and the Michaelis-Menten and inhibitor constants are measured from the slopes and intercepts of these plots. The nature of the inhibition is determined (e.g., competitive vs. noncompetitive).

KEY TERMS

Competitive inhibition	*Dixon plots*
Inhibition	*Inhibitor constants*
Initial velocity	*Lineweaver-Burk plots*
Michaelis-Menten kinetics	*Mixed-type inhibition*
Noncompetitive inhibition	*Product inhibition*
Uncompetitive inhibition	

BACKGROUND

Enzyme kinetics is the principal tool used to characterize enzymes and to determine the mechanism of catalysis. Our basic understanding of enzyme catalysis and the behavior of enzyme systems has been developed more by kinetics methods than any other approach. In 1902, Henri derived an equation describing the effect of substrate concentration on enzyme velocity and this equation was rediscovered 11 years later by Michaelis and Menten. This simple chemical equilibrium model of enzyme substrate interaction is intuitively attractive and most discussion of enzyme kinetics in introductory biochemistry textbooks begins with this approach. The conventional *Michaelis-Menten* equation describes a hyperbola; therefore, it is difficult to evaluate V_{max} and K_m from plots of velocity v vs. substrate concentration $[S]$, since v approaches V_{max} only as $[S]$ approaches infinity. Similarly, it is difficult to determine K_m without precise knowledge of V_{max}; $K_m = [S]$ at $v = V_{max}/2$.

Lineweaver and Burk suggested an alternative approach for analyzing the same data which affords a straight line, making such extrapolations more convenient and often more accurate before the age of high-speed computers. Although resisted by some,[1] this simple method has been a significant improvement in the analysis of enzyme catalysis and contributed much to the importance such analyses have played in the growth of modern biochemistry.[2]

The Michaelis-Menten equation can be inverted to give a linear equation:

$$V = \frac{V_{max}[S]}{K_m + [S]} \qquad (17\text{-}1)$$

Inverted and with the terms separated, the equation becomes

$$\frac{1}{V} = \frac{K_m}{V_{max}} \cdot \frac{1}{[S]} + \frac{1}{V_{max}} \qquad (17\text{-}2)$$

Lineweaver-Burk   Plots

A plot of $1/v$ vs. $1/[S]$ has a slope K_m/V_{max} and the intercept on the $1/[S] = 0$ axis (y-axis) is $1/V_{max}$ (see Figure 17-1). These kinetic parameters can be obtained from a carefully drawn plot of $1/v$ vs. $1/[S]$. Care should be taken in interpreting parameters obtained by visually extending a line through the data points. In such a plot not all points have been determined with equal precision. Small errors in V which are very likely at low $[S]$ levels are magnified when plotted as reciprocals whereas the more accurately determined velocities V determined at high substrate concentrations tend to cluster close to the $1/v$-axis and are not weighted as heavily if a line is "eyeballed" through all the points. A less subjective computer line-fitting program is recommended.[3] The concentrations of substrate used to generate a Lineweaver-Burk plot must range above and below the K_m-value; points in the neighborhood of K_m are more accurate than points for $[S] << K_m$ and points for $[S] >> K_m$ cluster too close to the axis and poorly define the slope.

Inhibition

Inhibition of enzyme activity is a major regulatory mechanism of living cells. Biochemists have long since learned to exploit enzyme inhibition to synthesize specific inhibitors; most drugs, antibiotics, and herbicidal or pesticidal poisons are cleverly designed enzyme inhibitors. The use of inhibitors is also one of the most important tricks in the repertoire of the enzymologist, interested in the mechanism of enzymes and their regulation. Inhibition studies tell us much about the mechanism of an enzyme-catalyzed reaction, including the structural requirements for a tightly binding or inhibitor, the order of substrate binding for multiple substrates, and the nature of the rate determining step in a multistep enzyme-catalyzed process. Enzyme inhibition falls into two categories: reversible and irreversible. The latter involves the permanent chemical modification of the enzyme.

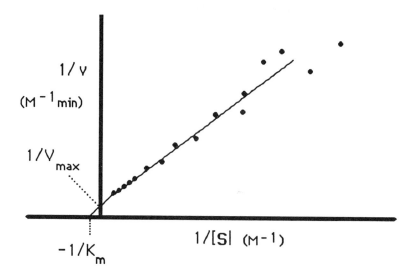

Figure 17-1 Lineweaver-Burk plot

Conceptually, the easiest form of reversible inhibitor to understand is the *competitive inhibitor* that associates with the free enzyme and prevents substrate binding. Enzyme cannot bind inhibitor and substrate simultaneously because they compete for the same site. Competitive inhibitors often are structurally similar to the substrate and bind to the active site by the same forces that operate with the substrate. Competitive inhibitors are often modified substrates or structural analogs that cannot undergo the enzyme-catalyzed reaction. Reaction products frequently are competitive inhibitors of their own biosynthesis (i.e., product inhibition is often by a competitive inhibition mechanism).

The Lineweaver-Burk equation can be modified to contain a term that takes into account the "dead-end" equilibrium between enzyme and inhibitor **I**.

$$E + S \underset{}{\overset{K_{eq}}{\rightleftharpoons}} ES \longrightarrow E + P$$

$$+\ I\ \big|\big|$$

$$K_I = \frac{[E][I]}{[EI]}$$

$$EI$$

In the formulation of the scheme shown above the dissociation constant for enzyme-substrate binding is written as K_{eq}. This term is not identical to the Michaelis-Menton constant K_m defined in chapter 14 (which was used in an expression derived according to the steady-state assumption), but we obtain the same expression for velocity in terms of substrate concentration if we make the assumption that substrate-enzyme binding is an equilibrium process. The Lineweaver-Burk expression in the case of competitive inhibition becomes

$$\frac{1}{v} = \frac{K_m}{V_{max}}\left(1 + \frac{[I]}{K_I}\right).\frac{1}{[S]} + \frac{1}{V_{max}} \qquad (17\text{-}3)$$

The presence of a competitive inhibitor has the effect of increasing the apparent magnitude of K_m (or K_{eq}) by a term that is directly proportional to the inhibitor concentration and inversely proportional to the enzyme-inhibitor complex dissociation constant. A plot of $1/v$ vs. $1/[S]$ for a series of different fixed inhibitor concentrations gives a family of lines intersecting at $1/V_{max}$ on the $1/v$-axis but with slopes increasing by $(1 + [I]/K_I)$.

The result that V_{max} is independent of inhibitor concentration indicates that the potential level of catalysis is not altered by a competitive inhibitor but that a higher level of substrate concentration is required in the presence of inhibitor to obtain it; the effect of inhibitor can be overcome by using additional substrate. An alternative plot of $1/v$ vs. $[I]$ gives a family of intersecting lines with slopes equal to $K_m/V_{max}K_I[S]$; this plotting procedure--a *Dixon plot*--can be used to evaluate K_I from velocities measured with a fixed substrate concentration and a range of inhibitor concentrations above and below K_I. The lines in the Dixon plot intersect at a point in the left quadrant with coordinants $(1/V_o,-K_I)$ and the inhibitor binding constant can be obtained by projecting this point to the $[I]$ axis (see Figure 17-2).

The Lineweaver-Burk plot for competitive inhibition with each line corresponding to a different fixed concentration of inhibitor is shown in Figure 17-2.

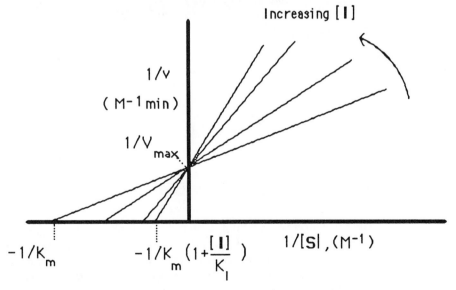

Figure 17-2 Lineweaver-Burk plots for various concentrations of a competitive inhibitor

The corresponding data can be plotted as a Dixon plot with all points on each line representing data obtained at the same substrate concentration, as shown in Figure 17-3.

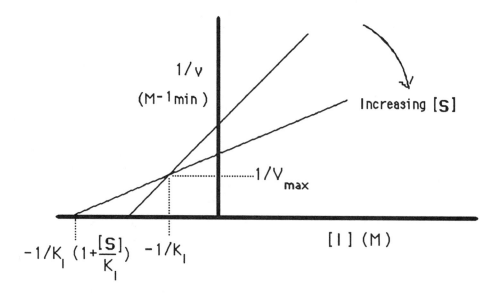

Figure 17-3 Dixon plots for various concentrations of a competitive inhibitor

The behavior of the lines of these plots is diagnostic of competitive inhibition, especially the intersection of all the lines at $1/V_{max}$ in the Lineweaver-Burk plot. The use of such plots to discriminate between the various types of reversible inhibition requires good-quality data with enough of a spread in V and $[S]$ to define the intersection points; the data must be plotted on a scale where the intercepts are clearly distinguishable from the origin (0,0).

Another often seen type of reversible inhibition is ***noncompetitive inhibition*** which corresponds to the scheme

$$E + S \underset{K_{eq}}{\overset{}{\rightleftharpoons}} ES \longrightarrow E + P$$

$$K_I = \frac{[E][I]}{[EI]} = \frac{[ES][I]}{[ESI]}$$

The noncompetitive inhibitor has no effect on substrate binding, so that S and I bind independently at different sites; however whenever I is bound, the enzyme has no catalytic activity. (Inhibition schemes where the presence of I bound to the enzyme changes the dissociation constant for substrate binding are said to be ***mixed-type inhibitions***. The inhibitor may prevent the proper orientation of the catalytic amino acid residues within the enzyme active site, alter their ionization, or in some other way destabilize the transition state. The Lineweaver-Burk expression for noncompetitive inhibition predicts a decrease in V_{max} by the term

$$\left(1 + \frac{[I]}{K_I}\right)^{-1} \tag{17-4}$$

but no change in K_m. This implies that unlike competitive inhibition, noncompetitive inhibition cannot be overcome by increasing the substrate concentration. The following equation is valid under the condition $K_m = K_{eq}$.[2,4]

$$\frac{1}{v} = \frac{K_m}{V_{max}}\left(1 + \frac{[I]}{K_I}\right)\frac{1}{[S]} + \frac{1}{V_{max}}\left(1 + \frac{[I]}{K_I}\right) \tag{17-5}$$

Lineweaver-Burk plots give a family of lines intersecting on the 1/[S] axis at $-1/K_m$ (Figure 17-4).

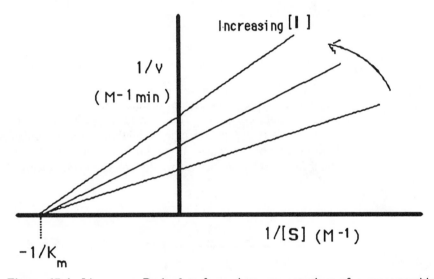

Figure 17-4 Lineweaver-Burk plots for various concentrations of a noncompetitive inhibitor

Data collected in this experiment for the inhibitors phosphate and fluoride should be plotted as shown in Figures 17-2 and 17-4, with lines drawn through the experimental points. Actual experimental data tends to fall off the line at low substrate concentration (large 1/[S] values) as shown in Figure 17-1 so that the theoretically most significant line will not be the conventional least-squares best fitting line, unless the line fit takes into account the reciprocal nature of the variables. There are software packages that take this into account.[3]

The alternative method of plotting data taken at fixed inhibitor and variable substrate, developed by Dixon, requires that several inhibitor concentrations be examined at each of a number of substrate concentrations. In principle the experiment can be planned so that the same data can be plotted either way. The Dixon plot for noncompetitive inhibition (Figure 17-5) readily permits estimation of K_I as the [I] axis intercept.

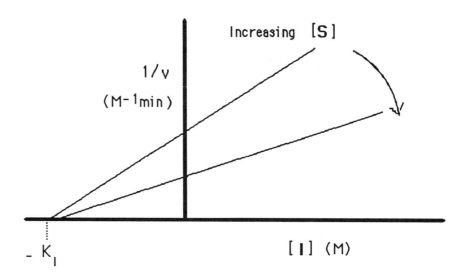

Figure 17-5 Dixon plots for various concentrations of a noncompetitive inhibitor

Uncompetitive inhibition occurs when the inhibitor must bind to the enzyme-substrate complex. This is a rare type of inhibition in single-substrate systems, but it is a common type of inhibition for a multisubstrate enzyme with ordered binding of substrates. In its simplest terms the scheme for an uncompetitive inhibitor for a bisubstrate enzyme would look like the following.

$$E + S \underset{\phantom{K_{eq}}}{\overset{K_{eq}}{\rightleftharpoons}} ES + S' \rightleftharpoons ESS' \longrightarrow E + P$$

$$+I \;\Big\Updownarrow\; K_I$$

$$ESI$$

The inhibitor **I** will be uncompetitive with respect to a given substrate if **I** binds only after the substrate binds. **I** could be an unreactive structural analogue of the second substrate **S'**. Uncompetitve inhibitors give a Lineweaver-Burk plot with a slope of K_m/V_{max}, but the $1/v$ intercept is multiplied by the term $(1 + [I]/K_I)$. Increasing [I] increases the intercept but not the slope, implying a family of parallel lines. For a single-substrate enzyme exhibiting uncompetitive inhibition, we have:

$$\frac{1}{v} = \frac{K_m}{V_{max}}\frac{1}{[S]} + \frac{1}{V_{max}}\left(1 + \frac{[I]}{K_I}\right) \qquad (17\text{-}6)$$

More complex expressions govern multisubstrate enzymes in which I is noncompetitive or uncompetitive depending on the order of substrate binding.[4] The plot for uncompetitive inhibition has a very characteristic pattern of parallel curves (Figure 17-6).

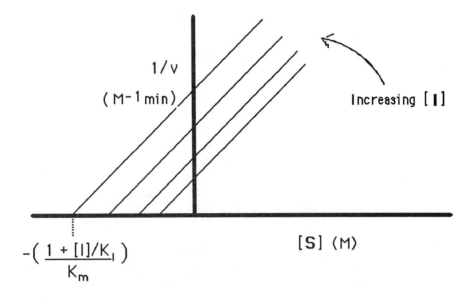

Figure 17-6 Lineweaver-Burk plots for various concentrations of a uncompetitive inhibitor

The Dixon plot is also a family of parallel lines with an $1/v$ intercept that decreases as [S] increases.

Product   Inhibition

Enzymes bind products as well as substrates; this is obviously the case for reactions with equilibrium constants less than 1, but it is equally true for highly favorable reactions. All products are inhibitors of the forward reaction since they bind to one of the forms of the enzyme and thereby backup the flow of the forward reaction. Product inhibition studies are an important part of the characterization of an enzyme catalyzed reaction. Different kinetic patterns are associated with different inhibition patterns; one obtains competitive inhibition when substrate and product bind to some enzyme form. Take for example the inhibition patterns of the two products P_1 and P_2 on the reactivity of the two substrates S_1 and S_2 that correspond to different enzyme schemes. Table 17-1 gives three representative patterns.

Table   17-1
PRODUCT INHIBITION PATTERNS

Mechanism	Product	Type of Inhibition[a]	Substrate
Ordered sequential bi-bi			
	P_1	Noncompetitive	S_1
	P_1	Noncompetitive	S_2
	P_2	Competitive	S_1
	P_2	Noncompetitive	S_2
Random sequential bi-bi			
	P_1	Noncompetitive	S_1
	P_1	Noncompetitive	S_2
	P_2	Noncompetitive	S_1
	P_2	Noncompetitive	S_2
Ping-pong bi-bi			
	P_1	Noncompetitive	S_1
	P_1	Competitive	S_2
	P_2	Competitive	S_1
	P_2	Noncompetitive	S_2

[a]Assuming unsaturated with the other substrate; see Ref. 7.

In an ordered sequential mechanism the two substrates add in order (S_1 before S_2) followed by the loss of the two products, also in order. In random sequential bi-bi the two substrates can add in either order, followed by the loss of the substrates, also in random order. In the ping-pong mechanism, P_1 leaves the enzyme after S_1 has bound (and been chemically transformed) but before substrate S_2 binds to the enzyme.

pH-Rate   Profiles

The ionizable groups within an active site must be in the proper ionic form for a catalytic event to occur. Moreover, substrates with ionizable groups must have the correct charge in order to bind to enzyme with maximum affinity. For these reasons pH will influence the rate of enzyme-catalyzed reactions. The pH dependence of an enzyme reaction is important information about the identity of the ionizable groups participating in substrate binding and in the catalytic event. The pK_a values of the ionizable groups can be estimated from the behavior of the velocity at saturating levels of substrate, V_{max}, as a function of pH. If one observes an upturn or downturn in a V_{max} vs. pH profile, the pH at the midpoint of either bending portion is often attributable to a single ionization constant pK_a (see Figure 17-7), especially if a plot of log V_{max} vs. pH is linear with unit slope (1 or -1) over this range. Wheat germ acid phosphatase is most active in the pH range 4.5 to 6.5; a plot of V_{max} versus pH gives a plateau over this range, dropping off at either extreme.[5] The pK_a given by the alkaline downward-bending portion of the V_{max} vs. pH plot is close to that expected for the ionization of a covalent phosphohistidine intermediate.

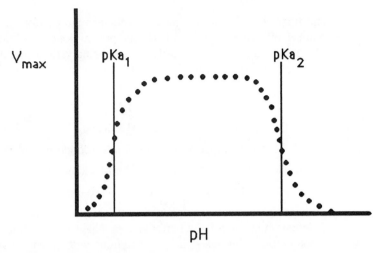

Figure 17-7   pH vs enzyme velocity profile

This pH dependence is taken as evidence of phosphoryl-imidazole enzyme intermediate, with the decomposition of this intermediate rate limiting at alkaline pH.[5,6]

The purpose of this experiment is to determine the V_{max} and K_m values for wheat germ acid phosphatase, either from a commercial source or from the preparation schemes given in Chapters 15 or 16, and to determine the K_I values for inorganic phosphate and fluoride ions in 0.1 M acetate buffer (pH 5) containing 0.01% Triton X-100. The hydrolysis of 1 mM p-nitrophenyl phosphate will be followed at 405 nm as described in Chapter 16. There have been previous studies of this enzyme in comparable systems.[5,6] Inhibition by the product, inorganic phosphate in 0.05 M acetate pH 5 containing 0.1 M NaCl and 0.01% Triton X-100, is competitive, as expected, but the inhibitor-enzyme dissociation constant changes little with pH. Fluoride acts as a noncompetitive inhibitor at least at pH 4.6 with 1 mM p-nitrophenyl phosphate in 0.11 M acetate buffer.[5] The pH dependence of the binding of inhibitors such fluoride anions has not been investigated fully.

EXPERIMENTAL SECTION

Students are expected to work in groups of three or four, with the preparation of solutions and the details of kinetic runs worked out in advance. The initial reaction velocity will be determined first as a function of enzyme concentration, at saturating substrate levels, then as a function of substrate concentration, at an enzyme concentration found convenient by the first set of kinetic runs. Velocity determinations as a function of substrate, with various concentrations of the two inhibitors phosphate and fluoride, will also be made. Each separate part of the project requires a large number of solutions, with all the component concentrations but one held constant. This is accomplished most easily by combining various proportions of stock solutions. Five 250-mL stock solutions of 0.01 M sodium acetate/acetic acid buffer (pH 5) containing 0.01% Triton X-100 are prepared for each group of students:

A. Contains just buffer and Triton X-100

B. Contains the above plus 1 mM substrate

C. Contains all that is in A plus 4 mM sodium phosphate

D. Contains all that is in B plus 4 mM sodium phosphate

E. Contains all that is in B plus 5 mM sodium fluoride

Solutions C-E should be adjusted to pH 5 with acetic acid after dissolving the salt. Solutions containing fixed buffer and variable substrate concentration are prepared by recombining solutions A and B; solutions with fixed phosphate concentration and variable substrate are made from various proportions of C and D, to a constant total volume, combined with a fixed volume of A. Similarly, solutions A, B and E allow for variable substrate with fixed fluoride concentration.

CAUTION: SODIUM FLUORIDE IS A POISON

If commercial enzyme is used, 20 mg of solid enzyme (0.5 U/mg) should be dissolved with gentle agitation in 10 mL of solution A and stored in an ice bath. This sample is enough for at least 24 students.

If freshly isolated enzyme is used, then the protein concentration should be adjusted to about 1.0 mg/mL for the following preliminary determination of the volume of enzyme and the reaction time required to give a final absorbance reading between 0.300 and 1.00, after quenching. The assay is initiated by adding 50 μL of enzyme solution, in buffer A, to 3 mL of solution B. The color change observed after 3 minutes, when the reaction is quenched with 250 μL of 1 M KOH, is measured with a colorimeter or spectrometer at 405 nm (see Chapter 15). The experiment is repeated for several additional enzyme solution volumes (1 to 100μL). The final absorbances are plotted as a function of volume enzyme solution and the smallest volume that gives a good change (0.3 < A < 1.0) within 3 minutes is determined. This volume of enzyme solution is used to initiate the reaction in three identical tubes to be run for 1, 2, and 5 minutes, respectively, and the final absorbances are determined. These absorbances must fall on a straight line in a plot of absorbance vs. time; otherwise, the enzyme volume added to initiate the reaction must be further reduced.

Determination of K_m

The first solution, A, is used to dilute carefully measured portions of B to prepare a series of ten 3-mL samples (in colorimeter tubes or cuvettes) containing substrate ranging from 0.01 to 1.0 mM concentration. These are then incubated at 25°C in a constant-temperature bath for 10 minutes and the reaction is initiated in each tube, in turn, by the addition of enzyme. It is not possible to initiate and then quench all the reactions simultaneously so they must be done one at a time or, if two students are working together, in staggered time order.

Care should be taken to have all the tubes run for exactly 5 minutes. This is not an easy task and a stopwatch or some sort of timer with a warning buzzer is necessary.

The absorbances at 405 nm are measured after the reactions are quenched and a plot of $(abs)^{-1}$ versus $1/[S]$ is prepared to determine linearity. If one or two of the absorbance values do not fall on a straight line defined by the other points, these points should be repeated. If more than two points appear off the line or if the plot is curved, then you must see your instructor.

Determination of Inhibitor Constants K_I

Much the same procedure is used to determine the inhibitor binding constants. Various portions of A, B, C, and D are combined to give three series of tubes with constant inhibitor concentration (0.5, 1.0, and 2.0 mM phosphate) but variable substrate concentrations (0.1 to 1 mM) so that the data for each series constitutes a linear Lineweaver-Burk plot. It is not necessary to use the same sample volume each time but the ratio of enzyme volume to sample volume must be kept constant. The inhibition by fluoride is assayed similarly; two series at fixed inhibitor concentrations (2.5 mM and 0.5 mM) with variable substrate are all that is required.

Data Analysis

The change in absorbance after 5 minutes is proportional to the reaction velocity. Rough data plots of $(Abs)^{-1}$ vs. $1/[S]$ should be prepared in the lab so that "bad points" can be redetermined at exactly the same bath temperature and pH. More sophisticated data analysis must be done outside of the lab employing a least-squares-fit with a calculator or with a computer.

There are a number of computer software packages available for enzyme kinetics. The K_m, V_{max} and K_I values must be reported with error ranges estimated from the degree of fit to Lineweaver-Burk plots. The types of inhibition caused by phosphate and fluoride ions are determined by calculating K_m, V_{max}, and K_I evaluated from plots corresponding to competitive and noncompetitive inhibition. The inhibition category that best fits the data will give the smallest error ranges. Note that all three parameters are positive numbers, intercepts that correspond to negative values must be the result of bad data or an improperly drawn line. Points for higher substrate values (ones nearer the origin in Lineweaver-Burk plots) are more reliable. Students should check to see if discarding the less reliable points improves the reasonableness of the binding parameters.

PROJECT EXTENSIONS

This is a lengthy experiment and gives considerable practice with obtaining data and graphical analysis. However additional work can be done on determining the effect of various Triton X-100 concentrations on enzyme activity. The pH dependence of fluoride inhibition can also bear further investigation.

DISCUSSION QUESTIONS

1. Rationalize the inhibition patterns described in Table 17-1, based on the preceding discussion of competitive and noncompetitive inhibition.

2. Why would you expect plots of $\Delta A/\Delta t$ (after 5 minutes) to be more curved for a concentrated enzyme or a more dilute enzyme?

3. Is it wise to work near saturating substrate concentrations when trying to discriminate between competitive and noncompetitive inhibition?

REFERENCES

1. H. Lineweaver and D. Burk, *J. Amer. Chem. Soc.*, **1934**, *56*, 658.

2. D. Burk, *Trends Biochem. Sci.*, **1984**, *9*, 202.

3. K. E. Neet in *Contempory Enzyme Kinetics and Mechanism*, ed. D. L. Purich, Orlando, Fla: Academic Press, 1983, pp. 284-287.

4. I. H. Segal, *Biochemical Calculations*, 2nd ed. New York: John Wiley & Sons, 1976, p. 257.

5. P. P. Waymack,"Isolation, Properties and Mechanism of an Isoenzyme of Wheat Germ Acid Phosphatase," Ph.D. thesis, Purdue University, 1978.

6. M. E. Hickey, P. P. Waymack, and R. L. Van Etten, *Archiv. Biochem. Biophys.*, **1976**, *172*, 439.

7. F. B. Rudolph in *Contemporary Enzyme Kinetics and Mechanisms*, ed. D. L. Purich, Orlando, Fla: Academic Press, 1983, p. 216.

Chapter 18

Chymotrypsin Burst Kinetics

The hydrolysis of p-nitrophenyl trimethylacetate is catalyzed by α–chymotrypsin. The kinetics of this process can be studied spectrophotometrically because the reaction liberates colored p-nitrophenoxide anion. An enzyme-bound trimethylacetyl intermediate accumulates during the reaction. The first mole equivalent of substrate reacts rapidly with the enzyme to yield p-nitrophenoxide ion and enzyme-bound intermediate. Subsequent utilization of substrate is slower since it depends on the slow breakdown of the intermediate to release free enzyme. The timewise dependence of colored product formation as a function of substrate concentration permits a detailed kinetic analysis of a complicated multistep reaction.

KEY TERMS

Acy-enzyme intermediate *Esterase activity*
Initial burst kinetics *Presteady-state kinetics*
Steady-state kinetics *Turnover number*

BACKGROUND

Although α-chymotrypsin is a proteolytic enzyme, it displays *esterase activity* in addition to its peptidase activity. With some ester model substrates it is a quite efficient esterase.[1] In the early 1950s, two British biochemists (B. S. Hartley and B. A. Kilby)[2] showed that p-nitrophenyl esters are substrates with the useful property that the colorless hydrolysis product p-nitrophenol is a weak acid ($pK_a = 7$) in equilibrium with the yellow p-nitrophenoxide anion with a maximum absorbance at 405 nm. The strongly colored product ($\varepsilon = 14{,}000$ L^{-1} m^{-1}) provides a convenient and sensitive means of following the reactivity of the colorless substrate near or above pH 7.

Another less expected discovery was that although the slow *steady-state* formation of reaction product could easily be monitored, extrapolating back to zero time gave a nonzero absorbance value that depended on substrate and enzyme concentration. The kinetics proved to be biphasic; there was an initial rapid liberation of p-nitrophenoxide followed by a slower release of this reaction product. Hartley and Kilby described the *presteady-state* behavior as an *initial burst*. Bender and coworkers[3,4] showed that this kinetic behavior depended on a second kinetically important intermediate after the initial Michaelis-Menten intermediate **ES**:

$$E + S \underset{}{\overset{K_s}{\rightleftharpoons}} ES \xrightarrow{k_2} ES' + P_1 \xrightarrow{k_3} E + P_2$$

The first-formed intermediate **ES** corresponds to the traditional Michaelis-Menton substrate-enzyme complex, while **ES'** is an acyl-enzyme intermediate and P_1 and P_2 represent p-nitrophenol and acetate, respectively.

By treating both the steady-state and presteady-state data by the method suggested by Bender,[4] K_s, k_2 and k_3 can be uniquely determined. For physiological substrates with turnover numbers of 100 s^{-1}, the presteady-state time interval is in the millisecond range and rapid reaction techniques are often required to detect acyl intermediates. With p-nitrophenyl trimethylacetate, the presteady-state time interval is on the order of 300 seconds, so that conventional recording spectrophotometers can be used. The rate of turnover of the acyl-enzyme intermediate following the initial burst is slower at lower pH and the acyl-enzyme intermediate can be isolated and studied.[5] The trimethylacetyl-enzyme derived at pH 5 has even been crystallized.[6]

Kinetic Analysis

Although this reaction proceeds through three consecutive steps, the system is amenable to kinetic analysis under certain conditions ([S]>>[E]) as the following derivation demonstrates. At high initial substrate concentrations, $[S]_0$>>[E] the formation of the Michaelis-Menten complex **ES** is a rapid equilibrium process characterized by the dissociation constant K_s.

$$K_s = \frac{[E]\ [S]_0}{[ES]} \tag{18-1}$$

The following conservation and differential equations are relevant:

$$[E]_o = [E] + [ES] + [ES']$$ (18-2)

$$\frac{d[P_1]}{dt} = k_2[ES]$$ (18-3)

$$\frac{d[P_2]}{dt} = k_3[ES']$$ (18-4)

$$\frac{d[ES']}{dt} = k_2[ES] - k_3[ES']$$ (18-5)

Equation (18-2) asserts that all enzyme is either free [E] or complexed with substrate [ES] or with its acyl moiety [ES']. The remaining three equations express the rate of formation of the two products in terms of the concentrations of complexed enzyme and the rate constants for individual steps in the kinetic scheme. Equation (18-5) points out that the turnover of the acyl-enzyme intermediate is the difference in the rate of its formation and the rate of its hydrolysis. Separating terms and integrating yields

$$\int_0^{[P_1]} d[P_1] = \int_0^t \frac{k_2[E]_o}{1 + K_s/[S]_o} dt - \int_0^t \frac{k_2 \frac{a}{b}(1-e^{-bt})}{1 + K_s/[S]_o} dt$$ (18-6)

giving an expression for $[P_1]$:

$$[P_1] = \frac{k_2([E]_o - \frac{a}{b})t}{1 + K_s/[S]_o} - \frac{k_2 a(1-e^{-bt})}{b^2[1 + K_s/[S]_o]}$$ (18-7)

Therefore $[P_1]$ has a time dependence of the form

$$[P_1] = At + B(1-e^{-bt})$$ (18-8)

This expression nicely describes the biphasic kinetics observed at some enzyme and substrate concentrations. At long reaction times, beyond the initial *burst*, $bt \gg 1$ and (18-8) reduces to the equation for a straight line that describes the steady-state region.

$$[P_1] = At + B$$ (18-9)

At short reaction times, the time dependence of $[P_1]$ deviates from the straight line by the difference

$$\Delta_t = Be^{-bt}$$ (18-10)

The kinetic behavior can be seen in Figure 18-1.

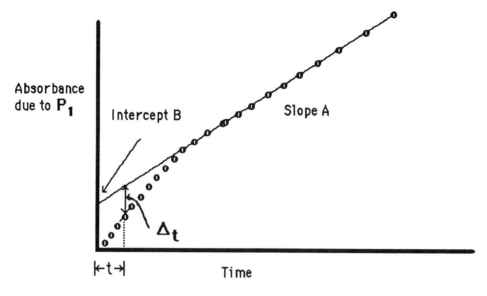

Figure 18-1 Determining Δt as a function of t during the initial burst

The term b can be evaluated from a plot of $\ln\Delta_t$ vs t because $\ln(Be^{-bt}) = \ln B - bt$ as seen in Figure 18-2.

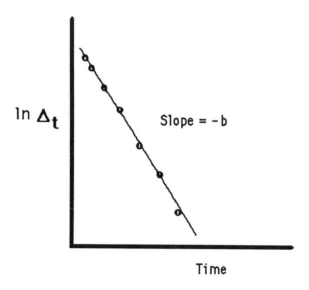

Figure 18-2 The term b can be valuated from a plot of ln Δ_t vs. t.

The terms A, B, and b have the meanings:

$$A = \frac{k_2 k_3 [E]_o [S]_o}{(k_2 + k_3)[S]_o + k_3 K_S} = \frac{k_{cat}[E]_o [S]_o}{[S]_o + K_m} \qquad (18\text{-}11)$$

$$B = \frac{[k_2/(k_2+k_3)]^2 [E]_o}{\{1 + K_m/[S]_o\}^2} \approx [E]_o \qquad (18\text{-}12)$$

$$b = \frac{(k_2+k_3)[S]_o + k_3 K_S}{K_S + [S]_o} \qquad (18\text{-}13)$$

where $k_{cat} = k_2 k_3/k_2 + k_3$ and $K_m = k_3 K_S/k_2 + k_3$.

The terms $k_2[S]_o/K_S$, k_3 and k_{cat} can be evaluated by determining A, B, and b for several initial substrate concentrations. Double-reciprocal plots are required. For example a plot of $1/A$ vs. $1/[S]_o$ gives an intercept and slope determined by the equation

$$\frac{1}{A} = \frac{1}{k_{cat}[E]_o} + \frac{K_m}{k_{cat}[E]_o} \frac{1}{[S]_o} \qquad (18\text{-}14)$$

A plot of $1/\sqrt{B}$ vs $1/[S]_o$ gives an intercept and slope according to (18-15). In the case of the hydrolysis of p-nitrophenyl trimethylacetate, where $k_3 K_S \ll (k_2 + k_3)[S]_o$, a plot of $1/b$ vs $1/[S]_o$ gives a slope of $K_S/k_2 + k_3 = K_m/k_{cat}$ because of equation (18-16).

$$\frac{1}{\sqrt{B}} = \frac{(k_2+k_3)}{k_2[E]_o} + \frac{K_m(k_2+k_3)}{k_2 \cdot \sqrt{[E]_o}} \cdot \frac{1}{[S]_o} \qquad (18\text{-}15)$$

$$\frac{1}{b} = \frac{[S]_o}{(k_2+k_3)[S]_o + k_3 K_S} + \frac{K_S}{(k_2+k_3) + k_3 K_S} \cdot \frac{1}{[S]_o} \qquad (18\text{-}16)$$

Now b is evaluated in the $[S]_o$ concentration range where $(k_2 + k_3)[S]_o \gg k_3 K_S$ since the expression above simplifies.

The purpose of this experiment is to determine the values of A, B and b for several values of substrate concentrations over the range 10^{-5} to 10^{-4} M. From these values $[E]_0$ can be evaluated from the value of B. The constants K_m/k_{cat}, k_3 and k_{cat} can be determined from the appropriate doublereciprocal plots. The values of the rate constant for this system at pH 8.2 have been reported by Bender (Table 18-1).[4]

Table 18-1
RATE CONSTANTS FOR CHYMOTRYPSIN

k_2	0.37 sec^{-1}
k_3	1.3×10^{-4} sec^{-1}
K_s	1.6×10^{-3} M
k_{cat}	1.3×10^{-4} sec^{-1}
K_m	5.6×10^{-7} M

[a]Source: Ref. 4

EXPERIMENTAL SECTION

This project requires a good recording spectrometer with a thermostated cell compartment, consequently, only part of a class can work on it at any one time. Students should work in pairs or groups of three preparing solutions, but each student must take several turns obtaining rate data. The enzyme solution is added with a automatic pipet and the reaction mixture is stirred with a small glass rod or paddle; then the cell compartment is closed and the recorder activated. Practice is required before a student is able to initiate the reaction and close the spectrometer cell compartment in time to obtain a reasonable burst effect.

A pH 8.5 TRIS buffer (0.01 M) is used for all rates; the enzymatic rates are almost constant in this pH region. Three milliliters of this buffer solution is placed in a cuvette and thermostated to 25°C. Then 10 to 100 μL of substrate solution is added. The substrate solution is made from 7.8 mg (3.5 mM) p-nitrophenyl trimethylacetate in 10.0 mL of pure acetonitrile. A 100-μL sample corresponds to approximately 1.1×10^{-4} M, which is quite close to the saturation limit, 1.5×10^{-4}. The spontaneous hydrolysis of the substrate is measured spectrophotometrically at 405 nm over the length of time expected for the reaction course (2000 s). This spectral change due to spontaneous hydrolysis must be subtracted from the absorbances observed in the enzyme-catalyzed process before any other calculations are made.

An enzyme-catalyzed reaction is initiated by addition of 10 μL of the enzyme stock solution to the buffered substrate solution.[7] The enzyme stock solution is made from 50 mg of α-chymotrypsin (2.0 mM) in 1.0 mL of pH 4.6 acetate buffer. The stock solution must be kept cold. The rate is monitored by the change in absorbance at 405 nm. Doubling $[E]_0$ at constant $[S]_0$ should double the intercept B and the slope A but should not affect b. Doubling $[S]_0$ at constant $[E]_0$ should double b but not affect B or A.

PROJECT EXTENSIONS

1. There is a long chain of calculations and graphs between the observed absorbance values and the calculated kinetic parameters. A calculation of the error in these parameters can be made by the propagation of errors method.[8]

2. Examine the pH dependence of the burst.[7]

REFERENCES

1. C. Walsch, *Enzymatic Reaction Mechanisms*, San Francisco: Freeman, 1979, p. 61.

2. B. S. Hartley and B. A. Kilby, *Biochem. J.*, **1954**, *56*, 288.

3. B. Zerner and M. L. Bender, *J. Amer. Chem. Soc.*, **1964**, *86*, 3069.

4. M. L. Bender, F. J. Kezdy and F. C. Wedler, *J. Chem. Educ.*, **1967**, *44*, 84.

5. A. K. Balls and H. N. Wood, *J. Biol. Chem.*, **1956**, *219*, 245.

6. E. Zeffren and P. L. Hall, *The Study of Enzyme Mechanisms*, New York: John Wiley and Sons, 1973, p. 174.

7. R. D. Allison and D.L. Purich, *Contemporary Enzyme Kinetics and Mechanism*, ed. D. L. Purich, Orlando, Fla: Academic Press, 1983, pp. 33-44.

8. B. Mannervik, *Contemporary Enzyme Kinetics and Mechanism*, ed. D. L. Purich, Orlando, Fla: Academic Press, 1983, pp. 75-94.

Chapter 19

Nucleic Acid Isolation, Purification, and Characterization

A general acount of nucleic acid structure, nomenclature, chemical reactivity, enzymology, purification and characterization is given as background for the projects described in Chapters 20 to 26.

KEY TERMS

Apurinic acid   Autoradiography
Chromatin   Chromosome
Complementary sequences   Deoxyribonucleic acid
Deoxyribonucleotide   Deoxyribose
Endonucleases   Exonucleases
Genes   Helix-to-coil transition
Intercalation   Ligases
Nick translation   Nucleic acid
Nucleoside   Nucleosome
Nucleotide   Oligonucleotide
Primary structure   Purines
Pyrimidines   Replication
Restriction endonucleases   Restriction fragments
Ribose   Ribonucleic acid
Self-complementary sequences   Transcription

GENERAL COMMENTS ON NUCLEIC ACIDS

Deoxyribonucleic acid (DNA) and *ribonucleic acid* (RNA) are constituents of all cells and they play the fundamental roles of preserving and expressing hereditary information respectively. DNA is the repository of genetic information within a cell. The genetic information is stored as the primary structure of the DNA molecule (i.e., the sequence of subunit bases that make up the DNA molecule). This information is expressed as individual units called *genes*. Most often there are only one or few copies of each gene within a cell. Many genes are associated together as parts of the sequence of a very large DNA molecule known as a *chromosome*. DNA is always associated with proteins within cells. In procaryotic cells the DNA exists as a single circular chromosome associated with diverse proteins, and in eukaryotic cells (cells with nuclei) this association involves a very specialized nucleoprotein structure known as *chromatin*. The fundamental unit of chromatin is an octameric complex of histone proteins wrapped with DNA; this structure is known as the *nucleosome*. There are usually a number of linear DNA molecules associated with proteins in the nucleus of a eukaryote (i.e., there are a number of chromosomes). RNA exists in a number of forms within a cell. Linear mRNA molecules are direct copies in RNA of the DNA base sequences of individual genes. These molecules are used as temporary copies required for the synthesis of proteins; the RNA sequence of each mRNA molecule determines the primary sequence of a particular protein. For this reason mRNA molecules are short-lived and are quickly broken down by ribonuclease (RNase) within cells. RNA also occurs as a structural element within ribosomes as rRNA and in ribonucleoprotein complexes associated with specialized functions such as transport and mRNA modification. RNA is also involved in activating individual amino acids for protein synthesis as tRNA. All these RNA structures have complex tertiary structures, fundamental to their biological activity.

In the following seven chapters are a series of experiments that involve either the isolation or the chemical modification of RNA and DNA and their characterization, mostly by electrophoretic means. The purpose of this chapter is to make some fairly general statements regarding the structure, stability and chemistry of nucleic acids, at least as far as they are relevant to the successful completion of these more advanced projects. We will focus first on the basic structural features of nucleic acids, especially the rather confusing nomenclature in this field, than on the factors governing the stability of nucleic acids, their chemistry and enzymology, and finally, on the use of electrophoresis, spectroscopy, and centrifugation to characterize nucleic acids.

NUCLEIC ACID STRUCTURE

Nucleic acids are linear polymers build up of subunits, which are composed of three types of chemical structures: bases, sugars, and inorganic phosphate. The backbone of the nucleic acid is an alternating linear polymer of sugars linked by phosphate diester linkages, so that the backbone is a polyanion with a charge equal to the number of phosphate groups. The neutral bases are also attached to the sugar group; one base for each sugar. There are two kinds of heterocyclic bases in nucleic acids: *purines* and *pyrimidines*. The structures of all the common bases in nucleic acids are given in Figure 19-1. The two purines adenine and guanine appear in both DNA and RNA; the two pyrimidines cytosine and thymine appear in DNA while uracil replaces thymine in RNA. These bases are designated A, G, C, T, and U, respectively. There are two sugars common to nucleic acids: *ribose*, which occurs in RNA, and *deoxyribose*, which occurs in DNA. The sugar atoms are numbered as indicated in Figure 19-2.

When a purine or pyrimidine base is attached to a ribose or deoxyribose, the result is a *nucleoside*. The phosphoric acid ester of a nucleoside (usually at the 5'-OH group) is a *nucleotide*. Nucleotide 5'-phosphates can form a second phosphate ester linkage to the 3'-OH of another nucleotide and the resulting phosphate diester is a *dinucleotide*. If this polymer is elongated by the formation of additional phosphate diester bonds to additional nucleotides, the result is an *oligonucleotide*.

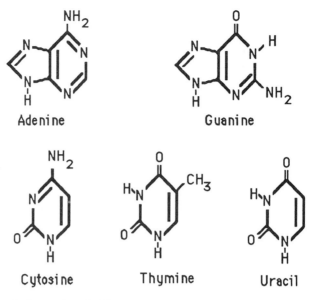

Figure 19-1 Purine and pyrimidine bases

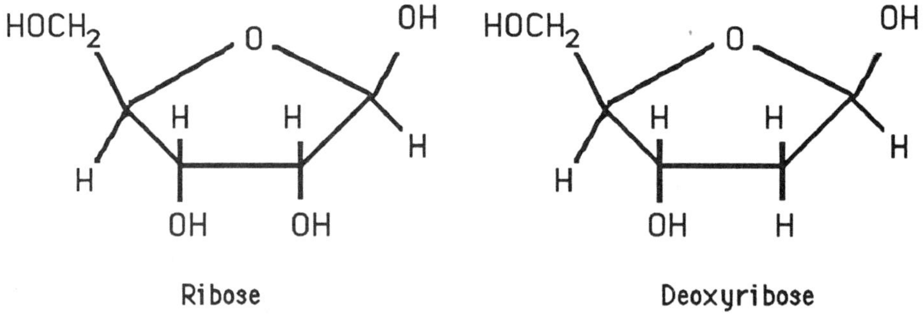

Figure 19-2 Ribose and deoxyribose

Phosphate diester linkages are often designated by inserting a "p" between the letter symbols for the connected nucleotides. When placed to the right of a nucleotide symbol the phosphate is attached to the 3' position of the ribose moiety; when placed to the left it indicates that the phosphate is attached to the 5' position. The oligonucleotide designated 5'-ApGpCpTp-3' contains the ribonucleotides adenosine, guanosine, cytidine, and ribosylthymidine linked in that order from the 3'-O of adenosine to the 5'-O of ribosylthymidine. Each of the phosphodiester phosphates between the 3'-O and 5'-O of linked nucleotides has a -1 charge, while the monoesterified 3' phosphate of ribosylthymidine has a -2 charge. It is important to remain aware that an oligonucleotide is a polyanion since much of its chemistry and physical properties stems from this fact.

The oligonucleotide described above is built up from ribonucleotides; the corresponding deoxyribonucleotides would have the names deoxyadenosine, deoxyguanosine, deoxyctidine, and thymidine. Note that ribosylthymidine is so named because the deoxynucleotide thymidine, which is a constituent in DNA but not RNA, is more common.

Nucleic acids share the same relationship to oligonucleotides as proteins do to peptides (i.e., they are longer versions of the same type of chemical compound). Nucleic acids are polymers made up of hundreds, thousands or even millions of nucleotides linked by phosphodiester linkages. DNA molecules are polymers of deoxyribonucleotides and RNA molecules are polymers of ribonucleotides, both connected in the 5'- to 3'- order. A nucleic acid is defined uniquely by the nucleotide sequence written in the 5'- to 3'- order; the same order in which a nucleic acid is synthesized during *replication* and read during *transcription* or *translation*. This sequence is known as the *primary structure*. DNA molecules are the primary repository of genetic information in cells and for this reason they can be very large. Table 19-1 gives the molecular weights and extended lengths of DNA from a number of biological sources as well as the molar proportions of each of the four bases in DNA from various sources.[1]

One can see from the table that the length of DNA increases markedly as one progresses up the classical phylogenic tree but that the base composition does not reveal any obvious correlation with the size of the DNA. A more extensive comparison of the composition of DNA from various sources reveals a very subtle but important regularity: Adenine and thymine are present in equimolar amounts and so are guanine and cytosine.[2]

This was the key to understanding that the bases in nucleic acid are often associated by hydrogen bonding to form *complementary base-pairs.* A is complexed with T by two hydrogen bonds and G with C by three.

Table   19-1

PROPERTIES OF DNA FROM VARIOUS SOURCES[a]

Source	Mol. Wt	Length	% A	% G	% C	% T
Bacteriophage $\phi\chi 174$	1.6×10^6	0.6 μM	24.3	24.5	18.2	32.3
E. coli	3.0×10^9	1.5 mm	26.0	24.9	25.2	23.9
Yeast	1.2×10^{10}	4.6 mm	31.3	18.7	17.1	32.9
Drosophila	4.3×10^{11}	20 mm	30.7	19.6	20.2	29.4

[a]Source ref. 1.

Oligonucleotides are complementary if the bases of one can pair up with the bases of the other. For example, ApApTpGp is complementary with CpApTpTp and ApGpCpTp is complementary with ApGpCpTp (itself). Oligonucleotides that can spontaneously pair with themselves are said to be *self-complementary.*

5'-ApApTpGp-3'   5'-ApGpCpTp-3'

3'-pTpTpApC-5'   3'-pTpCpGpA-5'

The pairs of flat bases are stacked one above the other so that the double-stranded structure is stabilized by hydrophobic as well as hydrogen bonding forces. Hence the physical chemistry of a nucleic acid is to be understood in terms of two complementary polyanions associating to form an extended base-paired structure, stabilized by hydrogen bonding and hydrophobic forces, despite the electrostatic repulsion between the chains. The base stacking of chiral nucleotides leads to an asymmetric helical structure: either a smooth right-handed double helix as in B-DNA or A-DNA or a zigzag left-handed double helix as in Z-DNA. The different forms of DNA have been characterized by a wide variety of physical methods but especially, X-ray diffraction structure studies.[3,4]

Factors Governing Nucleic Acid Stability

The equilibrium between double-stranded DNA (dsDNA) and single-stranded DNA (ssDNA) can be shifted to favor dissociation, by factors reducing hydrogen bonding or hydrophobic forces, and to favor association, by factors mitigating the electrostatic repulsion between chains. Temperature, solvent, and the concentrations of other ions are very important factors which must be taken into account in planning any experiment with DNA. Chain dissociation is a denaturation process that is often irreversible in a practical sense and it must be avoided if isolation of biologically active dsDNA is the point of the experiment. RNA forms intramolecular hydrogen bonds that account for the complex tertiary structures associated with tRNA or rRNA. The factors governing the stability of these structures has not been fully worked out,[3] but it is apparent that rather precise limitations on the torsion angles between sugar and base and within the sugar must be obeyed. The chemistry of RNA species is dependent on their conformations, and the loss of tertiary structure sometimes hastens subsequent degradations of these molecules by enzymes.

When working with either DNA or RNA, high temperatures must be avoided. Hydrogen bonds are constantly breaking down and re-forming; the rate of breakdown depends on the precise angle and distance between the donor and acceptor pair. Increasing the temperature decreases the number of hydrogen bonds between complementary bases, and when hydrogen bonds are no longer present to constrain the motion of bases, they can swing farther away from their stacked orientation and hence are much less likely to reform a hydrogen bond. This conformational change influences the conformation of neighboring nucleotides and increases the rate of hydrogen bond breaking for nearby bases so that the breakdown of base pairing is a cooperative phenomenon.[5] For a relatively short oligonucleotide (<100 base pairs) at low temperatures, the base pairs near the ends of the helix open and close easily, but structures with only a small percentage of base pairs are open. As the temperature increases the fraying of the ends becomes only slightly more prominent with very few internal base-pair hydrogen bonds broken, until a pseudophase transition temperature or "melting temperature" T_m is reached.[6] This is a change over a very narrow temperature range because of the zipper-like mechanism of hydrogen-bond breakdown, that depends on bases reorienting cooperatively. The *helix-to-coil transition temperature* T_m depends on the base-pair composition of the nucleic acid and upon the presence or absence of ions that can screen the electrostatic repulsion between phosphates on different chains. Nucleic acids with a higher percentage of GC base pairs have a higher melting temperature than nucleic acids with more AT base pairs; this is presumably due to the stronger hydrogen bonding in the former case.

Added salts stabilize dsDNA relative to ssDNA. At low concentrations (0.01 to 1 M) added cations shield the phosphate backbones and minimize the interchain repulsions that destabilize the dsDNA molecule. This results in an increase in the helix-to-coil transition temperature with increasing ionic strength.[7] The range of temperatures over which the helix-to-coil transition occurs is also narrower with added salt.

DNA is such a highly charged polyanion that the high charge density generates a steep local gradient of counterions.[8] Even at low concentrations of added salts, the local counterion concentration at the DNA surface is in the molar range. The denaturation of a nucleic acid reduces the charge density, which reduces the steepness of the local ion concentration gradient. Hence the helix-to-coil transition amounts to the release of ions to the bulk solvent. The change in charge density upon denaturation is relatively independent of salt but the change in entropy associated with the release of these ions to the bulk solvent is profoundly a function of salt concentration.[8] This is the origin of the salt effect.

For some kinds of experiments denaturation is specifically desired and protocols have been developed to ensure that it occurs. DNA denaturation can be provoked by high temperatures or very low electrolyte concentrations. RNA is completely unfolded because of the organic solvent formamide, which competes for hydrogen bonds with the bases. DNA and RNA can also be chemically degraded. Both undergo hydrolysis of the N-glycosidic bonds in acid. The purines come off easiest to form *apurinic* acids which are labile in acid; the phosphodiester bonds are preferentially cleaved at apurinic sites. RNA also undergoes hydrolysis of phosphodiester bonds in dilute alkali whereas DNA is much more stable in base. The 2'-OH of RNA is involved in the mechanism of base-catalyzed hydrolysis; the base removes the 2'-OH proton and the resulting nucleophilic 2'-O$^-$ alkoxide, then attacks the 3'-phosphate ester to form a cyclic intermediate which rapidly breaks down in base. This reaction will be employed in the project described in Chapter 20. Nucleic acids also are cleaved by a variety of inorganic and organic reagents, an important one is ferrous iron in the presence of peroxide. This reagent generates hydroxyl radicals which attack ribose C-H bonds indiscriminately, degrading nucleic acids.

ISOLATION AND CHARACTERIZATION OF NUCLEIC ACIDS

Three problems must be overcome in any DNA or RNA preparation:

1. The nucleoprotein complexes must be dissociated and the proteins removed from the nucleic acid.

2. Cleavage due to native DNase and RNase enzymes must be minimized.

3. Hydrodynamic shear of fragile DNA strands must be avoided.

There are now a large number of methods for accomplishing these ends and several are illustrated in the following chapters. For DNA the most common procedure was demonstrated in Chapter 10. The nucleoprotein complex is broken up with the detergent SDS (sodium dodecyl sulfate) and the deproteination completed by extraction with chloroform/*iso*-amyl alcohol. Deproteinated DNA in very cold 2 M NaCl is rendered insoluble by the addition of ethanol. The resulting DNA is a viscous filamentous mass which is spooled onto a glass rod. Water-saturated phenol is also used to deproteinate nucleic acids as shown in Chapter 20.

Detergents, chaotropic agents and chelating agents can be employed successfully to break up nucleic acid/protein complexes and in deactivating nucleases, as shown in Chapters 21 and 24. A chaotropic agent such as quanidinium thiocyanate disrupts hydrophobic interactions to break down protein tertiary structures. A chelating agent such as EDTA or citrate is necessary to remove the metal cations associated with DNA phosphate groups. It also removes the Mg^{2+} or Ca^{2+} required by nucleases. DNA cleavage due to DNase activity is also reduced by working rapidly and at low temperatures. Nucleic acids are also broken down by shearing forces and by trace metal ions, especially iron. Care must be taken to avoid shearing the very long strands of DNA, and techniques such as stirring and pipetting DNA through a capillary orifice, which can break the strands, must be avoided. All steps must be performed at 4°C with glass or plastic vessels; stainless steelware, including spatulas, must be avoided. Even minute traces of rust cause rapid degradation. A stainless steel tissue homogenizer should not be used if a Teflon-bladed one is available. All glassware should be rinsed with 1 mM EDTA before use.

Enzymic Degradation and Modification of Nucleic Acids

Enzymes catalyze the cleavage of phosphate diester bonds at either the 5' or 3' ends of a dinucleotide junction. DNA-cleaving enzymes (DNases) can be classified according to their specificity. Nucleases that cleave the terminal phosphodiester linkages of DNA from either end are known as *exonucleases,* while enzymes that cleave the phosphodiester linkages within a DNA molecule are *endonucleases.* Nucleases exhibit specificities either for single-stranded DNA (ssDNA) or for double-stranded DNA (dsDNA) although some nucleases will do both, especially at high concentrations. Endonucleases that introduce double- or single-stranded cuts in DNA, more or less at random, are widely used. *Deoxyribonuclease I*(DNase I) cleaves DNA to a complex mixture of mono- and polynucleotides with 5'-phosphate termini. In the presence of Mg^{2+} DNase I cleaves each strand of dsDNA independently introducing both double and single-strand cuts at statistically random positions along a dsDNA chain; in the presence of Mn^{2+}, double-stranded cuts to produce blunt ends predominate.[9] At low levels (several hundred thousand-fold dilutions of commercial enzyme stocks) and in the presence of mg^{++}, DNase I can be used to introduce occasional nicks in dsDNA. *E. coli DNA polymerase I* has both 5'-to3'-exonuclease and 5'-to3'-polymerase abilities and can translate, or move a nick introduced by low levels of DNase I. The polymerase binds at the nick and then moves down the chain replacing nucleotides with the nucleoside triphosphates included in the reaction mixture. If any of the latter are radiolabeled then this method of *nick translation* can be used to radiolabel a DNA chain.[10]

Nuclease S1 from *Aspergillus oryzae* degrades ssDNA to yield 5'-phosphoryl oligonucleotides. Double-stranded nucleic acids (dsDNA, dsRNA, and DNA·RNA) are cleaved more slowly. This enzyme cleaves at nicks or mismatched regions in the presence of the required cofactor Zn^{2+} at pH 4.5. Single-stranded breaks can be introduced into dsDNA by mechanical and hydrodynamic forces. Such nicked dsDNA can be cleaved at the nick sites by digestion with S1. We will use this enzyme in this project to explore the single-stranded nicks introduced into dsDNA by mechanical shearing. The S1 cleavage pattern of nicked and nick-free DNA will differ because the former will be cut into shorter fragments by S1.

DNase I and a related enzyme, *micrococcal nuclease*, can be used to study the binding of proteins to specific locations along a DNA chain. The bound protein shields the DNA from the nuclease so that even high levels of the latter do not cleave any of the phosphate diester bonds sheltered by the protein; exhaustive double-stranded cleavage will result in small oligonucleotides except for the protein sequence specifically protected by the protein. This fragment is said to be the "footprint of the protein on the

DNA" and the DNA cleavage technique is known as *footprinting*.[11] The footprint polynucleotide can be isolated and sequenced. This is a principal method for studying the binding mechanism of proteins that bind to specific DNA sequences because the resulting footprint will have a unique or nearly unique sequence. DNase digestion of all but a specific protein-bound region of DNA has been used to isolated specific protein-binding regions of DNA, including the repressor binding region of the *lac* operon. Although helical dsDNA appears at first glance to be a molecule of almost monotonous repetitiveness, sequence-specific proteins can bind and interact with unique regions with quite remarkable tenacity and selectivity. Our knowledge of the interactions between protein and nucleic acid is still very limited, but we know that such interactions occur in the regulation of DNA replication and expression and are of the greatest importance to life. Ferreting out the details of protein-nucleic acid interactions is one of the greatest challenges to modern biophysical chemistry.

Among the most impressive examples of specificity in protein-DNA binding are the very specific DNA cleavage patterns caused by *restriction endonucleases*. Restriction endonucleases have been found in nearly every microorganism examined and are known to catalyze double-stranded breaks in DNA to yield *restriction fragments*. The restriction enzyme cleavage pattern is specific for a given DNA and enzyme and can be used to characterize plasmids and viral DNA. The techniques are relatively easy[12] provided that the enzymes are stored very cold and with stabilizing reagents such as glycerol. Care must also be taken to ensure that the ionic strength is adjusted for maximum activity; each enzyme has specific buffer requirements.The use of restriction enzymes to characterize a plasmid by restriction mapping is demonstrated by the experiment described in Chapter 23.

There are also enzymes that can modify nucleic acids without cleaving them. Polynucleotide kinases can transfer the γ-phosphate from ATP (or other nucleotide triphosphate) to the 5'-end of DNA. The phosphorylation reaction requires magnesium ions and a sulfhydryl reducing agent. If $[\gamma\text{-}^{32}P]$ ATP is used this reaction can be used to radio-label DNA as illustrated in Chapter 26. *Ligases* are used to catalyze the formation of a phosphodiester bond between the free 5'-phosphate group of one oligonucleotide and the 3'-OH end of another. T4 DNA ligase requires ATP as a cofactor, while *E. coli* or *B. subtilis* ligase requires NAD.

Electrophoretic Characterization of Nucleic Acids

The electrophoretic migration of DNA through agarose gels depends on the following parameters:

1. **Molecular weight.** Linear dsDNA migrates through the gel in an end-on orientation, and within limits the migration rate is inversely proportional to the logarithm of the molecular weight.[13,14] Figure 19-3 illustrates mobility as a function of molecular weight. Very long DNA molecules $(MW > 10^7)$ can migrate efficiently only after adopting certain conformations.[15]

2. **Agarose concentration (% gel)** Increasing the agarose concentration will retard the electrophoretic mobility of a DNA fragment; there is a linear relationship between the logarithm of the electrophoretic mobility (μ) and the gel concentration (g):

$$\log \mu = \log \mu_o - K_r\, g \qquad (19\text{-}1)$$

where μ_o is the free electrophoretic mobility and K_r is the retardation coefficient.

Agarose is a complex polysaccharide derived from agar. When dissolved in hot aqueous buffer and allowed to cool, it forms a soft but mechanically rigid matrix of highly hydrated carbohydrate in buffer. The electrophoretic mobilities of DNA in agarose gels depend on the average pore size of this matrix, which is a function of the percent agarose in the gel. Table 19-2 illustrates the separation range for linear DNA molecules as a function of the amount of agarose in the gel.[13],[14] The molecular sieving effect is clearly evident.

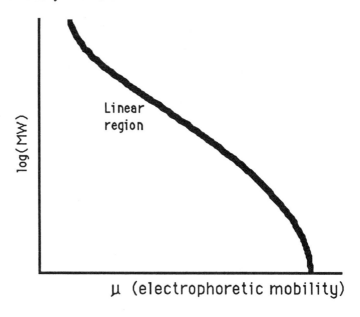

Figure 19-3 Electrophoretic Mobility and DNA size

Table 19-2
DNA FRACTIONATION ON AGAROSE GELS

Agarose Gel (% w/w)	Separation Range (kb)
3.0	0.07-0.1
2.0	0.1- 3
1.5	0.2- 4
1.2	0.4- 6
0.9	0.5- 7
0.7	0.8- 10
0.6	1 - 20
0.3	5 - 60
0.1	Up to 800 kb

3. **DNA Conformation** The ability of a DNA molecule to pass through the highly hydrated matrix of an agarose gel depends on both the size and shape of the DNA molecule.[16] Long rigid rodlike native double-stranded dsDNA runs considerably slower than collapsed, and hence more compact, single-stranded ssDNA in agarose gels. Closed circular, nicked circular, and linear dsDNA all migrate through agarose at

different rates, but their relative order is a complicated function of gel percentage ionic strength and voltage. The concentration of ethidium bromide, used to visualize DNA gels, while they are being run, is also important. Supercoiling in the closed circular form increases the compactness of this form, so that it can migrate faster than linear DNA; however, ethidium bromide binding reduces the negative supercoiling of closed circular dsDNA and will lower its mobility below that of linear dsDNA. At very high levels of ethidium bromide (>> 0.5 μg/mL) bound dye introduces positive supercoiling so that closed circular DNA again moves faster than linear dsDNA. Concentrations of ethium bromide higher than 0.5 μg/mL also introduce positive supercoiling in negatively supercoiled circular dsDNA. At such high concentrations of the cationic dye, the charges of all forms of DNA are reduced and linear and nicked circle dsDNA migrate more slowly. For most preparations of linear dsDNA, the ethidium bromide concentration is in the range 0.1 to 0.5 μg/mL. Ethidium bromide is a planar polyaromatic compound that inserts itself or **intercalates** between stacked bases in dsDNA; see Figure 19-4.

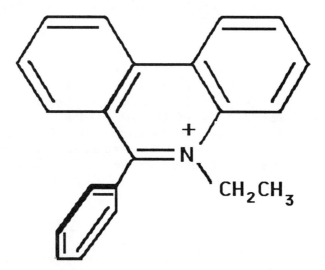

Figure 19-4 Ethidium cation is a planar species

Bound dye has a considerably enhanced fluoresence compared to dye in buffer alone; bound dye absorbs UV light (300 to 360 nm) and emits in the red-orange region (590 nm). Unlike protein polyacrylamide gels, which must be fixed and stained after the gel is run, ethidium bromide can be added to the agarose gel electrophoresis buffer so that the progress of the electrophoresis can be directly monitored. Continuous monitoring is not recommended because UV irradiation does destroy DNA, but brief UV illumination once or twice near the end of a run can be used to ensure maximum running time and resolution without sample running off the end of the gel.

Electrophoretic characterization of RNA is best done under denaturing conditions. An unambiguous correlation between molecular weight and electrophoretic mobility can be obtained only if the secondary structure of the RNA is abolished before and during electrophoresis. Single-stranded RNA contains hairpin loops that are stabilized by hydrogen bonds. Such structures may contain hidden breaks which cannot be seen until the fragments dissociate as seen in Figure 19-5.

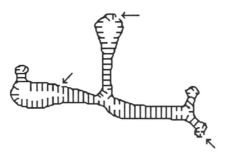

Figure 19-5 RNA with "hidden breaks"

There are a number of denaturing gel systems used to measure the molecular weight of RNA and to separate RNA of different sizes:

1. Electrophoresis in 4% polyacrylamide gels in 98% formamide.[17]

2. Electrophoresis through agarose gels after denaturation of the RNA with glyoxal and dimethyl sulfoxide.[18]

3. Electrophoresis with an agarose gel containing 1 M formaldehyde.[19, 20]

In the first method the RNA is completely unfolded because of the organic solvent formamide, which is infinitely miscible with water, can compete for hydrogen bonds with both water and with the bases. Hydrophobic interactions that promote base stacking in native RNA are also lost in this solvent. This method has two disadvantages: The formamide must be deionized before use to remove formate and ammonium ions, and it is a very poisonous compound. In the second method the RNA solution is incubated at 50°C for 60 minutes with excess glyoxal in water-DMSO. The glyoxalated RNA cannot reform RNA secondary structure below pH 8.0.

$$\underset{\text{H-C}}{\overset{\text{O}}{\|}}\,\underset{\text{C-H}}{\overset{\text{O}}{\|}} \qquad \text{Glyoxal}$$

Undegraded rRNA gives two prominent low-mobility bands corresponding to 28S and 18S ribosomal RNA. Degradation leads to a higher-mobility fraction and the replacement of these bands by a prominate band farther down the gel. In the third method, which is suggested for the experiment described in Chapter 21, formaldehyde reacts with the primary amino groups of A, G, and C bases to prevent the formation of base-pair hydrogen bonds. The mRNA is denatured by brief heating to 95°C in a loading buffer containing formaldehyde and formamide then applied to an agarose gel containing formaldehyde. The loading buffer and agarose gel pH is maintained with the buffer MOPS (3-[N-morpholino]propanesulfonic acid), which is a tertiary amine that does not react with formaldehyde. Because formaldehyde is a poisonous gas, the gel must be run in a fume hood. The molecular weight of RNA run on denaturing gels is determined by comparing its mobility to that of known standards, just as with DNA (see Chapter 10). Convenient RNA markers are 28S and 18S eukaryotic rRNA, which are 5.1 and 2.0 kb, respectively. Total RNA samples will contain prominent amounts of these species.

DNA restriction fragments which have been fractionated according to molecular weight in agarose gels can be transferred to a solid support of nitrocellulose or chemically activated paper in such a way as to retain the original pattern. This blotting method was developed by Southern[21] and the resulting pattern is called a "Southern blot." (An analogous procedure for transfering RNA is humorously refered to as "Northern blots.") In most instances the DNA attached to the solid support is then hybridized to ^{32}P-labeled DNA or RNA and autoradiography is used to locate any bands complementary to the probe. The Southern blotting method will be demonstrated in Chapter 26. Chemically modified ABM paper contains covalently attached aminobenzyloxymethyl groups which are activated with nitrous acid to form diazobenzyloxymethyl paper (DBM paper) just prior to use. The nucleic acid bases become covalently bonded to the paper through diazo linkages to the benzyloxymethyl groups. The diazo linkage is a covalent attachment that is not readily reversible.

Because the DNA fragments are covalently bound to the DBM paper, small fragments are retained efficiently. This is not the case with nitrocellulose which is best used only with large fragments or glyoxylated DNA.

Autoradiography is very sensitive and gives a high-resolution, non-destructive way of locating radio-labeled nucleic acids in polyacrylamide gels or on paper or nitrocellulose supports. The strongly β-emitting isotope ^{32}P gives very good autoradiographs on X-ray film from the procedure described in the Experimental Procedure section. The film is sandwiched between two calcium tungstate phosphor screens (shiny sides toward film) and the sandwich is placed next to the DNA sample bound to DBM paper. The whole thing is taped to a backing sheet and wrapped and taped tightly with more backing and stored at -70°C. The events recorded on film are the long-wave length fluorescence emissions that occur when β-particles strike the screen.[22] The film sensitivity is increased 10-fold by the use of two fluorescent screens. The exposure is carried out at -70°C, if possible, in order to prolong the period of fluorescence. The film is developed with Kodak liquid X-ray developer (5 min.), a 3% acetic acid stop bath (1 minutes), and Kodak rapid fixer (10 minutes).[23]

A gel stained with 0.5 μg/mL ethidium bromide can be conveniently photographed with the gel in its tray. A hand-held UV lamp illuminating the gel from above and photographing the gel from reflected and emitted light is an inexpensive approach, but the results are less than optimum. Illumination from below, so that the photograph is taken with transmitted light, is better. Several commercial systems are designed along the scheme pictured in Figure 19-6. A homemade light box with a 366 filter and a GE germicidal bulb can be assembled with little difficulty. A camera with Polaroid type 55 (50 ASA) or type 56 (3000 ASA) film is quite suitable. While the former film requires several seconds to a minute or two of exposure time, it gives both a positive and negative copy and excellent contrast. The latter film requires a shorter exposure time but gives more background. The camera must be mounted on a tripod or some other rigid support.

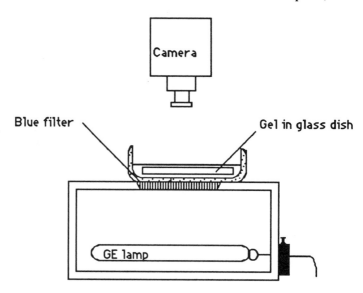

Figure 19-6 Photographing gels with transillumination

Spectroscopic and Centrifugative Characterizations of Nucleic Acids

The concentration of nucleic acid can be estimated from the UV absorbance at 260 nm. The $E^{1\%}$ is 200 at 260 nm at pH 7.4 which roughly corresponds to an absorbance of 1 for a nucleotide concentration of 0.1 mM. The purity of nucleic acid can be assessed either spectroscopically or electrophoretically. RNA preparations with a UV-absorbance ratio $A_{260\,nm}/A_{280\,nm}$ of 2.1 or better are nearly free of protein contamination.

The classical way of separating types of DNA with different base composition is sedimentation through a CsCl gradient. This is also an important way to characterize DNA since GC base-pairs are less hydrated than AT base pairs and the G+C content can be determined on the basis of the buoyant density.[24] The G+C composition can be calculated from the density ρ of the DNA by the formula:

$$\%(G{+}C) = \frac{\rho - 1.660}{0.98} \times 100 \qquad (19\text{-}2)$$

Enhanced separations of DNA based on base composition can be obtained by introducing sequence specific ligands like actinomycin D or netropsin into the CsCl gradient.

DNAs with the same base composition and molecular weight can also be fractionated based on differences in conformation, especially in the presence of ethidium bromide, which reduces the density of the DNA. The amount of ethidium bromide that intercalates depends on the conformation of the DNA. Superhelical DNA binds less ethidium bromide than does linear DNA and the change in buoyancy upon ethidium bromide binding can be used to determine the superhelicity of DNA.[24] Plasmid isolation by centrifugation in a CsCl-ethidium Bromide gradient is part of the project described in Chapter 22.

REFERENCES

1. R. L. P. Adams, J. T. Knowler and D. P. Leader, *Biochemistry of the Nucleic Acids*, London: Chapman and Hall, 1986, pp. 13-14.

2. E. Chargaff, *Essay on Nucleic Acids*, Amsterdam: Elsevier/ North Holland, 1963.

3. W. Saenger, *Principles of Nucleic Acid Structure.*, New York: Springer-Verlag, 1984.

4. R. E. Dickerson in *Unusual DNA Structures*, ed. R. D. Wells, New York: Springer-Verlag, 1988, pp. 287-306.

5. D. Pörschke, *Mol. Biol. Biochem. Biophys.*, **1977**, *24*, 191.

6 D. Pörschke and M. Eigen, *J. Mol. Biol.*, **1971**, *62*, 361.

7. R. J. Britten and D. E. Kohne, *Science*, **1962**, *161*, 529.

8. M. T. Record et al., *Ann. Rev. Biochem.* **1981**, *50*, 997.

9. S. Moore, in *The Enzymes*, Vol. 14, ed. P. D. Boyer, New York: Academic Press, p. 121.

10. R. G. Kelly, N. Cozzarelli, M. P. Deutscher, I. R. Lehman and A. Kornberg, *J. Biol. Chem.*, **1970**, *245*, 39.

11. W. Gilbert and B. Maxam, *Proc. Natl. Acad. Sci. USA*, **1973**, *70*, 3581.

12. S. Linn and W. Arbor, *Proc. Natl. Acad. Sci. USA*, **1968**, *59*, 1300; A. Hall, in *Experiments in Molecular Biology*, ed. R. J. Slater, Clifton, N. J.: Humana Press, 1986, pp. 29-38.

13. C. Aaij and P. Borst, *Biochem. Biophys. Acta,* **1972**, *269*, 192.

14. R. B. Helling, H. M. Goodman, and H. W. Boyer, *J. Virol.*, **1974**, *14*, 1235.

15. P. Serwer and J. L. Allen, *Biochemistry*, **1984**, *23*, 922.

16. P. G. Sealey and E. M. Southern in *Gel Electrophoresis of Nucleic Acids, a practical approach*, eds. D. Rickwood and B. D. Hames, Washington, D. C.: IRL Press, 1982, pp. 39-76.

17. P. Thomas, *Proc. Natl. Acad. Sci. USA*, **1980**, *77*, 5201.

18. J. C. Pindar, D. Z. Staynov, W. B. Gratzer, *Biochemistry*, **1974**, *13*, 5373.

19. S. L. C. Woo, J. M. Rosen, C. D. Liarkos, Y. C. Choi, H. Bush, A. R. Means, B. W. O'Malley and D. L. Robberson, *J. Biol. Chem.*, **1975**, *250*, 7027.

20. L. G. Davis, M. D. Dibner and J. F. Battey, *Basic Methods in Molecular Cloning*, Amsterdam: Elsevier/ North Holland, 1986, pp. 130-141.

21. E. M. Southern, *J. Mol. Biol.*, **1975**, *98*, 503.

22. T. Maniatis, E. F. Fritsch and J. Sambrook, *Molecular Cloning, a Laboratory Manual*, Cold Spring Harbor, N. Y.: Cold Spring Harbor Laboratory, 1982, p. 470.

23. R. J. Roberts, P. A. Meyers, A. Morrison and K. Murray, *J. Mol. Biol.*, **1976**, *102*, 157.

24. D. Rickwood and J. A. A. Chambers, in *Centrifugation, a Practical Approach*, 2-ed. ed. D. Rickwood, Washington, D.C.: IRL Press, 1984, pp. 110-114.

Chapter 20

Isolation of Wheat Germ rRNA and Ribonucleotides

In this experiment, wheat germ ribosomal ribonucleic acid (rRNA) is obtained by a phenol extraction. The rRNA is hydrolyzed to the 2'- and 3'-nucleotides by alkali and the nucleotides separated by anion-exchange chromatography. The nucleotides are characterized by their elution order and quantitated by their UV spectra.

KEY TERMS

Nuclease denaturation
rRNA

Anion exchange
Phenol extraction

BACKGROUND

A typical eukaryotic cell contains about 10^{-5} µg of RNA, with 80 to 85% of that being high-molecular-weight ribosomal RNA (28S,18S, and 5S rRNA) and the remainder, mostly low-molecular-weight transfer RNA (tRNA, 10 to 15%). Present in smaller amounts are a heterogeneous mixture of messenger RNA (mRNA, 1 to 2%) and small nuclear RNA (snRNA). The phenol extraction method of Kirby is the most widely used method to isolate a mixture of various kinds of RNA.[1] Nucleic acids are polyanions which are often tightly bound to cationic proteins. Cells can be pulverized in the presence of buffered phenol, which serves to denature proteins, including most nucleases, as well as disrupt cellular membranes. The phenol extraction procedure frees nucleic acids from proteins and membranes and gives a nucleic acid extract without producing the extremes of acid, alkali, or heat that will destroy nucleic acids. The denaturation of nucleases is also a key to the success of this method. The resulting "crude RNA" can be freed from proteins by partitioning the extract between water-saturated phenol and water, with the nucleic acids ending up in the aqueous phase. The two-phase system of phenol and water contains RNA in the aqueous phase and denatured proteins (including nucleases) dissolved in the phenol layer. Interfacial solid containing protein and carbohydrate is frequently present. After centrifugation, the upper aqueous layer is removed from the phenol and interfacial material and shaken with ether to remove dissolved phenol. The nucleic acid solution is then separated into a fraction containing large ribosomal and messenger RNA molecules and one containing more soluble low molecular weight species. Enough NaCl is added to bring the solution to 1 M and precipitate the higher-molecular-weight species (mostly rRNA), leaving the low-molecular-weight RNA (mostly tRNA) in solution. The high RNA to DNA ratio of wheat germ makes removal of DNA unnecessary. The rRNA is recovered by centrifugation and further purified by reprecipitation with ethanol. The ethanol-precipitated form can be stored in the cold for several days.

The rRNA can be hydrolized in dilute alkali. The 3'-to5'-phosphate ester linkage is broken by the nucleophilic action of the ionized 2'-OH group. A 2',3'- cyclic phosphate intermediate then affords a roughly equal mixture of nucleoside 2'- and 3'- phosphates as shown in Figure 20-1. The reaction product mixture contains oligonucleotides, 2'- and 3'-nucleotides, and nucleosides.

Oligonucleotides and nucleotides bind tenaciously to a polystyrene anion-exchange resin at alkaline pH. Nucleosides bind less tightly. A large component of the binding force is due to nonionic hydrophobic interactions, which can be substantially reduced by adding an organic cosolvent to the elution buffer. The order of recovery from a polystyrene anion-exchanger reflects the role of both ionic and nonionic forces in binding ribonucleotides to the resin. With 20% ethanol cosolvent at pH 10, the nucleosides come off an anion-exchange column in the order C, A, U, and finally G. The corresponding ribonucleotide (2'- and 3'- phosphates) come off next as four sets of double peaks with the 2'-isomer immediately preceding the 3'-isomer in each case. The elution order is Cp, Ap, Up and Gp. Oligonucleotides are eluted from the column with 40% ethanol. The relative abundance of each component can be determined by the UV absorbance for each peak.

Elution volumes are minimized if the eluting buffer contains ethanol and salt which disrupts hydrophobic and ionic binding. The method of Asteriades et al. [2] employs a linear gradient of 0.1 to 0.5 M NH_4Cl (pH10.0) in aqueous ethanol to elute the bases from an AG 1-X4 column. For a 100-cm column, 200 1-mL fractions are collected and the nucleotides in each fraction are characterized and quantitated by their UV absorption spectra.

The strong-base anion-exchanger AG-1 has quaternary ammonium cationic groups attached to a styrene-divinylbenzene copolymer lattice. This resin is equivalent to Dowex-1 resins but with a narrower particle size range. AG-1 resin comes in different cross-linkages, each with its characteristic and fractionation properties. For example, AG-1-X2 (2% cross-linker) is used for high-molecular-weight substances such as peptides and proteins, and AG-1-X8 (8% cross-linker) is for ion-exchange separation of low-molecular-weight anions, including organic carboxylates and cyclic nucleotides. The more cross-linked resins do not swell as much as the less cross-linked resins at low ionic strength.

EXPERIMENTAL SECTION

Isolation of rRNA

Students are expected to work in pairs during the extraction steps. Fresh uncooked wheatgerm (10 g) is soaked in 60 mL of 0.1 M potassium phosphate buffer (pH 7) and after about 15 minutes, 60 mL of 88% phenol (v/v) is added. The flask is tightly stoppered and stirred with a magnetic stirring bar for 30 minutes at room temperature.

Alternatively, the stoppered flask is shaken (but not inverted) once a minute for 30 minutes.

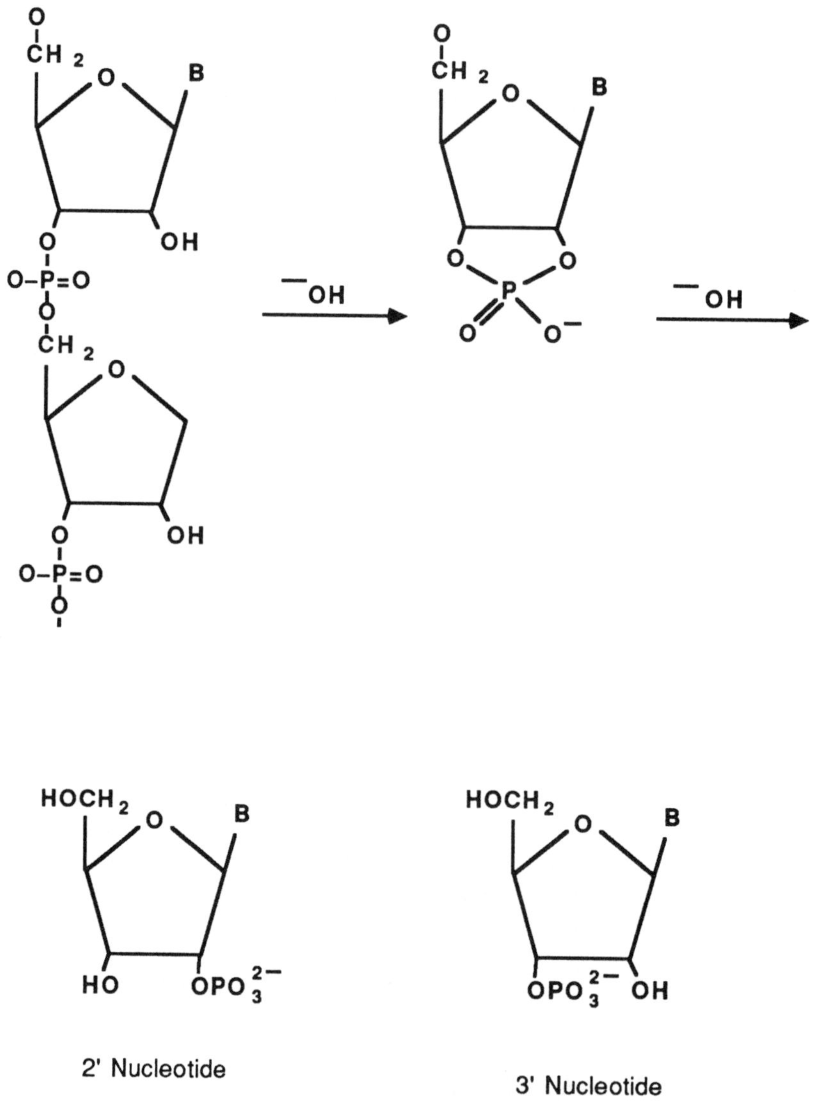

Figure 20-1 Ribonucleotides released upon rRNA hydrolysis

CAUTION: PHENOL WILL BURN SKIN UPON CONTACT.
WEAR GLOVES AND SAFETY GLASSES AT ALL TIMES.
WIPE UP ALL SPILLS AT ONCE!

The wheat germ suspension is transferred to polyallomar centrifuge tubes (not polycarbonate) and centrifuged for 30 minutes at 5000 x g. The resulting liquid should contain a clear yellow oily lower phase (phenol), interfacial solid material (protein), and a cloudy aqueous upper phase. The upper phase is carefully removed with a Pasteur pipet, with care to exclude the interfacial material. The aqueous phase is vigorously extracted with an equal volume of ether in a small separatory funnel. The resulting emulsion requires nearly an hour to clear at room temperature and this time can be spent preparing solutions for the hydrolysis and chromatographic steps. Once the layers are separate, the lower (aqueous) layer is recovered and enough solid NaCl is added to bring the salt concentration to 1 M. The suspension is allowed to stand overnight in the cold room to precipitate rRNA.

The rRNA can be recovered by centrifugation at 8000 x g for 10 minutes at 15°C. The supernatant is discarded and the solid pellet washed with an equal volume mixture of 1 M NaCl and ethanol. The precipitated RNA is redissolved in a minimum amount of water and two volume-equivalents of cold 95% ethanol are added and the suspension is stored overnight at -20°C. The precipitate is recovered by centrifugation at 8000 x g for 15 minutes at 4°C. The resulting white pellet is primarily rRNA, which can be dissolved in sterile water. Insoluble material (turbidity) can be removed by prolonged centrifugation at 15,000 x g.

The concentration of rRNA can be estimated from the UV absorbance at 260 nm; an absorbance of 1.0 roughly corresponds to a nucleotide concentration of 0.1 mM (see page 214).

Alkaline Hydrolysis

Enough ethanol precipitated rRNA is dissolved in 0.3 mL of 0.25 M NaOH to give a 1% RNA solution. The solution is heated in a sealed glass tube in a 50° oven overnight. The hydrolysate is diluted to 1.0 mL with water before application to the column.

Anion Exchange Chromatography

The strong-base anion-exchanger AG-1 in the chloride form is allowed to swell by standing overnight in 0.5 M NH_4Cl (pH 10) with 20% ethanol. The resin may be freed from fine particles by first suspending the resin in buffer and allowing it to settle for 1 hour. Any floating resin should be discarded and the water layer should be clear. The glass column should be filled with buffer and enough slurried resin added to give a 100 x 0.5 cm resin bed. The column must be gently shaken or vibrated with a rubber hammer (or a rubber stopper on a glass rod) during the packing process. The solvent delivery system is attached to the top of the column.

The chromatographic solvent is prepared from freshly weighed NH_4Cl added to water. The pH of the solution is adjusted to 10 with concentrated NH_4OH (use hood).

Ethanol is added to give a final concentration of 20% (v/v). The solvent should be filtered and degassed by aspiration prior to use.

The sample application requires the following additions in the indicated order:

> 0.4 mL of 1 M NaOH
>
> 1.0 mL of 20% ethanol/water
>
> 1.0 mL of sample hydrolysate in water
>
> 1.0 mL of 20% ethanol-water

The elution is carried out slowly (10 mL/h) with 200 mL of 20% ethanol with a linear gradient of 0 to 0.5 M N NH_4Cl (pH 10). The samples are collected with a fraction collector.

Spectroscopic Characterization

The nucleotide in each tube can be identified from the absorption at 260 and 280 nm and from the wavelength of maximum absorbance. The concentrations can be estimated from the extinction coefficients given in Table 20-1.

Table 20-1
UV-ABSORPTION SPECTRA OF 2' AND 3'-RIBONUCLEOTIDES[a]

Base	λ_{max} (nm)	$\epsilon_{max} \times 10^{-3}$	$\epsilon_{260} \times 10^{-3}$	$\epsilon_{280} / \epsilon_{260}$
A	259	15.3	15.3	0.15
C	270	9.2	7.6	0.9
U	262	10.0	9.9	0.3
G	252	13.4	11.4	0.7

[a]Source: Ref. 3.

PROJECT EXTENSION

Extraction of cyclic nucleotides cAMP and cGMP involves homogenization of the source tissue, deproteination of the extract, and separation of the nucleotides on a 1-mL AG 1-X8 column (200 to 400 mesh). The neutralized deproteinated sample is applied to a column that has been equilibrated with 10 mL of 0.1 M formic acid. The column is washed with 10 mL of the same buffer, then eluted with 20 mL of 2 N formic acid follwed by 20 mL of 4 N formic acid. The cAMP comes off with about 10 mL of 2N formic acid and the cGMP after about 25 mL total elution volume.[4-6]

DISCUSSION QUESTIONS

1. Why is RNA more easily hydrolyzed in alkali than is DNA?

2. Why was the wheat germ not pulverized prior to extraction?

REFERENCES

1. K. S. Kirby, *Methods in Enzymology*, **1968**, *12B*.

2. G. T. Asteriades, M. A. Armbruster, and P. T. Gilham, *Anal. Biochem.*, **1976**, *70*, 64; J. J. Sninsky, G. N. Bennett and P. T. Gilham, *Nucleic Acid Res.*, **1971**, *1*, 1665.

3. K. Burton in *Data for Biochemical Research*, 2nd ed., ed. R. M. C. Dawson, Oxford: Clarendon Press, 1969, pp. 169-178.

4. K. Nakazawa and M. Sanao, *J. Biol. Chem.,* **1974**, *249*, 4207.

5. E. H. Schwarzel, S. Bachman, and R. A. Levine, *Anal. Biochem.*, **1977**, *78*, 395.

6. N. Krishnan and G. Krishna, *Anal. Biochem.,* **1976**, *70*, 18.

Chapter 21

Isolation of Eukaryotic mRNA

Intact ribonucleic acid (RNA) is extracted from fresh rat livers by homogenization in 4 M guanidinium thiocyanate with 0.1 M mercaptoethanol. The RNA was isolated free of protein by repeated ethanol precipitation at -20°C. Optional purification by ultracentrifugation through a CsCl gradient and further purification by selection of poly(A$^+$) tails on oligo(dT)-cellulose is also described. The RNA is characterized by UV absorption spectroscopy and electrophoresis in formaldehyde/agarose under denaturing conditions.

KEY TERMS

Affinity chromatography   Chaotropic agents
CsCl gradients   Ethanol precipitation
Oligo(dT)cellulose   RNase deactivation
Ultracentrifugation

BACKGROUND

The preparation of intact messenger ribonucleic acid is difficult because of the action of nucleases which are liberated upon tissue homogenization. In many cells high concentrations of ribonucleases (RNases) are reserved in secretory granules and normally sequestered from RNA. Upon disruption of the cell, RNA and RNase are mixed and the former are degraded. Rutter and coworkers[1] showed that enzymatic degradation of RNA is minimized by homogenization in a 4 M solution of the potent protein denaturant guanidinium thiocyanate plus 0.1 M β-mercaptoethanol. Guanidinium chloride and thiocyanate are potent *chaotropic agents* that reduce hydrophobic interactions and disrupt protein tertiary structures, dissociate protein nucleic acid complexes, and disintegrate cellular structures. Guanidinium thiocyanate is an especially strong protein denaturant because both the cation and anion disrupt hydrophobic bonds between amino acid side chains.[2] RNA is usually bound to proteins within a cell, and this agent dissociates the nucleoprotein complex. Neither ion disrupts RNA structure. The reducing agent β-mercaptoethanol breaks disulfide bonds, destabilizing the tertiary structures essential for enzyme activity. Even the extremely robust enzyme RNase, which can survive brief boiling is deactivated by these agents when cells are homogenized by this procedure; all liberated RNase is promptly denatured before significant RNA degradation can occur.

The liberated RNA is freed of insoluble particulate matter by low-speed centrifugation and the RNA precipitated from the homogenate by the addition of acetic acid and ethanol followed by storage overnight at -20°C. The RNA precipitates from cold aqueous ethanol solutions and can be recovered by centrifugation.

$$H_2N^+{=}C-NH_2 \qquad \overset{\displaystyle NH_2}{|}$$

Guanidinium Cation

Even when freed from cellular debris, isolated RNA can undergo the ravages of RNase. This enzyme is an ubiquitous contaminate of almost all laboratory glassware that has been washed by most conventional methods. Use of contaminated glassware can lead to RNA degradation. Dilutions of denatured RNase results in renaturation and the recovery of RNase activity. For this reason sterile conditions must be used throughout this experiment. Sterile disposable plasticware is generally free of RNase and can be used without pretreatment. But this is an expensive alternative for a teaching laboratory, and considerable attention must be paid to the storage and distribution of such plasticware to insure that it remains sterile. Laboratory glassware must be soaked in a 0.1% solution of diethylpyrocarbonate, then baked overnight at 180° or for 4 hours at 250°C before use to destroy any residual RNase. This compound is a nonspecific inhibitor of RNase, but since it also reacts with adenine residues of RNA it must be removed by prolonged baking before the treated glassware is used.[3] All items of glassware to be used for experiments with RNA should be marked distinctively and stored in a designated oven or dust-free cupboard. Such glassware should be stored at room temperature only when sealed with sterile cotton or sterile plastic wrap to prevent internal contamination. Gloves should be worn at all stages of the preparation and all transfers of dry chemicals should be done with baked spatulas. Freshly glass-distilled water must be used and all buffers must be sterilized in the autoclave.

In this experiment the mRNA will be partially purified by repeated precipitations with ethanol. An alternative purification[4,5] by ultracentrifugation through a CsCl gradient is also described. This method gives excellent recovery of quite pure RNA if the capacity of the gradient has not been exceeded. The partially purified mRNA is layered onto CsCl solution and run overnight at 174,000 x g. The proteins remain in the upper aqueous layer, DNA bands partway down the tube, and RNA forms a jellylike pellet. About 250 mg of RNA requires 5 mL CsCl solution.

Messenger RNA represents only a small fraction of the total RNA of a eukaryotic cell, but since most eukaryotic mRNA contain 3'-poly(A) "tails" while all other RNA is not polyasenylated, this form of RNA can be selected from the bulk of RNA by an affinity chromatography method. One pass through an oligo(dT)cellulose column removes most rRNA and tRNA and two passes yields more than 95%

poly(A$^+$)mRNA.[5-9] The very tight and specific coordination between the poly(A)tails and the column poly(dT)-residues is broken down by decreasing the ionic strength; the desired mRNA fraction is washed from the column with low salt buffer. One gram of column material binds 40 to 100 O.D. units of poly(A).

The purity of RNA can be assessed either spectroscopically or electrophoretically. RNA preparations with a UV-absorbance ratio $A_{260}/A_{280 nm}$ of 2.1 or better are nearly free of protein contamination. The RNA concentration can be estimated from an $E^{1\%}_{1cm}$ of 200 at 260 nm.

Electrophoretic characterization is best done through an agarose gel containing 1 M formaldehyde.[5,6,10] Formaldehyde reacts with the primary amino groups of A, G, and C bases and prevents the formation of base pair hydrogen bonds. The mRNA is denatured by brief heating to 95°C, a loading buffer containing formaldehyde and formamide then applied to the agarose gel containing formaldehyde. The loading buffer and agarose gel pH is maintained with the buffer MOPS (3-[N-morpholino]propanesulfonic acid), which is a tertiary amine that does not react with formaldehyde. Because formaldehyde is a poisonous gas, the gel must be run in a fume hood. The molecular weight of RNA run on denaturing gels is determined by comparing its mobility to that of known standards, just as with DNA (see chapter 10). Convenient RNA markers are 28S and 18S eukaryotic rRNA which are 5.1 and 2.0 kb, respectively. Total RNA samples will contain prominent amounts of these species.

EXPERIMENTAL SECTION

This experiment involves the humane sacrifice of a small animal; this aspect of the experiment can be done by the teaching or support staff, but the liver must be removed immediately. Liver tissue from a meat market is not a reliable source of mRNA. The remainder of the experiment should be done by pairs of students.

One rat, which has fasted for 24 hours is painlessly sacrificed and the liver quickly removed, trimmed of veins and fat, weighed, and then dropped into 20 mL of cold (10°) guanidinium thiocyanate stock solution (see below) in a square polyethylene bottle. The liver is immediately homogenized for 1 minute at full speed with a Brinkmann Polytron or similar tissue homogenizer. Insoluble particulate matter is quickly removed by centrifugation for 10 minutes at 8000 x g, also at 10°C. The supernatant is a crude lysate that can be purified by repeated ethanol precipitations or by CsCl ultracentrifugation; both methods are given below.

Extraction   Buffer

The extraction buffer is prepared from the following stock solution:

Guanidinium thiocyanate stock solution (stable for 30 days)

Guanidinium thiocyanate (Fluka)	50 g
Sodium N-lauroylsarcosine	0.5 g
2-Mercaptoethanol	0.7 mL
Antifoam A 30% (Sigma)	0.33 mL

All is dissolved, with warming and stirring, in 25 mM sodium citrate, pH 7.0, to a final volume of 100 mL. Insoluble material is removed by filtration and the pH adjusted to 7.0 with a small amount of 1 M NaOH.

CAUTION: GUANIDINIUM THIOCYANATE AND
2-MERCAPTOETHANOL ARE POISONOUS.
AVOID CONTACT WITH SKIN OR INHALATION OF VAPORS.
USE A HOOD.

The RNA within the guanidinium thiocyanate extract is then recovered by either ethanol precipitation or by CsCl ultracentrifugation then purified further by oligo(dT)-cellulose chromatography.

Ethanol Precipitation Method

Enough 1 M acetic acid (approximately 0.5 mL) is added to the lysate to bring the pH down to 5, then 15 mL (0.75 volume equivalent) of absolute ethanol is added. After thorough shaking, the solution is stored at -20°C overnight. The resulting precipitate (mostly RNA) is sedimented by centrifugation for 10 minutes at 6000 x g at -10°C (or as near that temperature as possible) and all but the firm pellet are discarded. The pellet is resuspended in 10 mL of guanidinium chloride solution (pH 7) by gentle shaking and warming to ensure complete dispersion of the pellet. The RNA is reprecipitated by adding 0.25 mL of acetic acid and 5 mL (0.5 volume equivalent) of ethanol. After three hours at -20°C, the RNA is sedimented by centrifugation as before. The pellet is dispersed in 5 mL ethanol at room temperature; vigorous shaking is required. The RNA is recovered by centrifugation for 5 minutes at 6000 x g. Ethanol is removed from the pellet by a gentle stream of nitrogen and the RNA dissolved in sterile water (10 mg of RNA per 500 μL of water) and used for oligo(dT)cellulose chromatography

CsCl Ultracentifugation Method

Fresh lysate (5 mL) is carefully layered on top of 5 mL of a 5.7 M CsCl/30 mM sodium acetate pH 6 buffer in a 10-mL ultracentrifuge tube, and the sample is sedimented overnight at 175,000 x g at 20°C.

CsCl buffer:

CsCl	95.97 g
Sodium acetate	0.408 g
Water to 100 mL	

Cold CsCl solutions precipitate. The supernatant is removed from the soft pellet by aspiration (see Chapter 11) and the pellet is resuspended in 200 μL of 0.3 M sodium acetate buffer, pH 6, and transferred to a 1.5 mL Eppendorf tube. The centrifuge tube is rinsed with 100 μL of buffer.

The RNA suspension is overlayered with 750 μL of very cold ethanol briefly shaken and stored in the freezer or on dry ice for 10 minutes. The RNA is pelleted by microcentrifugation for ten minutes, the supernatant is decanted, and the pellet resuspended in 500 μL of ethanol. The RNA is pelleted again and ethanol is removed (as above) by a gentle stream of nitrogen. The RNA in sterile water (10 mg of RNA per 500μL of water) is used immediately for oligo(dT)cellulose chromatography.

Oligo(dT)cellulose Chromatography

A column of commercially available oligo(dT)cellulose (BRL or P.L. Biochemical) is prepared by pouring 0.5 mL of the cellulose, as a slurry in water, into an autoclaved Pasteur pipet with the tip packed with glass wool. The column is washed with 2 to 3 mL of 0.1 M NaOH, followed by repeated washings with sterile distilled water and then several times, until the pH of the eluent drops below 8. The column is equilibrated with several volumes of washing buffer. An RNA sample, containing 10 mg RNA in 500 μL of water in a sterile Eppendorf tube, is combined with an equal volume of loading buffer and denatured by warming to 61 to 65° for 1 to 2 minutes, followed by rapid quenching in an ice-water bath. The sample is applied to the column followed by 0.5 mL of washing buffer. The liquid eluting from the column is briefly heated as before and applied again to the column, followed by 5 mL of washing buffer. The liquid eluting a second time is collected as 1-mL fractions and diluted with 20 mL of water. The optical densities at 260 and 280 nm are recorded. The material coming off the column at high salt contains very little poly(A)RNA and can be discarded or saved for some other project. The column is then washed with 2 to 3 mL of the low-salt eluting buffer to collect poly(A)RNA in 0.5-mL fractions.

The oligo(dT)cellulose chromatography buffers, all at pH 7.4 (must be mixed and autoclaved just prior to use).

Loading buffer:	Washing buffer:	Elution buffer:
40 mM TRIS-HCl	20 mM TRIS-HCl	10 mM TRIS-HCl
1 M NaCl	0.1 M NaCl	no salt
1 mM EDTA	1 mM EDTA	1 mM EDTA
0.1% SDS	0.1% SDS	0.05% SDS

The poly(A) samples are not diluted before taking optical density measurements. A plot of optical density at 260 nm versus eluent volume is prepared and the poly(A) fractions are pooled and precipitated in 0.15 M sodium acetate with a 2.5-fold volume of very cold ethanol. The desired RNA is precipitated by centrifugation and the liquid decanted and replaced with 500 μL of cold 80% ethanol. After a second centrifugation, the supernatant is discarded and the RNA pellet is dried in a stream of nitrogen (as above). Poly(A)-RNA comprises only about 3 to 5% of the total RNA isolated in the first part of this experiment, so the pellet will be quite small. The solid RNA can be stored at -70°C for a month. The column can be reused if reequilibrated with 0.1 M NaOH, water, and washing buffer as before.

Formaldehyde/Agarose Gel Electrophoresis[5,6,10]

A 1% gel is prepared by combining 3 g of agarose in 255 mL of hot (> 80°C) sterile distilled water with 30 mL of MOP concentrated buffer. The gel is allowed to cool to 60° in a fume hood; then 16.2 mL of 37% formalin solution is added and mixed by swirling. Ten microliters of ethidium bromide stock solution (10 mg/mL) is also added and a submarine-style gel (see Chapter 10) is poured in the hood.

CAUTION: FORMALDEHYDE IS VERY POISONOUS
GLOVES MUST BE WORN WHEN HANDLING FORMALDEHYDE
GELS AND SOLUTIONS.

(For very precise work a nonsubmarine-style gel is used and the electrophoresis buffer is circulated between compartments with a peristaltic pump to neutralize pH changes at the electrodes. These precautions will not be used in this experiment.[6]

Formaldehyde Gel loading buffer:

750 μL of formamide (poisonous)

150 μL of MOPS buffer concentrate

250 μL of formalin solution (37% formaldehyde)

250 μL of water

2 mg of bromophenol blue

0.2 g of Ficoll (add last with vigorous shaking)

MOPS buffer concentrate for formaldehyde/agarose gel (1 Liter):

MOPS	200 mM	46.26 g
Sodium acetate	0.05 M	6.8 g
EDTA	10 mM	20 mL of 0.5 M stock

A 2-μg sample of poly(A)RNA or 10 μg of crude RNA is added to 20 μL of electrophoresis loading buffer (above) and heated to 95° for two minutes to denature RNA, then applied to one of the lanes of the gel. Each group of students working on this project will use one lane on the same formaldehyde gel with two lanes (middle and right-hand side) reserved for marker RNA's. The electrophoresis running buffer is MOPS buffer diluted 20-fold. The gel is run as described in chapter 10 and photographed under UV transillumination (see Chapter 19) for a permanent record.

PROJECT EXTENSIONS

1. A rapid Quanidium Thiocyanate/CsCl Gradient Method has been reported.[13]

2. RNA can also be run in a denaturing agarose-urea gel with citrate buffer.[12]

3. To instructors reluctant to sacrifice warm animals in a teaching experiment, we point out that a number of other biological sources of eukaryotic mRNA are available. For those working in older buildings, we mention the isolation of intact RNA from pulverized cockroach heads.[14]

4. The mRNA fraction isolated by this method can be used for cDNA preparation or for *in vivo* translation but it is essential to remove any remaining SDS before carrying out these reactions. A second round of alcohol precipitation at at least -20° will be required.

5. RNA may be transfered immediately after electrophoresis from agarose gels to nitrocellulose filters. The gels must not be stained with ethidium bromide or soaked for prolonged periods in buffer; otherwise no additional treatment of the gel

is necessary.[11] The gel is placed in contact with the nitrocellulose filter and blotted as described in Chapter 26. DNA-RNA cross-blotting can also be done.

5. An interesting plant mRNA isolatrion procedure that employs liquid scintillation counting to monitor the extraction of total nucleic acids from pea seedlings, incubated in ^{32}P-orthophosphate, and subsequent preparation of mRNA by oligo (dT) cellulose chromatography has been reported.[15] The proportion of the total radioactivity found in the polyA-rich fraction, retained by the column, is used to calculate the proportion of mRNA in the total nucleic acid.

DISCUSSION QUESTIONS

1. Given an $E^{1\%}_{1cm}$ = 200 value for RNA, how much does 100 O.D. units of RNA weigh?

2. Why is a peristaltic pump used to recirculate the running buffer between the two tanks for precision electrophoresis with formaldehyde/agarose gels? Hint: what do you know about the chemistry of formaldehyde in aqueous solutions exposed to air?

3. In the absence of a reducing agent the reaction of formaldehyde with primary amines is reversible. Why must an RNA containing formaldehyde/agarose gel be soaked in several changes of a sodium chloride, sodium citrate pH 7 buffer before its use in blotting experiments?

4. Self-complementary antiparallel sequences in RNA can pair to form short duplex structures. Computer analysis of an RNA sequence can identify potential duplex forming sequences and these sequences have been used to make educated guesses about the tertiary structure of biologically important RNA molecules such as t RNA or rRNA. The cloverleaf structures assigned to tRNAs before X-ray crystal structures were based on possible hydrogen bonding between sequences of RNA and such structures proved reasonably good guesses of the tertiary structure. Is there a fundamental reasons why the same approach cannot be applied to rRNAs?

5. There are a number of ribonucleases with sequence specific endonuclease activity:

enzyme	product
Ribonuclease T_1	3'-Gp, 5'-OH
Ribonuclease T_2	3'-Ap, 5'-OH
Ribonuclease U_2	3'-(G or A)p, 5'-OH
Ribonuclease A	3'-(U or C)p, 5'-OH

Describe the elution pattern for cleavage products of 5'-ApGpCpUpApGp-3' with each of the above enzymes followed by the ion-exchange procedure described in this experiment.

REFERENCES

1. J. M. Chirgwin, A. E. Przybyla, R. J. MacDonald and W. J. Rutter, *Biochemistry*, **1979**, *18*, 5294.

2. R. A. Cox, *Methods Enzymol.*, **1968**, *12B*, 120.

3. D. D. Blumberg, *Methods Enzymol.*, **1987**, *152*, 20.

4. D. Grierson in *Gel Electrophoresis of Nucleic Acids, a practical approach*, eds. D. Rickwood and B. D. Hames, Washington D. C. : IRL Press, 1982, p. 16.

5. L. G. Davis, M. D. Dibner and J. F. Battey, *Basic Methods in Molecular Cloning*, Amsterdam: Elsevier, 1986, pp. 130-141.

6. B. Perbal, *Practical Guide to Molecular Cloning*, New York: Wiley-Interscience, 1984, pp. 385-397.

7. H. Aviv and P. Leder, *Proc. Natl. Acad. Sci. USA*, **1972**, *69*, 1408.

8. M. Edmonds, M. H. Vaughn, Jr. and H. Nakazato, *Proc. Natl. Acad. Sci. USA*, **1971**, *68*, 1336.

9. P. T. Gilham, *J. Am. Chem. Soc.*, **1980**, *86*, 4982.

10. P. Thomas, *Proc. Natl. Acad. Sci. USA*, **1980**, 77, 5201.

11. J. C. Pindar, D. Z. Staynov, W. B. Gratzer, *Biochemistry*, **1974**, *13*, 5373; B. W. O'Malley, and D. L. Robberson, *J. Biol. Chem.*, **1975**, *50*, 7027.

13. T.H. Turpen and O. M. Griffith, *BioTechniques*, **1986**, *4*, 13.

14. T. H. Turpen and O. M. Griffith, *BioTechniques*, **1986**, *4*, 11.

Chapter 22

Isolation of pBR322 Plasmid DNA

This experiment involves the growth of *E. coli* strain RR1 and its transformation with the plasmid pBR322, which confers ampicillin resistance to the host bacterium. Brief heating in a TRIS-CaCl$_2$ transforming buffer is required to permit some plasmids to penetrate into bacteria. Successfully transformed bacteria are selected by using ampicillin-containing media, which poisons nontransformed bacteria while permitting transformed ones to survive. One surviving colony is used to establish a large-scale culture. The antibiotic protein synthesis inhibitor chloramphenicol inhibits host chromosome replication, but not pBR322 plasmid replication so that the copy number of this plasmid is increased by adding chloramphenicol solution (in alcohol) to an established culture. The resulting bacteria, containing up to a 1000 plasmids per cell, are concentrated by centrifugation and lysed in a small amount of SDS buffer containing lysozyme. The large chromosomal DNA is fragmented under such conditions and is irreversibly denatured by brief heating while the smaller circular plasmid is not. The viscous solution is quickly cooled and the chromosomal DNA pelleted by centrifugation (10,000 x g); the plasmid DNA remains in the supernatant and can be used in restriction mapping as described in Chapter 23 without further purification. Purification by ultracentrifugation through a cesium chloride/ethidium chloride gradient at 50,000 x g is an optional project extension.

KEY TERMS

Chimeric plasmids
Cloning vectors
Density-gradient centrifugation
Plasmid copy number
Stringent and relaxed control

Chloramphenicol amplification
Competent cells
Insertional inactivation
Plasmids

BACKGROUND

Plasmids are supercoiled double-stranded circular molecules of DNA that exist in a bacterial cell as stable, self-replicating, autonomous extrachromosomal genetic elements. They are much smaller than the host chromosome, ranging in size from 1000 to greater than 200,000 base pairs. The genes within plasmids can be expressed and they often code for enzymes advantageous to the host bacterium, including enzymes that deactivate, or in some other way, confer resistance to antibiotics. Some plasmids are part of nature's repertoire for natural genetic recombination. Under natural conditions, such plasmids (types F and R) are transmitted to new hosts by bacterial conjugation. If the plasmid is an R type containing genes specifying resistance to antibiotics, this advantage will be conferred to the new host as well. Antibiotic resistance is useful in designing plasmid *cloning vectors*, which are used to "carry" inserted foreign DNA for the purpose of producing cells with new and interesting properties, most often the ability to produce a useful protein product. A common procedure is to use a plasmid that has genes specifying resistance to two antibiotics. One of the genes is used to identify bacteria that have picked up the plasmid and the second serves as a site used to insert foreign DNA. If foreign DNA is inserted within this second gene, the resulting plasmid will loose resistance to that antibiotic. A bacterium with a *chimeric plasmid* (i.e., one containing foreign DNA) can be selected by its resistance to one antibiotic, but loss of resistance to the second. This selection process is known as *insertional inactivation.*

Figure 22-1 illustrates this principle. The plasmid cloning vector containing antibiotic resistance genes A^R and B^R, and the foreign DNA are digested with a restriction endonuclease that cuts the plasmid at a unique site within the antibiotic resistance gene B^R. The two DNA pieces are treated with DNA ligase and then used to transform A-sensitive host bacteria so that some may establish colonies in a medium containing the antibiotic A. All the colonies that grow in the presence of A contain plasmid DNA, but not all contain the foreign DNA since some plasmid may have recircularized without insertion of the foreign piece. These will have retained their resistance to both A and B while chimeric transformants will grow in the presence of A but not B.

The plasmid pBR322 is a human made *E. coli* cloning vector constructed *in vitro* from components spliced from natural plasmids; for example, pBR322 contains a tetracycline resistance gene (*Tc*) from the natural plasmid pSC101 as well as an ampicillin gene (*Ap*) from another source. The intact plasmid contains 4363 base pairs. Numbering of the sequence begins within the unique Eco R I site: the first T in the sequence GAATTC is designated number 1 and the numbers increase around the molecule in the direction Tc to Ap. A number of enzymes cut this plasmid only once or twice offering unique splicing sites in both antibiotic resistance genes. The reader is referred to Figure 22-2 for a map of the unique restriction sites of this plasmid.

This project employs an *E. coli* strain RR1 devised by Genentech to carry pBR322. The experimental section includes a protocol for introducing this plasmid into *E. coli* strain RR1 bacteria by the Ca^{2+} transformation method described below, but the instructor may elect to start with strain RR1 already carrying the plasmid, in which case the project amounts to producing a 100 mL volume culture of plasmid containing bacteria, increasing the plasmid copy number with chloramphenicol amplification (see below), harvesting the bacteria by centrifugation and isolating the plasmid DNA separately from the bacterial chromosomal DNA.

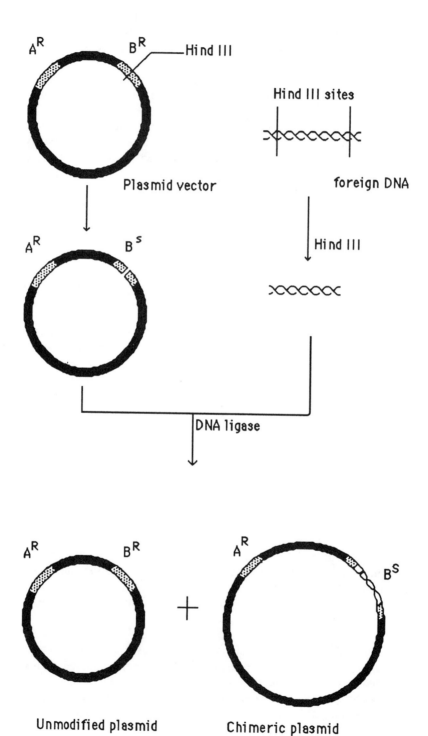

Figure 22-1   Insertional inactivation

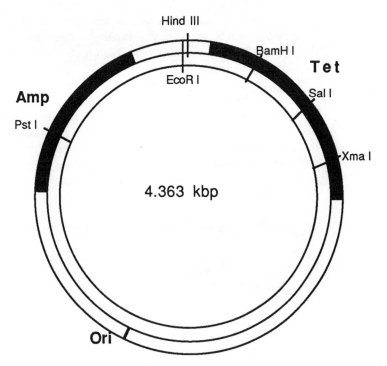

Figure 22-2   Gene map of pBR322

Ca Ion Transformation Method

Rapidly growing cells are made ***competent***, that is permeable to DNA by incubation with DNA in a cold TRIS buffer containing $CaCl_2$. The DNA is taken up from this buffer and this uptake is facilitated by a brief heat shock. This technique of $CaCl_2$ transformation was first applied to plasmids by Cohen and coworkers in 1973[1] and it is now the most common and most important experimental method for transferring a plasmid from one strain to another. Cells that have acquired permeability to foreign DNA are called competent cells. Cells retain this ability for only a short while; they seldom remain competent for more than 1 or 2 days. For most protocols only about 10^5-10^7 successful transformants are obtained per microgram of intact pBR322 plasmid which works out to one DNA molecule in approximately 10,000 successfully transformed. Bacterial cells that have taken up the pBR322 plasmid are resistant to ampicillin and are selected for by adding ampicillin to the growth medium. Cells not carrying this plasmid will be sensitive to this antibiotic and will not form stable colonies.

Chloramphenicol Plasmid Amplification

Plasmids are maintained in a cell in a stable and characteristic ***copy number*** from generation to generation. The replication of plasmid DNA requires the same set of stable enzymes used to replicate the host chromosome, but the copy number is regulated by a plasmid-encoded inhibitor that is a negative regulator of the initiation of replication. Some plasmids are present only as one or a few plasmids per cell; such plasmids are said to be under ***stringent control***. Plasmids under ***relaxed control*** have copy numbers as high as 200. High-copy-number plasmids exhibit chloramphenicol amplification of the number of plasmid copies per cell. Chloramphenicol is an inhibitor of protein synthesis, and its presence markedly reduces the concentration of unstable proteins, which must be continuously synthesized to remain at high levels.

Host chromosomal DNA synthesis requires unstable initiation proteins (e.g., dnaA protein), whereas plasmid replication employs only stable host enzymes therefore, in the absence of protein synthesis, chromosomal DNA will not be synthesized, but plasmids will. Moreover, the copy number regulator coded by the plasmid is an unstable protein which will not be at high enough concentration to inhibit plasmid replication until there are many more copies of the plasmid than required in the absence of chloramphenicol. In the presence of chloramphenicol the plasmid copy number can reach 1000 or more.

Separating Plasmid from Host Chromosomal DNA

Plasmid DNA exists as supercoiled circles and such small intertwined double strands are only reversibly denatured at high temperatures or in alkali. The much larger host chromosome DNA molecule is usually fragmented upon extraction from the bacterium and is irreversibly denatured at temperatures above the dsDNA-to-ssDNA transition or at alkaline pH levels that break interchain H-bonds. Denatured random coil ssDNA can be pelleted by centrifugation, while intact plasmid DNA remains in the supernatant. This difference permits the rapid separation of plasmid from chromosomal DNA. The experimental section contains two procedures for harvesting plasmids from *E. coli* using lysozyme to cleave the bacterial cell wall and either SDS or Triton X-100 to solubilize the cell membrane. The instructor will make clear to the student which method is to be followed. The method of Holmes and Quigley[2] for cell lysis and for separating plasmid and chromosomal DNA employs the differential precipitation of chromosomal DNA upon heating. The cells are suspended in a Tris-EDTA-NaCl buffer containing high levels of lysozyme and 0.5% of the nonionic detergent Triton X-100. The enzyme solution must be prepared just before use and the pH must be above 8 to ensure rapid cell lysis. The cell suspension is heated in a boiling water bath for 40 seconds, then cooled in an ice bath and centrifuged for 20 minutes at 12,000 x g. The supernatant contains reasonably pure plasmid. The method of Birnboim and Doly[3] as reported by Maniatis et al[4] uses lysozyme to lyse cells in NaOH-SDS solution for about 10 minutes, then the surfactant, chromosomal DNA, and other cellular debris is precipitated in potassium acetate buffer. The potassium salt of dodecyl sulfate is insoluble in water. The plasmid remains in the supernatant after centrifugation at 20,000 x g for a half-hour, then is precipitated with isopropanol at room temperature and collected by centrifugation.

EXPERIMENTAL SECTION

The following protocols all require sterile techniques. All glassware, centrifuge tubes, and pipet tips must be autoclaved. Culture flasks must be autoclaved, then the flask necks sterilized by flame. The media are prepared according to the sterile techniques and formulations outlined in Chapter 10.

Several hours before the laboratory begins, 10 mL of Luria-Bertani LB medium is prepared in a sterile culture flask according to the formulation given in Chapter 10. A small culture of *E. coli* strain RR1 is prepared from a single colony from an agar stock plate using a flame-sterilized metal transfer loop. The loop is not cooled in air but by touching the agar plate in a spot away from colonies. A single colony is scraped up and shaken in the LB broth, which is quickly covered with sterile cotton and placed in a 37°C shaker bath.

If these cells already contain pBR322 plasmid, this stock colony should be scaled to 100 mL and used in the plasmid amplification protocol given on page 235.The following protocol will be followed in laboratories where plasmid incorporation by $CaCl_2$ transformation is done first. In this case a sample of pBR322 must be on hand. (In the author's laboratory, students run through the plasmid preparation twice, the first time starting with plasmid-containing bacteria; the plasmid is isolated and used to transform a second culture, which is then a source of plasmids for extensive restriction mapping as described in Chapter 23.)

First Laboratory Period

$CaCl_2$ Transformation with pBR322

It is important that cells are growing logarithmically to achieve optimal tranformation. The cell density must correspond to about 0.4 O.D. units at 550 nm to ensure a high level of competent cells. When this point is reached (3 to 4 hours) the cells are gently pelleted at 2500 x g for 5 minutes at room temperature and the pellet suspended in 5 mL of 50 mM $CaCl_2$, 10 mM TRIS HCl (pH 7.4), with occasional gentle shaking for 20 minutes.

After the $CaCl_2$ buffer is added, the cells are stored on ice with only very gentle agitation since competent cells are very fragile. The cells can be pelleted again at 2500 x g for 5 minutes at 4°C and mixed with 1 μg of plasmid DNA (0.2 mg/mL) in 2 mL of 50 mM $CaCl_2$, then suspended in 5 mL of ice cold buffered 50 mM $CaCl_2$ and stored on ice in a clean tube for a second half hour. The cells are heat shocked by heating the tube in a 42° warm water bath for <u>exactly</u> 2 minutes then transferred to 5 mL of prewarmed 37°C LB media and placed in the constant-temperature 37°C shaker bath for about 45 minutes. After the cells have had a chance to grow and express the ampicillin resistance gene, various aliquots (10, 25, 50, 100, and 250 μL) are removed with sterile pipette tips and spread with a sterile glass rod on agar plates containing ampicillin. The plates are placed in a 37°C incubator, agar side down, for about 20 minutes, until the medium is dry, then stored overnight at 37°C, agar side up. (The plates are prepared in advance from the following protocol: One liter of LB broth-- no glucose--is used to dissolve 15 g of agar and sterilized by autoclaving. After the solution is cooled to 50°C, 1 mL of sterile 50 mg/mL ampicillin stock solution in water, sterilized by filtration through a 0.45-μm filter, is added. The plates must be poured before the agar solution cools to the gel point.)

Second Laboratory Period (One Hour)

Amplification of Plasmid DNA

A single ampicillin-resistant colony is removed from the agar plate with a sterile wire loop and shaken off in 100 mL of sterile LB broth containing ampicillin (100 μL of ampicillin stock solution, 50 mg/mL), which is then shaken at 37°C for several hours. When the optical density reaches 0.5 at 550 nm, the plasmid copy number can be amplified by the addition of 0.5 mL of chloramphenicol solution (35 mg/mL in 100% ethanol) and incubation for an additional 12 to 16 hours.

Third Laboratory Period (Three Hours)

Isolation of pBR322 Plasmids

Triton/Heat Method of Lysis

The cells are harvested by centrifugation at 4000 x g for 10 minutes and washed once with 100 mL of 10 mM TRIS, 0.1 mM EDTA, pH 8.0, and pelleted as before. The pellet is suspended in 15 mL of the same TRIS, EDTA buffer, but this time also containing 10% sucrose and combined with 2 mL of freshly prepared lysozyme. and allowed to sit on ice for 10 minutes The lysozyme solution is prepared just before use by dissolving 20 mg of hen egg white (HEW) lysozyme in 2 mL of cold water. The clear solution is decanted away from any undissolved sediment. The lysozyme cleaves the peptidoglycan linkages of the *E. coli* cell walls and the sucrose cushions the resulting spheroblasts. Two milliliters of 0.5 M EDTA are added to chelate calcium and magnesium ions to deactivate DNases and 5 mL of 1% Triton X-100 (in water) is added. It takes some time to dissolve Triton so plan ahead. The solution is warmed to 37°C for several minutes, until lysis occurs, which is accompanied by a marked increase in viscosity and visible bacterial debris. The chromosomal DNA is sedimented at 12,000 x g for 20 minutes at 4°C. The supernatant contains both intact and nicked plasmids as well as linear fragments of host chromosome.

SDS/NaOH Method of Lysis

The following procedure can be run on the same 100 mL volume scale (or larger) as the preceding method; however, it can also be conveniently scaled down to a "mini-prep" for partial purification of plasmid DNA from a single colony.[5] The mini-prep is as follows:

A single bacterial colony is picked up with a sterile wire loop and suspended in 500 μL of 25 mM TRIS/50 mM glucose/10 mM EDTA buffer in a 15-mL Corex centrifuge tube. To this is added 6 mg of lysozyme and 50 mg of glucose, and the mixture is incubated at room temperature for 5 minutes. Then 1.2 mL of 1% SDS in 0.5 M NaOH is added and the reaction mixture is allowed to sit on ice, with occasional mixing for five minutes. The SDS and chromosomal DNA is precipitated by adding 900 μL ice-cold 3.0 M potassium acetate/acetic acid solution (pH 5.5) solution and centrifugation at 12,000 x g for 10 minutes at 4°C. The clear supernatant containing plasmid is transfered to a new Corex tube and combined with an equal volume of isopropanol and stored in the freezer until the next experiment. The white pellet is mostly potassium dodecylsulfate and can be discarded.

PROJECT EXTENSIONS

1. The plasmid DNA obtained in this experiment can be characterized by restriction endonuclease digestion (Chapter 23) and by determining the thermal denaturation temperature (Chapter 25).

2. The plasmid DNA can be further purified by ultracentrifugation through a cesium chloride/ethidium bromide gradient.[5] This is a convenient option for laboratories with an ultracentrifuge. The plasmid suspension is concentrated and freed from protein by two extractions with isoamyl alcohol/chloroform, as described in Chapter 10, or water-saturated phenol/chloroform, followed by an extraction with

just chloroform to pick up dissolved phenol. After each extraction the aqueous layer is transferred to a new clean tube. The plasmid is then precipitated with ethanol at dry ice temperatures and redissolved in TRIS-EDTA buffer in a thin walled polyallomar sealable ultracentrifuge tube. Solid cesium chloride is added up to a concentration of 1 g of CsCl/mL then 0.8 mL of ethidium bromide solution (10 mg/mL in water) is added for every 10 mL of CsCl solution. The CsCl solution is overlayered with light paraffin oil, the tube sealed, and the entire works centrifuged at 45,000 x g for 36 hours or at 225,000 x g overnight at 20°C. A purple fuzzy layer on top of the water layer is precipitated protein; two additional bands corresponding to DNA are visible under either UV or visible light. The uppermost bluish-white band is nicked plasmid and chromosomal DNA fragments and the lower band contains intact plasmid molecules. A pellet of RNA is formed at the bottom of the tube (see Chapter 21). The use of a soft centrifuge tube permits the direct removal of the plasmid layer. The outside of the tube is cleaned without shaking up the layers and a piece of cellophane tape is applied to the tube to serve as a gasket, preventing leaks. A 21-gauge syringe needle is inserted into the tube through the tape about 1 cm below the band and is inclined upward toward the band (see Figure 22-3). The needle bevel is up. The plasmid band is withdrawn from the tube as a gentle flow when air is allowed to enter the top of the tube through a second needle. The drops are collected in an Eppendorf microcentrifuge tube or with a 5-mL syringe or a Pasteur pipet. The CsCl is removed from the purified plasmid by dialysis at room temperature against a 1000-fold volume of TRIS-EDTA (pH 8.0) buffer.

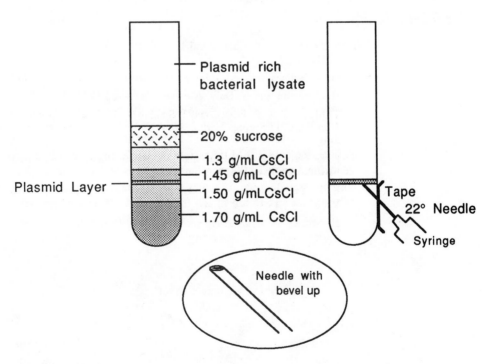

Figure 22-3 Use of a needle to remove the plasmid band from a centrifuge tube

3. The phage in the lysate fraction can be concentrated by polyethylene glycol precipitation in high salt (as in Chapter 10) and recovered by extraction of the pellet with low salt buffer.[6]

4. Variations on all these isolation and transformation methods abound.[7]

QUESTIONS TO PONDER

1. What is the purpose of the EDTA in all buffers for dissolving DNA?

2. The copy number of a particular plasmid is determined by the concentration of a plasmid-encoded inhibitor of plasmid DNA synthesis. How would you imagine the concentration of this inhibitor to affect the copy number? How would the concentration of this inhibitor depend on the number of plasmid copies (i.e., the number of copies of the inhibitor gene)? Why would protein synthesis inhibition lead to lower inhibitor concentration and hence higher copy numbers?

3. Supercoiled plasmid sediments to a lower band than does nicked plasmids or broken pieces of chromosomal DNA, even if they are larger than the plasmid. Explain.

4. In agarose electrophoresis, the plasmid obtained in this preparation often appears as a broad band appearing much farther down the gel than one would expect given its molecular weight. Brief treatment with Eco RI before electrophoresis results in a narrower, much slower moving band. Explain.

REFERENCES

1. S. Cohen, A.C.Y. Chang, and L. Hsu, *Proc. Natl. Acad. Sci. (USA)*, **1973**, *69*, 2110.

2. D. S. Holmes and M. Quigley, *Anal. Biochem.*, **1981**, *114*, 193.

3. H. Birnboim and S. Doly, *Nucleic Acid Res.*, **1979**, *7*, 1513.

4. T. Maniatis, E. F. Fritsch and J. Sambrook, *Molecular Cloning, a Laboratory Manual*, Cold Springs Harbor, New York: Cold Springs Harbor Laboratory, 1982, p. 366.

5. L. G. Davis, M. D. Dibner, and J. F. Battey, *Basic Methods in Molecular Biology*, New York: Elsevier, 1986, pp. 102-108.

6. K. R. Yamamoto, B. M. Alberts, R. Benzinger, L. Lawthorne, and G. Treiber, *Virology*, **1970**, *40*, 734.

7. H. Miller, *Methods Enzymology*, **1987**, *152*, 145.

Chapter 23

Nucleases and Restriction Enzymes

Plasmid pBR322 DNA is cleaved by the restriction endonucleases BamH I, Hae II, Hae III, Hind III, EcoR I, and Pst I, as well as by a mixture of EcoR I and Pst I to demonstrate the specificity of restriction enzymes on the same small DNA molecule. The relationship between electrophoretic mobility and linear fragment size is emphasized, but the importance of superhelicity on the mobility of circular DNA is also demonstrated. The circular double-stranded replicating form (RF-DNA) of ϕX174 exists in both relaxed and supercoiled forms that migrate differently during agarose electrophoresis. The agarose gel electrophoresis gel patterns can be visualized either by ethidium staining or by autoradiography after ^{32}P labeling by nick translation. As an optional experiment the identity of an unknown circular DNA molecule from three possibilities (ϕX174, SV40, or pBR322) can be discovered by comparing the restriction cleavage patterns observed with the restriction enzymes Hind III and Hae III with those predicted by published restriction maps. As another option to demonstrate the susceptibility of dsDNA to "nicking" by mechanical shearing forces, calf thymus DNA is cleaved with the ssDNA cleaving enzyme S1 nuclease, both with and without prior passage through a syringe to introduce single- and double-stranded breaks. The cleavage pattern of mechanically fractured and S1-treated calf thymus DNA are compared by agarose gel electrophoresis on 1.2% gels stained with ethidium bromide.

KEY TERMS

Endonucleases   *Equivalence point*

Exonucleases   *Isoschizomers*

Nick translation   *Nucleases*

Restriction mapping   *Restriction nucleases*

Restriction recognition sites

BACKGROUND

The ability to cut DNA at predetermined positions is a necessary precondition for genetic engineering, as well as any investigation of the ultimate sequence of complicated genomes. Although some human-made, chemically synthesized reagents have been devised to afford specific cleavages of DNA,[1] by far the most commonly used reagents are enzymes isolated from micro-organisms. Almost all the techniques employed to investigate the structure and regulation of genes depend on *nucleases.* These enzymes catalyze the cleavage of phosphate diester bonds at either the 5' or 3' ends of a dinucleotide junction.

DNA cleaving enzymes (DNases), which cleave the terminal phosphodiester linkages of DNA from either end are known as *exonucleases,* while enzymes that cleave the phosphodiester linkages within a DNA molecule are *endonucleases*. The latter type of enzyme is widely used to introduce double- or single stranded cuts in DNA, more or less at random. *Deoxyribonuclease I* (DNase I) is such an enzyme. In the presence of Mg^{2+}, DNase I cleaves each strand of dsDNA independently, introducing both double- and single-stranded cuts at statistically random positions along a dsDNA chain; in the presence of Mn^{2+}, double-stranded cuts yielding blunt ends predominate.[2] At low levels (several hundred thousand-fold dilutions of commercial enzyme stocks) and in the presence of Mg^{2+}, DNase I can be used to introduce occasional single-stranded "nicks" in dsDNA. *E. coli DNA polymerase I* has both 5'-to-3'-exonuclease and 5'-to-3'-polymerase abilities and in the presence of all four deoxyribonucleotides (dATP, dGTP, dCTP, and TTP), this enzyme can replace bases near a single-stranded nick introduced by low levels of DNase I. The polymerase binds at the nick and then moves down the chain in the 5'- to-3' direction, replacing old nucleotides with the new nucleotides included in the reaction mixture. If any of the latter are radiolabeled, this method can be used to radiolabel a DNA chain.[3] The original broken phosphodiester linkage is replaced with a new one synthesized by the attack of a 3'-OH on the α-phosphate of the incoming nucleoside triphosphate and the nick is moved along the DNA hence the term *nick translation* . A ligase is required to finally seal the nick. The sequence of events is illustrated in Figure 23-1.

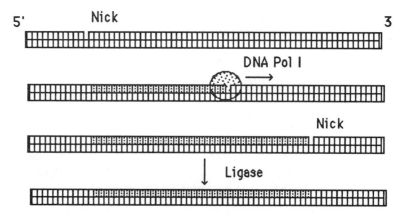

Figure 23-1 Nick translation is used to introduce ^{32}P-labeled nucleotides into a DNA chain

Restriction Enzymes

Among the most impressive examples of specificity in protein-DNA binding are the very specific DNA cleavage patterns caused by *restriction endonucleases*. Restriction endonucleases have been found in nearly every microorganism examined and are known to catalyze double-stranded breaks in DNA to yield *restriction fragments*. The restriction enzyme cleavage pattern is specific for a given DNA and enzyme and can be used to characterize plasmids and viral DNA. The techniques are relatively easy.[4]

There are three kinds of restriction endonucleases; all play a role in bacterial systems to guard against foreign DNA. Complex endonucleases that are associated with a DNA methylase in a complex host restriction/modification system are designated type I and III endonucleases. They are large oligomeric proteins capable of both cleaving and modifying DNA; they require ATP, Mg^{2+}, and S-adenosyl-L-methionine (AdoMet). Type I has a very complex recognition site and the enzyme cleaves at nonspecific sites 400 to 7000 base pairs from this site, whereas type III is a simpler complex that cleaves DNA only 25 to 27 base pairs from specific recognition sites. Foreign DNA entering a cell will be cleaved if not modified in the proper pattern. These enzymes play a role in protecting a cell from phage infection.[5] Type II restriction endonucleases are small monomeric or dimeric proteins requiring only Mg^{2+} to cleave DNA directly at specific recognition site so that the site of cleavage is always the same. It is this type of restriction enzyme that is so useful to molecular biologists. *Restriction recognition sites* are four to six base pairs in length and usually have twofold (dyad) symmetry. Cleavage at the center of the sequence gives a blunt-end cut, whereas cleavage to either the left or right of center gives cohesive "sticky" termini with protruding ssDNA residues. Such off-center cleavage of DNA generates short complementary single-stranded termini which can base-pair with each other or with any other fragment produced by the same enzyme.

To date, approximately 400 restriction endonucleases have been identified. A nomenclature has been developed based on an abbreviation of the name of the micro-organism from which the enzyme is isolated: the first initial of the genus and the first two initials of the species followed by a roman numeral, indicating the order of discovery.[6] For example, Hae II is purified from *Haemophilus aegypticus*. (The very well known endonucleases EcoR I and EcoR II were isolated from *E. coli* long before consensus on nomenclature was obtained.) Some of the 400 enzymes share recognition-site specificities. Restriction enzymes that recognize identical sequences are known as *isoschizomers*.

The restriction enzymes used in this experiment are given in Table 23-1, along with the specific cleavage pattern of their recognition site and the number of cleavage products produced by their reaction with four types of dsDNA. It can be seen that enzymes with a six-base-pair recognition site cleave less frequently than do enzymes with smaller recognition sites. If all base sequences are chemically random, then, for dsDNA containing A = T = G = C, the probability that any one type of base will occur at a particular lattice position will be 1/4. The probability **p** of finding a specific sequence of **n** bases can be estimated as

$$p = (1/4)^n \qquad (23\text{-}1)$$

For a tetrameric sequence an average of one site for every 256 base pairs will be cleaved; for a pentameric sequence one site every 1024 base pairs; and for a hexameric sequence one site every 4096 base pairs would be expected on average. These statistical predictions are only partially reflected in the actual cleavage patterns of the four DNA specimens given in the table. The number of fragments and their molecular weights can be used to characterize an unknown DNA sample.

Table 23-1
RESTRICTION ENZYME RECOGNITION SITES
AND THEIR VIRAL AND PLASMID DNA CLEAVAGE PRODUCTS

Enzyme	Recognition Sequence[a]	*Number of Cleavage Sites*			
		Lambda	φX174	pBR322	SV40
BamH I	G^GATCC	5	0	1	1
EcoR I	G^AATTC	5	0	1	1
Hae II	PuGCGC^Py	48	8	11	1
Hae III	GG^CC	>50	11	22	19
Hind III	A^AGCTT	6	0	1	6
Pst I	CTGCA^G	28	1	1	2

[a]Recognition sequences are given from 5' to 3' for one strand only; the point of cleavage is indicated by ^. For example the BamHI recognition sequence is

5' G^GATCC 3'
3' C CTAG^G 5'   which yields

5'G   GATCC 3'
3' CCTAG   G 5'

Working with restriction enzymes is relatively easy if care is taken to ensure that all solutions and glass or plasticware are sterile and that the enzymes, which are fragile at room temperature, are stored cold under the appropriate conditions. Commercial samples of restriction endonucleases are in concentrated form, stabilized by bovine serum albumin (BSA) and 50% glycerol. These stock solutions remain stable if they are stored in the freezer (-20°C or colder) and removed only briefly just before use; the enzyme vial is removed from the freezer and immediately put into ice water so that the temperature never rises appreciably above 0°C. A fresh sterile pipette tip or a fresh piece of fine plastic tubing attached to the needle of a Hamilton microliter syringe is used to withdraw the required aliquot immediately before use. Usually, 0.1 to 1 μL of the commercial enzyme is sufficient to digest 10 μg of DNA in an hour. The enzyme stock solution must be returned to the freezer right away. The specificity of many restriction endonucleases is reduced by the presence of more than 5 % glycerol[7] so the enzyme stock solution must be diluted at least by a factor of ten by the digestion buffer. A typical digestion buffer contains TRIS, Mg^{2+}, NaCl, 2-mercaptoethanol and BSA. All restriction enzymes require Mg^{2+} and most work best within the pH range 7.2 to 7.6 and at 37°C. Each restriction enzyme is most active at an optimal ionic strength. All reaction buffers contain 10-50 mM TRIS-HCl (pH 7.5), 10 mM $MgCl_2$ and 100 μg/mL bovine serum albumin, but various amounts of NaCl are required. Three typical 10 x stock buffers containing no salt, low-salt (0.5 M NaCl), and high-salt (1.0 M NaCl) are used. These stock solutions are diluted with water to give reaction buffers with the salt and buffer concentrations required for each enzyme. Buffer and salt concentration profoundly influences the specificity of restriction enzymes, and most enzymes operate only over a fairly narrow electrolyte concentration range. The optimum buffer and NaCl concentration and temperature are given in Table 23-2.[14]

Table 23-2
EFFECT OF TEMPERATURE, BUFFER AND NACL CONCENTRATION ON RESTRICTION ENZYME ACTIVITY

Enzyme	Optimum Temp. (°C)	Optimum [TRIS-HCl] (mM)	Optimum [NaCl] (mM)	Reference for Recognition Sequence
BamH I	37	10	50	8
EcoR I	37	50	100	9
Hae II	37	10	50	10
Hae III	37	10	50	11
Hind III	37-55	10	50	12
Pst I	37	10	100	13

Substitution of manganese for magnesium, inappropriate ionic strength or pH, and the presence of an organic cosolvent (glycerol or ethanol) can reduce specificity. For example, EcoR I, in 50 mM NaCl, 5mM $MgCl_2$, 100 mM TRIS-HCl (pH 7) recognizes and cleaves the hexanucleotide sequence GAATTC, while in 2 mM $MgCl_2$, 20 mM TRIS-HCl (pH 8.5) the recognition specificity is reduced to the tetranucleotide sequence AATT.[15] Even in the same buffer a particular enzyme may cleave some sites faster than others. There is a 10-fold difference in the rate of cleavage of different lambda phage DNA sites by EcoR I and a 14-fold difference by Hind III.[16] Greater amounts of some restriction enzymes are required to cleave the supercoiled plasmid pBR322 than for linear lambda DNA. When DNA is to be cleaved with two or more restriction enzymes, the digestions can be carried out simultaneously provided that the enzymes both work at the same ionic strength and temperatures. Alternatively, the enzyme that requires the lower ionic strength is used first, then the salt concentration is increased so that the second enzyme can be used.

Restriction endonuclease activity is defined by an *equivalence point* rather than a rate term as used conventionally to define enzyme activity. For the enzymes used here, one unit of activity is the amount of enzyme required to digest 1.0 μg of lambda phage DNA completely in 60 minutes in a 50-μL volume at appropriate temperature and buffer conditions.

Restriction Enzyme Mapping

The number and size (mobility) of the fragments produced by a particular enzyme are like a fingerprint of the DNA molecule. This is the idea of *restriction mapping*, which can be used to characterize and identify small DNA molecules. One digests the DNA of interest with a series of restriction enzymes and the products of the digestion are resolved by analytical electrophoresis on agarose. The DNA is visualized by ethidium bromide staining or by autoradiography if the DNA has been radiolabeled by nick translation before the digestion step. DNA can also be digested with mixtures of two or more endonucleases; the resulting pattern reflects the combined activity of the several enzymes. For example, both EcoR I and Hind III cleave pBR322 into a single large piece because they each cleave the dsDNA circle once. But since the enzymes cut DNA at different sites, their combined effect is to yield two pieces. The fragment profile for four or five enzymes used separately and in combination can be used to generate restriction map that uniquely characterizes a small plasmid. Table 23-3 lists the fragments and their sizes for a number of circular DNA molecules that can be used as unknowns to be identified by restriction mapping.

In the first experiment described in this chapter, the specificity of restriction endonucleases is demonstrated by cutting pBR322 DNA with a variety of restriction enzymes (BamH I, EcoR I, Hae II, Hae III, and Pst I) acting alone and EcoR I and Pst I acting together. The cleavage patterns can be shown by agarose electrophoresis with ethidium bromide staining.   An optional scheme for nick translation and autoradiography is also included. The instructor may choose to give each student a circular DNA unknown to identify by restriction mapping.

Table   23-3
RESTRICTION FRAGMENTS OF φX174 RF, SV40,
AND pBR322 FROM HIND III AND HAE III CLEAVAGE

	Fragment Sizes (Base Pairs)
φX174 RF	
Hind III	5386 (not cleaved)
Hae III	1353, 1078, 872, 603, 310, 281, 271, 234, 194, 118, 72
SV40	
Hind III	1768, 1169, 1118, 526, 447, 215
Hae III	1661, 752, 540, 372, 329, 308, 300, 299, 227, 179, 49, 45, 33, 30, 30, 14, 9, 6
pBR322	
Hind III	4363 (cleaved once)
Hae III	587, 540, 504, 458, 434, 267, 234, 213, 192, 184, 124, 123, 104, 89, 80 64, 57, 51, 21, 18, 11, 7

Supercoiled Circular DNA

One feature that complicates any analysis of the mobility of circular DNA is supercoiling. The difference in electrophoretic mobility of relaxed and supercoiled φX174 RF circular DNA and the effect of a single cut by Pst I is demonstrated in the second experiment described in this chapter. (The RF designates the double-stranded replicating form of this viral DNA.) A nicked circle and an intact non-supercoiled dsDNA circle have the same shape and flexibility and migrate through an agarose gel at the same rate, but supercoiling increases the mobility of DNA.[17] A single twist introduced into an intact circle increases the mobility of DNA. Each additional twist increases the mobility an additional increment. A nicked circle cannot become supercoiled. A compact supercoil does not always migrate as the highly flexible linear form, which can change orientation as it migrates through the gel matrix. The difference in mobility of supercoiled and linear DNA of the same molecular weight depend on the gel percentage and the concentration of the intercalating dye, ethidium bromide, but both migrate about the same rate and both much faster than a relaxed or nicked circle. The behavior of commercial φX174 RF indicates the amount of supercoiled and relaxed (nicked) circular forms. As an optional experiment the student may distinguish between nicked and merely relaxed forms by *brief* digestion with S1 nuclease (see below), which would open nicked circles to give faster migrating linear dsDNA.

Mechanical Shearing of DNA

Finally, the effect of hydrodynamic shearing forces in producing both double- and single-stranded cuts in calf thymus DNA is demonstrated by comparing the agarose electrophoresis patterns of this DNA before and after several passages through a small gauge syringe needle. Calf thymus DNA is a heterogenous mixture which will show up on a gel as a smear ranging from high- to medium-molecular-weight mobilities; the more mobile portions are due to smaller fragments set free by random double-stranded brakes. Single-stranded breaks can be introduced into dsDNA by mechanical and hydrodynamic forces. Such nicked dsDNA can be cleaved at the nick sites by digestion with S1. We will use this enzyme in this project to explore the single-stranded nicks introduced into dsDNA by mechanical shearing. *Nuclease S1* from *Aspergillus oryzae* degrades ssDNA to yield 5'-phosphoryl oligonucleotides. Double-stranded nucleic acids (dsDNA, dsRNA, and DNA·RNA) are cleaved more slowly. This enzyme cleaves at nicks or mismatched regions in the presence of the required cofactor Zn^{2+} at pH 4.5. The S1 cleavage pattern of nicked and nick-free DNA will differ because the former will be cut into shorter fragments by S1.

Single-stranded nicks do not set fragments free but digestion of syringe-treated calf thymus DNA with S1 nuclease will convert single-stranded nicks into double-stranded brakes which will be revealed by the presence of additional higher-mobility fragments. Calf thymus DNA so treated will show up on an agarose gel as a smear in the moderate-to high-mobility range (smaller pieces). Figure 23-2 illustrates the gel pattern observed for the above three studies.

EXPERIMENTAL SECTION

The DNA solutions, the nuclease stop solution and most buffer stock solutions (see below) should be made by the laboratory staff to ensure high quality and to enable the first three parts of the experiment to be completed in a single 4-hour laboratory period. As a necessary precaution, all glassware and plasticware must be sterilized by autoclaving; this includes microcentrifuge tubes, centrifuge tubes, and disposable pipette tips.

Siliconized glassware is not necessary for these experiments, but such a precaution is not a bad habit to get into. Research-quality experiments, especially sequencing of very small amounts of DNA, can be jeopardized by the loss of nucleic acids by absorption on untreated glass surfaces. Reaction of glassware with a small quantity (1 to 2 mL) of dichlorodimethylsilane in a vacuum desiccator followed by baking at 180°C for 2 hours will coat the glass surface and minimize such loss.

CAUTION: DICHLORODIMETHYLSILANE
IS A VOLATILE POISON.

All water used to prepare buffers and gels should be glass distilled just before the experiment. All reagents should be of high quality and stored under sterile, cold conditions. Pipette tips and microcentrifuge tubes should never be reused.

Students working in pairs should prepare agarose for a 1.2% gel at the beginning of the laboratory period according to the procedure described in Chapter 10. Electrode buffer must also be prepared either by students or by stockroom staff.

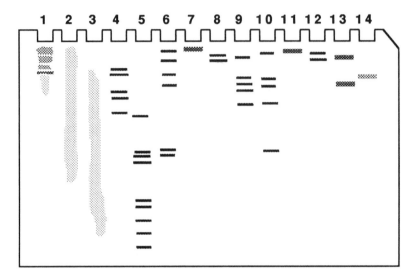

Key to gel patterns:
1. Calf Thymus DNA – heterogeneity due to mechanical cleavage
2. Calf Thymus DNA – cleaved with S1 nuclease without pretreatment
3. Calf Thymus DNA – cleaved with S1 nuclease after pretreatment
4. Lambda DNA – cleaved with BamH 1
5. Lambda DNA – cleaved with Hae III
6. Lambda DNA – cleaved with Hind III
7, 11. Lambda DNA – not cleaved
8, 12. Lambda DNA – cleaved with Xho I
9. Lambda DNA – cleaved with EcoR I
10. Lambda DNA – cleaved with both Xho I and EcoR I
13. ØX174 DNA – not cleaved (note relaxed and supercoiled circles)
14. ØX174 DNA – cleaved with Xho I (linear form only)

Figure 23-2 Gel pattern for restriction enzyme and nuclease digestion studies described in this chapter

A number of steps employing DNA radiolabeled by nick translation are included in the following protocols. These steps should only be attempted in an area designated for radiochemicals, and since ^{32}P is a high-energy β–emitter, **extreme caustion must**

be employed. The laboratory must be monitored with a Geiger -Müller counter and all workers must wear film dosimetry badges. All work must be done behind plexiglass radiation safety shields, and all personnel must wear plastic gloves and protective goggles.

Restriction Enzyme Reactions

Coordination of all the pairs of students running this experiment is necessary to avoid repeated thawing and refreezing of frozen restriction enzymes. Such enzymes are very expensive and must be treated with the utmost care to avoid denaturation or contamination. Consequently, students must budget their time so that all are ready for the restriction enzymes at approximately the same time and all are ready to add their samples to the agarose gel at the most convenient time.

Three 10x stock buffers should be prepared by trained stockroom personnel:

1. **Low salt buffer**:

   100 mM TRIS-HCl, pH 7.4   1 mg/mL BSA

   100 mM MgCl$_2$ 10 mM 2-Mercaptoethanol

   0.5 M NaCl

2. **High salt buffer**:

   100 mM TRIS-HCl, pH 7.4   1 mg/mL BSA

   100 mM MgCl$_2$ 10 mM 2-Mercaptoethanol

   1.0 M NaCl

3. **High salt, high TRIS buffer**:

   500 mM TRIS-HCl, pH 7.4   1 mg/mL BSA

   100 mM MgCl$_2$ 10 mM 2-Mercaptoethanol

   1.0 M NaCl

The buffers are placed on ice. An additional solution used to stop restriction digests must also be on hand.

4. **Nuclease stop solution**:

   500 mM EDTA, pH adjusted to 7.5

In addition, agarose gel electrophoresis electrode, gel and loading buffers must be prepared as described in Chapter 10. Plasmid DNA isolated according to Chapter 22 can be used. The plasmid DNA solution should contain about 1 to 5 μg/μL if ethidium bromide visualization is used; lower levels of radiolabeled DNA can be used (see below).

To run a small volume restriction digest, the following are combined in a 1.5 mL microcentrifuge tube:

   1 μL of the appropriate 10x buffer

   0.5 to 5 μL of the thawed restriction enzyme

   1 μL of DNA sample (0.1 to 5 μg)

   enough water to give 10 μL of solution

The reaction mixture is incubated for one hour at 37°C, then quenched by adding 5 μL of **nuclease** *stop buffer* or by heating to 60°C for several minutes. Avoid overheating, which will dissociate the dsDNA fragments.

The reported enzyme activity of the commercial sample of restriction enzymes should be used to calculate the required amount of enzyme to digest 5 μg of DNA in 60 minutes. The appropriate amount of restriction enzyme is transfered from the ice-cold stock vial with a micropipet. Commercial restriction enzymes range in concentration from 8000 to 40,000 units/mL and are sold in 300 to 10,000 unit aliquots.

The DNA digest products are combined with 15% Ficoll, 0.1% SDS and 0.1% bromophenol blue as described in Chapter 10, and each is loaded into a separate sample well of the agarose gel. Accurate notes must be taken to sort out the gel lanes after staining with ethidium bromide.

CAUTION: USE GLOVES WHEN USING ETHIDIUM BROMIDE.

It is recommended that two lanes contain molecular weight standards (Hind III digest of lambda phage). Since these lanes are easy to spot, if they are arranged asymmetrically (e.g., second and middle lanes) one can orient all other lanes with respect to them. The protocol for staining and viewing gels is also described in Chapter 10. The eight Hind III digest fragments span a wide range of polynucleotide lengths (given as kilobase pairs):[18]

23.7, 9.5. 6.6, 4.3, 2.2, 1.9, 0.59, and 0.15

The last two are sometimes hard to see on a gel visualized with ethidium bromide. Commercially available molecular weight standards employing ^{32}P-labeled restriction fragments are available (NEN). Labeled base-pair length standards can be prepared by 5'-end labeling protocols supplied with nonlabeled fragments by BRL or New England Bio-Labs; see also Chapter 26.

Nick Translation Labeling

There are several steps in nick translation labeling: The DNA (0.02 to 0.2 μg/μL) is nicked with diluted DNase I, labeled with DNA polymerase, the polymerase reaction is stopped, and the labeled DNA deproteinated and separated from labeled nucleotides by gel filtration. If necessary, the nick-translated DNA is resealed with ligase. Nick translation reagent kits are available from a number of biochemical suppliers (BRL, New England Biolabs, NEN), and if one employs such a kit, the directions for that kit must be followed. (DNase is a frequent contaminate of commercial preparations of DNA polymerase, but since the amount of nuclease activity varies widely from source to source, the protocols associated with each kit may be slightly different.) The following general procedure is to be followed if a kit is not used.

1. Commercial DNase I is used to prepare a stock solution containing 1 mg/mL enzyme in 0.15 M NaCl with 50% glycerol. This solution is stored in the freezer. A portion is removed and diluted to a concentration of 0.05 μg/μL) in 50 mM TRIS-HCl, pH 7.4/10 mM $MgCl_2$ with 1 mg/mL BSA just prior to use.

2. A 10x nick translation buffer containing

500 mM TRIS-HCl (pH 7.4)
50 mM $MgCl_2$

      1 mM Dithiothreitol

      0.5 mM of dCTP, dGTP, TTP (in each)

is used to dissolve 0.5-1.0 µg of DNA to be labeled.

3. Using appropriate safety precautions, one must transfer 50 to 300 µCi 5'-[α-^{32}P] -dATP (300 Ci/mmol, 10 µCi/µL), 2 µL 10x nick translation buffer and 1 µL diluted DNase I and 19 µL water to an Eppendorf microcentrifuge tube.

THIS OPERATION MUST BE DONE BEHIND A PLEXIGLASS RADIATION SAFETY SHIELD.

ALL WORKERS MUST WEAR PROTECTIVE CLOTHING, GOGGLES, AND GLOVES.

The author recommends that the transfer of radioactive material be done by the laboratory instructor or someone familiar with the hazards associated with high-energy β–emitters; see Chapter 6.

4. Two units of DNA polymerase are added and the reaction mixture incubated at 15°C for 90 minutes. The reaction mixture should not be allowed to warm up.

5. The reaction is stopped by adding 0.1 volume nuclease stop solution and heating the solution to 70° for 5 minutes. The resulting DNA can be precipitated by adding 5 µL of tRNA (500 µg/mL) , 0.5 volume of 7.5 M ammonium acetate, and 2.5 volumes of ethanol. Further purification by gel filtration through a Pasteur pipet containing Bio-Gel P-60 (100-200 mesh) will remove the incorporated nucleotides.

After precipitation the DNA specific activity should be around 1×10^6 cpm/µg.

    Agarose gel electrophoresis patterns of DNA labeled this way may be observed with autoradiography of the frozen gel; see Chapter 19. A minimum exposure for film detection is 10 dpm/ mm,^{2} but a good exposure may require 100 times this amount. The author recommends Kodak X-OmatAR film developed with GBX developer (Sigma).

Supercoiling of φX174 Circular DNA

The commercial RF form of φX174 (BRL Laboratories) is more than 90% in the superhelical form at the time of preparation but will relax in the presence of even a trace of DNase. Student groups should be provided 1-µg samples prepared under sterile conditions from a fresh commercial stock. These can be digested with 0.1 volume of the dilute DNase solution discribed above, all diluted to 10 to 15 µL with water. The resulting material is combined with an equal volume of loading buffer and applied to an agarose gel. A comparison of relaxed and nonrelaxed phage DNA samples should reveal the effect of supercoiling.

Mechanical Shearing of Heterogeneous DNA

Calf thymus or chicken erythrocyte DNA solution (2 µg/µL) is divided into three portions. The first is reserved as a control to estimate the heterogeneity of DNA before syringe or S1 treatment. The second is treated with 1 unit of S1 nuclease per microliter of DNA solution and incubated for 15 minutes at 37°C. A 1-mg portion of commercial calf thymus DNA, for each pair of students, should be cut free from the swatch of DNA fibers with a glass knife (sharp shard of glass) and placed in 500 μL of 30 mM sodium acetate (pH 4.6) containing 50 mM NaCl, 1 mM $ZnSO_4$ and 5% glycerol and allowed to incubate with occasional agitation at 5 to 10°C for two days to dissolve. (Alternatively, chicken erythrocyte DNA from Chapter 24 may be used.) As the DNA begins to go into solution, it forms a quite viscous lump and some agitation is required to ensure its complete dissolving.

The remaining third portion of DNA solution is passed repeatedly through a fine-gauge syringe needle. The syringe-treated DNA is divided into two portions; one is treated with S1 as above and the other is not. The enzyme reaction in each case is quenched after the same period (15 minutes) by the addition of enough Nuclease stop solution, to bring the EDTA concentration to 10 mM. All four DNA samples are stored in an ice bath until time to load the agarose gel for electrophoresis.

PROJECT EXTENSION

Students may produce restriction maps of an unknown circular DNA molecule with Hind III and Hae III and identify the unknown from among the three possibilities (φX174-RF, SV40 and pBR322) using the information in Table 23-3. A 5 μg aliquot of unknown DNA should be divided into two portions and digested with the two enzymes under the conditions recommended in Table 23-2. The product fragments are analyzed by electrophoresis on 1 % agarose with ethidium bromide stain. The smaller fragments may not show up and fragments with only a minor difference in number of base pairs may not be resolved separately.

DISCUSSION QUESTIONS

1. If plasmid X is cleaved by EcoR I into two fragments and by Hind III into three fragments, would you expect the two enzymes working together to cleave X into five fragments?

2. Can two restriction enzymes act at the same site?

3. Can a restriction enzyme cut at two sites with different base sequences?

4. How would you imagine supercoiling would affect the rate of cleavage by restriction endonucleases?

5. Is there a phosphate left at the 3' end of a restriction fragment "sticky end"? What is left at the 5' end?

6. Alkaline phosphatases cleave off terminal phosphate groups from oligonucleotides. How does one use this enzyme to ensure that a plasmid cloning vector does not reform after restriction endonuclease cleavage?

7. Given the activity of S1 indicated by the supplier and the dilutions indicated in the above protocol how much time would be required to cleave 10% of all phospho-diester binds in 20 μg/ mL ssDNA?

REFERENCES

1. P. G. Schultz and P. B. Dervan, *J. Amer. Chem .Soc.* **1983**, *105*, 7748.

2. E. Melgar and D. A. Goldthwaite, *J. Biol. Chem.*, **1968**, *243*, 4409.

3. R. G. Kelly, N. Cozzarelli, M. P. Deutscher, I. R. Lehman and A. Kornberg, *J. Biol. Chem.*, **1970**, *245*, 39.

4. J. E. Brooks, *Methods Enzymology*, **1987**, *152*, 113; R. J. Roberts, *Nucleic Acids Res.*, **1985**, *13*, r165.

5. S. Linn and W. Arbor, *Proc. Natl. Acad. Sci. (USA)*, **1968**, *59*, 1300; S. E. Luria, *Cold Spring Harbor Symp. Quant. Biol.*, **1953**, *18*, 237.

6. H. O. Smith and D. Nathans, *J. Mol. Biol.*, **1973**, 81, 419.

7. J. G. Chirikjian and J. George in *Gene Amplification and Analysis*," Vol. I, ed. J. G. Chirikjian, New York: Elsevier/North Holland, 1981, p. 74.

8. R. J. Roberts, G. A. Wilson and F. E. Young, *Nature*, **1977**, *265*, 82.

9. J. Hedgpeth, H. M. Goodman and H. W. Boyer, *Proc. Natl. Acad. Sci. (USA)*, **1972**, *69*, 3448.

10. C.-P.D. Tu, R. Roychoudhury and R.Wu, *Biochem. Biophys. Res. Commun.*, **1976**, *72*, 355.

11. S. Bron and K. Murray, *Mol. Gen. Genet.*, **1975**, *143*, 25.

12. R. Old, K. Murray and G. Roizes, *J. Mol. Biol.*, **1975**, *92*, 331.

13. N. L. Brown and M. Smith, *FEBS Letters*, **1976**, *65*, 284.

14. T. Maniatis, E. F. Fritsch and J. Sambrook, *Molecular Cloning, a Laboratory Manual*, Cold Spring Harbor, New York: Cold Spring Harbor Laboratories, 1982, pp. 98-106.

15. B. Polisky, p. Greene, D. E. Garfin, B. J. McCarthy, H. M. Goodman and H. W. Boyer, *Proc. Natl. Acad. Sci. (USA)*, **1975**, *72*, 3310.

16. K. Nath and B. A. Azzolina in *Gene Amplification and Analysis*, Vol. I, ed. J. G. Chirikjian, New York: Elsevier/North Holland, 1981, p 113.

17. W. Keller, *Proc. Nat. Acad. Sci. (USA)*, **1975**, *72*, 4876.

18. E. H. Szybalski and W. Szybalski, *Gene*, **1979**, 7, 217.

Chapter 24

Chicken Erythrocyte Chromatin and Histone Protein Preparation

In this experiment, fresh chicken blood is diluted with saline EDTA/citrate buffer and the erythrocytes freed from plasma by centrifugation. The packed erythrocytes are lysed in Triton X-100 sucrose buffer to release nuclei, which are washed by repeated centrifugations through a sucrose buffer at 6000 x g. The nuclei pellet is divided into two portions. The chromatin in one portion is digested with low levels of micrococcal nuclease to liberate nucleosomes. Sucrose gradient isolation of nucleosomes is followed by their characterization by "particle gel electrophoresis." An optional procedure for extracting DNA from chromatin is also given. The second portion of packed erythrocytes is lysed. The histones in the second portion are isolated by either sulfuric acid extraction or fractionation on a hydroxyapatite column. The histones can be further fractionated by gel exclusion chromatography on Bio-Gel P-60. All fractions can be characterized by acetic acid-urea or SDS polyacrylamide gel electrophoresis. A number of advanced electrophoretic methods are introduced to characterize modified histones.

KEY TERMS

Acid-Urea gels	*Chromatin*
Chromatosome	*Core particle*
High density chase liquid	*Histones*
Hydroxylapatite	*Nucleosome ladder*
Nucleosome	*Particle gels*

BACKGROUND

Procaryotic DNA can be isolated conveniently from *E. coli* and student preparations from this source, such as that described in Chapter 10, suffice for demonstrating the properties of uncomplexed short dsDNA molecules (4 x 10^8 base pairs). Eukaryotic dsDNA is different in a number of significant ways, so that the isolation of such material involves special problems. It is a longer molecule (10^{11} base-pairs) and it is more tightly associated with proteins. Eukaryotic DNA occurs in a special structure of nucleoprotein fibers in the cell nucleus known as *chromatin*. The amount of protein is equal to or greater than the amount of DNA in chromatin. and a large number of proteins are involved. The most abundant type are highly basic *histone* proteins, but a large number of non-histone proteins are involved in chromatin as well. The DNA is condensed by a series of packaging mechanisms involving protein structures derived from histones and other structures. The DNA from a eukaryotic cell can be on the order of 1 m in length yet packed into a 10-μM nucleus, while maintaining accessibility during transcription and remaining tangle-free during replication. The initial coiling of DNA involves 146 bp of DNA wrapped one-and-three-quarter turns around an octomeric complex of histones. This complex of protein and DNA is known as a *core particle*.[1] The protein aggregate contains two molecules each of H2A, H2B, H3 and H4 type histones.[2] H2A and H2B are moderately lysine-rich (10-14 mole %), whereas H3 and H4 are arginine-rich (13 mole %) histones. In most somatic cells a very lysine-rich (27 mole %) histone H1 is overlaying the DNA strands wrapping the core particle. The entire particle, including two full turns of DNA (166 bp DNA) and the H1 protein is known as a *chromatosome*. Chromatosomes are arranged together on the DNA like beads on a string, with strands of linker DNA stretching between adjacent chromatosomes. The chromatosome plus linker DNA is termed a *nucleosome*. The length of the linker and hence the repeat frequency of nucleosomes is variable between species and even within tissues of the same species. In chicken erythrocytes the nucleosome repeat frequency is 212 bp, which is common for an inactive tissue; more active tissues have shorter repeat frequencies as short as 165 bp. Chicken erythrocyte chromatin is different in another important way as well. Inactive nucleated erythrocytes from birds, fish, and amphibians contain another type of very lysine-rich (24 mole %) histone, H5, which partially replaces H1.

Chromatin and Eukaryotic DNA

The structural details of chromatin were first probed by nuclease digestion studies.[3] The elementary subunit of chromatin structure, the nucleosome can be freed from chromatin by micrococcal nuclease digestion. Micrococcal nuclease makes double-stranded breaks in DNA. The large nuclease molecule preferentially cuts in the more accesible linker regions between histone-DNA complexes, first to release a particle containing 170 to 240 bp DNA (depending on the linker) as well as dimers and higher oligomers of this particle, formed by the failure to cut the linker regions between some nucleosomes. Continued digestion gives a more stable particle trimmed down first to 166 bp and then to 146 bp DNA. This structure, the "nucleosome core particle," can be separated from heavier forms by centrifugation through a sucrose density gradient. Electrophoresis through a 5 % polyacrylamide "particle gel" of the product from a brief micrococcal nuclease digestion of solubilized chromatin or nuclei will give a *nucleosome ladder* with slower-moving dimers and trimers trailing behind single nucleosomes.[4] An analysis of the size of the DNA fragments released by this approach indicates that the spacing between nucleosomes is about 200 bp, depending on the species.

The traditional source of eukaryotic chromatin and DNA has always been calf thymus gland. However, veal is becoming a luxury food item, and this organ is increasingly difficult to obtain, and even when available the cost can be prohibitive. Chicken erythrocytes are a good alternative source of eukaryotic DNA. Unlike mammalian erythrocytes, chicken red blood cells contain chromatin and are a convenient source of DNA[5] and histone proteins.[6] Fresh chicken blood can readily be obtained from a poultry slaughterhouse or frozen from commercial sources. In the nucleosome-preparation portion of this experiment, chicken erythrocytes are diluted with saline EDTA/citrate buffer, freed from plasma by centrifugation and lysed by freezing. The liberated chromatin is separated from cellular debris by centrifugation and the pellet redissolved in the digestion buffer.

As an optional experiment DNA can be extracted from the chromatin complex if the latter is broken up with the detergent SDS (sodium dodecyl sulfate) and the deproteinated DNA in 2 M NaCl is rendered insoluble by the addition of ethanol. The resulting DNA is a viscous filamentous mass which is spooled onto a glass rod. DNA cleavage due to DNase activity is reduced by using EDTA and by working rapidly and at low temperatures. Moreover, natural DNA sources rich in DNA but low in DNase activity should be used; chicken erythrocytes fit this criterion. All steps must be performed at 4°C with glass or plastic vessels; all stainless steelware, including spatulas, must be avoided. Even minute traces of rust cause rapid degradation. A stainless steel tissue homogenizer should not be used if a Teflon-bladed one is available. All glassware should be rinsed with 1 mM EDTA before use.

Histones

Histones are highly basic proteins complexed with DNA in the cells of eukaryotes. The simplest subunit of the basicity of histones permits their extraction from chromatin with dilute sulfuric acid.[7] Differences in their molecular weights and net charge allow histone fractionation either by gel-filtration on Bio-Gel P-60 or by hydroxylapatite chromatography. This experiment employs either acid extraction of histones from nuclei followed by Bio-Gel P-60 gel filtration chromatography or hydroxylapatite column chromatography of sheared chromatin to separate the histones into three groups which are further fractionated by Bio-Gel P-60 gel-filtration. The gel-filtration method developed by van der Westhuyzen[8] separates whole histone into four homogenous fractions, H1, H5, H2A, and H4 with the much larger protein H1 eluting at the void volume. The separation order does not correspond to the monomeric protein molecular weights but reflects the differences in aggregation of the proteins at the salt concentration employed (50 mM NaCl). At high elution volumes a fifth fraction containing both H3 and H2B is obtained, but this mixture can be cleanly separated by gel-filtration on Sephadex G-100. The gel filtration approach is quite time consuming but offers the advantage of a clean separation.

Hydroxylapatite is a special hydrated form of calcium phosphate widely used as a column packing material in preparative biochemistry. The molecular separation on hydroxylapatite depends on specific interactions between negative charges on the macromolecule and Ca^{2+} in the hydroxylapatite matrix, but interactions between positively charged groups on the macromolecule and phosphate groups also play a role. For this reason the retention on a hydroxylapatite column is not an exclusive function of any one parameter such as molecular weight, molecular shape, charge density or iso-electric point. Consequently, hydroxylapatite chromatography nicely complements most other chromatographic separation techniques. Sheared chromatin binds to hydroxylapatite in 0.1 M potassium phosphate, pH 6.7, and the DNA component remains bound while three histone groups (H1 + H5, H2A + H2B and H3 + H4) are eluted from the column by progressively higher KCl concentrations.[8] DNA will remain bound up to 2 M KCl. The separation of histones from DNA in solutions of

The histones in the various fractions can be identified by their electrophoretic migration on acetic acid-urea-20% polyacrylamide gels.[9],[10] Panyim and Chalkley have shown that the mobility increases for the series H1, H3, H2B, H2A, H4 with the five major histone types involved. Histone bands for H1, H3, and H4 may appear further subdivided because of histone breakdown due to endogenous proteolytic enzymes and because of natural sequence variant forms or lysine acetylation. The proteolytic activity may be suppressed by carrying out the preliminary steps of nuclei and chromatin isolation with 0.2 mM phenylmethylsulfonyl fluoride (PMSF) in the collection and lysis buffers. Histones also occur as modified forms due to *in vivo* acetylation and phosphorylation. Both modifications alter the net charge of the histone protein molecule so that modifications can lead to electrophoretically distinctive forms. An alternative series of acid-urea-polyacrylamide electrophoretic methods, which capitalize on the tight binding of histones and the nonionic detergent Triton X-100, have been used to resolve the differences in natural variants and acetylation products.[11] These are discussed as possible project extensions.

Histone proteins can also be separated by SDS-polyacrylamide slab gel electrophoresis. Because these proteins are so cationic to begin with, the binding of SDS does not fully mask their charges and their mobility on an SDS gel does not depend on molecular weight alone; the charge differences show through and the mobility increases in the order H1, H3, H2B, H2A, and H4.[12] The acetylated forms of each histone migrate separately. The presence or absence of non-histone proteins in these preparations can be determined by SDS electrophoresis as well.

EXPERIMENTAL SECTION

Chicken blood is obtained from freshly slaughtered chickens. Clotting is minimized by diluting the fresh blood with an equivalent volume of ice-cold 0.075 M NaCl, 0.025 M Na_2EDTA/0.01 M sodium citrate containing 0.2 mM phenylmethylsulfonyl fluoride (PMSF, Sigma P-7626, predissolved in ethanol), within 1 minute after draining the bird.

CAUTION PMSF IS A POTENT POISON!

The blood should be poured into the buffer and mixed by swirling. Flasks of diluted fresh blood should be transported to the laboratory in an ice chest. All subsequent operations should be carried out in the cold room (4°C) and the centrifuge should be refrigerated (4°C). The blood suspension should be filtered through two layers of cheese cloth, then centrifuged for 10 minutes at 4000 x g and the plasma discarded. The erythrocytes should be suspended in additional buffer to restore the volume to the original value, and then centrifuged as before. The packed erythrocytes are then divided into two portions. The first portion is used for nucleosome preparation, the second for histone purification. A wide variety of methods for purifying histones are described here, including some advanced electrophoretic methods suggested as project extensions. The instructor and student must decide ahead of time on the scope and methods to be employed in isolating and characterizing the components of chromatin.

Chromatin   Preparation

Freshly packed erythrocytes are gently resuspended in an equivalent volume of sucrose buffer (0.25 M sucrose, 0.010 M $MgCl_2$, 0.5 mM PMSF, and 0.05 M TRIS-HCl, pH 7.5). The suspension is added stepwise to a twofold volume of the same sucrose buffer, also containing 0.5% Triton X-100, and stirred overnight. (The use of a 20-fold volume of Triton buffer can reduce the lysis time to 20 minutes, but the resulting additional centrifugations will take longer.) The nuclei are released as the non-ionic surfactant Triton X-100 causes lysis of the erythrocyte cell membranes but not the nuclear membranes. The nuclei are obtained as a soft dark pink pellet after centrifugation at 2000 x g for 10 minutes; repeated centrifugations from fivefold volumes of sucrose buffer (no Triton) until the pellet is pale pink or white affords nuclei clean enough for histone preparations. Contaminating cellular debris are discarded with the supernatant.

Isolated chromatin can be partially digested by micrococcal nuclease to give cuts only in that portion of the DNA that is not tightly bound to histones. When some of this "linker region" is cut away, pieces of DNA of a characteristic length heterogeneity are produced corresponding to one, two, three and higher numbers of nucleosomes linked together. A "nucleosome ladder" can be obtained by electrophoresis of the DNA isolated after such partial digestion.[13] The nuclei from about 10 g of cells should be suspended in 15 mL of 10 mM TRIS-HCl (pH 6.85), 1 mM $CaCl_2$, 1.5 mM $MgCl_2$, 0.2 mM PMSF, and 0.25 M sucrose. The sample is then warmed to 37°C in a water bath for 10 minutes with gentle swirling, and enough micrococcal nuclease is added to bring the final concentration up to 25 units/mg DNA (for complete digestion). An enzyme stock solution in the foregoing buffer of about 70 units/μL is convenient. The digestion is carried on for 10-15 minutes, then terminated by adding enough EDTA to bring the concentration up to at least 2 mM EDTA; Ca^{2+} is necessary for nuclease activity and the chelating agent ties up this metal ion.

The nuclei are lysed by adding additional EDTA up to 8 mM at 0° and the cell debris spun down at 10,000 x g for 5 minutes. The nucleosomal particles can be isolated by centrifugation through a sucrose gradient (5 to 20% sucrose in 8mM EDTA, 10 mM TRIS-HCl pH 7.5) and removed from the centrifuge tube by replacement with the very dense and chemically inert chase liquid Fluorinert FC-40 (Sigma F 9775) in the apparatus shown in Figure 24-1. Such an apparatus can be constructed from readily available laboratory supplies and the chase liquid is reusable. (A high density chase liquid can be used to fractionate the centrifugate from density gradient studies[14] including that described in Chapter 31.)

The isolated nucleosomes are characterized by 5% polyacrylamide gel electrophoresis in TRIS-borate buffer at 5°C. Migration is from negative to positive at 50 mA constant current with bromophenol blue tracking dye in a separate lane.

Crude Histone   Preparations

Acid Extraction Method

An older method for the fractionation of histones according to their solubility in dilute sulfuric or hydrochloric acid is useful if large amounts of histone are to be prepared. The DNA is usually discarded in this preparation since it is extensively apurinated and cleaved by hydrolysis anyway. In the acid extraction method the nuclei are resuspended in 2 liters of 0.4 N sulfuric acid and stirred at 4°C for 4 hours. The supernatant after centrifugation at 2000 x g for 15 minutes is saved and the pellet is re-extracted with 1 in 2 liters of 0.4 N sulfuric acid and stirred at 4°C for 4 hours. The supernatant after centrifugation at 2000 x g for 15 minutes is saved and the pellet is re-extracted with 1 liter of 0.4 N acid. The second extract is pooled with the supernatant from the first extraction and dialyzed against distilled water. The dialyzed extract is lyophilyzed to total dryness to afford 0.5 g of a histone mixture per liter of blood. The crude histone mixture can fractioned by gel filtration chromatography (see below).

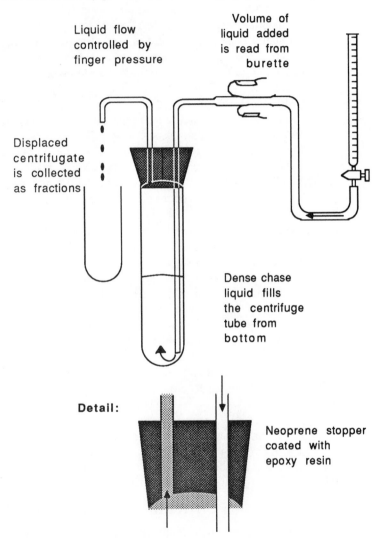

Figure 24-1 The use of a high density chase liquid

Hydroxylapatite Method

This method requires a much more expensive column chromatographic step but it avoids the use of moderately concentrated strong acids employed in the preceding method. The histones are also fractionated more completely, making the gel-filtration step unnecessary for some purposes. The nuclear pellet is largely chromatin and can be used directly with this histone fractionation procedure. The pellet is resuspended in 1 mM TRIS-HCl, pH 8, with 1 mM Na_2-EDTA. The chromatin is sheared for 90 seconds at maximum homogenizer speed and then centrifuged at 16,000 x g for 30 minutes to remove any insoluble material. The chromatin content of the supernatant was determined by diluting 100 μL of the supernatant with 1 mL of 2 M NaCl/5 M urea and recording its UV-absorbance spectrum from 220 to 300 nm. The DNA concentration can be calculated by measuring the absorbance at 260 nm; the absorbance for 1 mg/mL DNA is 20 O.D units.

Sheared chromatin at approximately 1 mg/mL is dialyzed overnight at 4° against 0.63 M KCl, 0.1 M potassium phosphate (pH 6.7) and 0.1 mM dithiothreitol. The nucleoprotein may precipitate and redissolve during the dialysis. The dialyzed chromatin is then centrifuged at 16,000 x g for 30 minutes to remove any insoluble material, and applied to a hydroxylapatite column. For large-scale preparations, a volume of clear chromatin solution containing approximately 10 mg of DNA in about 10 mL 0.63 M KCl buffer is added to a hydroxylapatite column (1.0 x 20 cm, Bio-Gel HTP), which is first washed with 20 mL of the 0.63 M KCl buffer, then washed with 150 mL of 0.93 M KCl, 0.1 M potassium phosphate and 0.1mM dithiothreitol. Next a 150-mL linear gradient from 0.93 to 1.20 M KCl is applied, followed by an additional 75 mL of the 1.20 M KCl buffer. A gradient mixer with a large volume of 1.20 M KCl in the nonstirred reservoir is used. The gradient mixer is removed and a final 100 mL of 2 M KCl, 0.1 M potassium phosphate and 0.1 mM dithiothreitol is used to elute the H3 + H4 proteins. Fractions (10 mL) are collected and their UV absorbance measured at 230 and 260 nm. The histones absorb at 230 but not at 260. During the initial wash, H1 and the nonhistone proteins (HMG 14 and 17) are eluted. At higher salt levels H5 followed by H2A + H2B come off the column. There may be considerable overlap between H5 and H2A + H2B. Fractions containing significant quantities of the desired protein are concentrated by Centriprep-10 concentration (Chapter 11) and characterized by acetic acid-urea-PAG electrophoresis (see below).

The hydroxylapatite column is easily prepared but the following considerations must be kept in mind. Bio-Gel HTP powder is weighed out and diluted with 6 parts 0.1 M potassium phosphate buffer with gentle swirling. This mixture should not be stirred vigorously, as this will destroy the hydroxylapatite crystals. When fully hydrated, Bio-Gel HTP occupies about 2 to 3 mL per gram dry weight. The slurry is allowed to settle for 10 minutes before the fines are decanted. Half the column is filled with the 0.63 M KCl, 0.1 M KH_2PO_4, and 0.1 mM dithiothreitol (pH 6.7) buffer, and the settled hydroxylapatite is poured into a wide-mouth funnel at the top of the column. The column is allowed to settle under gravity, then two bed volumes of buffer are passed through the column. The top of the bed should not be allowed to dry out, as it absorbs CO_2 to form a crust on the top of the bed.

Histone Purification

Bio-Gel P-60 Gel Filtration

The H2A + H2B histone mixture eluted from the hydroxyapatite column in 0.93 to 1.20 M KCl can be separated by gel filtration on Bio-Gel P-60. The histone mixture is dialyzed against 0.02 M HCl/0.05 M NaCl/0.02% NaN_3 overnight. Then the contents of the dialysis bag is brought to 8 M in urea and 1% in 2-mercaptoethanol; this is best done by the laboratory assistant several hours before students begin the application of this sample to a 2 cm x 60 cm P-60 (100-200 mesh) column. The column is eluted with 0.02 M HCl, 0.05 M NaCl, 0.02% NaN_3 (no urea) at a flow rate of 15 to 30 mL per hour. The isolated fractions containing histones can be concentrated by Centriprep concentration using a Centriprep-10 concentrator which retains all molecules larger than 10,000 D.

Other Chromatographic Methods

Gel exclusion chromatography on Bio-Gel P-60 with 0.02 M HCl/0.05 M NaCl separates whole histone into four nearly homogenous fractions H1, H5, H2A, H4, and a fifth fraction containing H2B and H3.[8] This mixture can be separated on sephadex G-100 with 0.05 M sodium acetate (pH 5.1) and 5 mM sodium bisulfite to suppress H3 oxidation. A mixture of H2A and H4 resulting from overlap of these two elution bands can be separated on a P-10 column with 0.01 M HCl (no NaCl).[15] Care must be taken when exposing Bio-Gel media to solutions more acidic than pH 2, as the amide groups are susceptible to hydrolysis at low pH. The gel must be rinsed between runs and stored with neutral buffers containing azide to suppress microbial degradation.

Electrophoretic Characterization of Histones

Acid-Urea Gels

The purity of histone preparations can be assessed by gel electrophoresis on high-density polyacrylamide gels made up with acetic acid and urea. The original formulation of Panyim and Chalkley[9,10] employing 2.5 M urea, 0.9 N acetic acid (pH 2.7), and 15% polyacrylamide is useful for routine examination of histone preparations. No stacking gel is used, but long gels are required to resolve histones H2A, H2B, and H3 or to reveal modifications such as lysine acetylation. However, this method requires a time-consuming pre-electrophoresis step to remove all extraneous charged species. The gels are prepared by combining two parts A, one part B, and five parts C, where:

> solution A contains 60.0 g of acrylamide and 0.4 g of BIS made up to 100 mL with water,
>
> solution B contains 43.2% (v/v) acetic acid and 4% (v/v) TEMED in water,
>
> solution C contains 40 mg of ammonium persulfate added to 20 mL of 4 M urea.

Solutions B and C are combined and degassed before addition to freshly degassed solution A. Polymerization requires 2 hours for optimum resolution. Because these cationic proteins migrate toward the cathode [the(-) pole], a cationic tracking dye such as benzene azo α-naphthylamine or methyl green must be used. The histone banding pattern, from top to bottom, is H1,H3, H2B, H2A, and H4. The gel is best stained with amido black (0.1% dye in 25% MeOH/10% HOAc).

Triton Gels

The addition of the nonionic detergent Triton X-100, which complexes with the more hydrophobic histones permits the resolution of histones differing by only a single amino acid residue or by a single covalent modification.[11,16] The histone banding pattern is different with 2.5 M urea, 0.9 M acetic acid, and 0.4% Triton X-100 (i.e., from top to bottom: H2A, H3, H4, H2B, and H1).[18]

PROJECT EXTENSIONS

1. The complete characterization of histone mixtures is possible with a discontinuous 8 M urea/1 M acetic acid/8 mM Triton X-100 system developed by Bonner.[16, 17] The resolving gel contains 15 % acrylamide and 0.2% BIS buffered with acetic acid and ammonium hydroxide (pH 4.0). The gel is made in 20-mL batches according to the formulation.

7.5 mL of 60% acrylamide

1.2 mL of 2.5% BIS

1.8 mL of glacial acetic acid

0.15 mL of TEMED

0.09 mL of conc. NH_4OH and 14.4 g of urea

All are made up to 27.4 mL with water and combined with 0.6 mL of 25% (w/v) Triton X-100, then degassed. The polymerization is initiated by adding 2.0 mL of 0.004% riboflavin and exposing the poured gel slab to light. The photopolymerization process is different from the method described previously (see chapters 7 and 8).

The stacking gel is prepared in 20-mL batches from:

1.1 mL of 60 % acrylamide

1.3 mL of 2.5% BIS (note that the acrylamide-to-BIS ratio has changed)

1.2 mL of glacial acetic acid

0.1 mL of TEMED

0.06 mL of conc. NH_4OH and 9.6 g of urea

After the above mixture is made up to 18.7 mL with water and degassed, 1.3 mL of riboflavin solution is added. No Triton X-100 is used in the stacking gel. To avoid oxidations of histone H3 Cys side chains, which leads to artifactual heterogeneity, the histone samples must be prepared and stored with 0.25 % (w/v) 2-mercaptoethanol or dithiothreitol.

2.   Bonner has also developed a two-dimensional discontinuous gel system (see also Chapter 29) whereby the foregoing urea-acetic acid-Triton X-100 gels, even after staining, can serve as the first dimension against a 6 M urea, 1 M acetic acid, 0.15 % cetyltrimethylammonium bromide gel.[16] The histones appear in the first dimension in the banding order (from top to bottom) H2A, H2, H3, and H2B and H4. In the second dimension the histones are spread out along a diagonal with modified histones and other proteins appearing off the diagonal.

3. An acetic acid/urea/polyacrylamide gel with a transverse Triton X-100 gradient is able to differentiate between core histones and other protein species when crude histone preparations are analyzed.[18]

4. Chicken erythrocytes are a good source of eukaryotic DNA. The packed erythrocytes for DNA preparation are lysed by freezing (12 to 24 hours in a freezer) and thawed at 4°C. Several hours are required to thaw a large sample, and project planning must allow time for this. The broken cells are stirred in a threefold volume equivalent of saline EDTA/citrate (as above but with no PMSF) and centrifuged for 15 minutes at 4000 x g. The supernatant is discarded. This operation is repeated several times until the residue is free from hemoglobin. The washed nuclei yield a white to pink-colored paste and can be stored as such overnight or used immediately. The nuclei paste is suspended in one volume of citrate buffer solution and disrupted by <u>brief</u> homogenization (with a Teflon tissue homogenizer or a Brinkman polytron) then centrifuged for 1 hour at 1800 x g. The supernatant is discarded. (This step may repeated a second time.) The crude nucleoprotein pellet is suspended in a threefold volume equivalent of cold 2 M NaCl and the mixture allowed to stand at 4°C for 24

hours. Insoluble material that has precipitated overnight is removed by centrifugation for 30 minutes at 1900 x g. Two volumes of absolute ethanol are slowly added to the supernatant at the rate of 150 mL per hour. The flask must be stirred gently but continuously. The resulting DNA fibers can be spooled onto a glass rod, lifted from the supernatant, and washed with 75% (v/v) ethanol. The washed fibers are then dissolved in 25 mL of 0.14 M NaCl, 0.15 M sodium citrate buffer (pH 7.1). This is combined with 8 mL of 5% sodium dodecylsulfate (SDS) in 45% (v/v) ethanol. The solution is stirred for 1 hour then left overnight at 4°C. Next day the solution is clarified by centrifugation at 21,000 x g for 1 hour and two volumes of absolute alcohol are again added slowly to the supernatant. The fiber is lifted and washed.

The purified DNA can be dissolved in the 0.14 M NaCl/0.015 M sodium citrate and an absorbance spectrum carried out in this solvent. The optimum absorbance is around 0.5 at 260 nm and the DNA concentration should be adjusted to this level. (The reader is directed to Chapter 19 for more details of the spectroscopic characterization of DNA.)

DISCUSSION QUESTIONS

1. What is your explanation of the markedly different migration order for histones caused by Triton when added to acid-urea gels? The migration order (fastest to slowest) in 2.5 M urea/0.9 m acetic acid is H4, H2A, H2B, H3 and H1 but in 2.5 M urea-0.9 M acetic acid with 0.4% Triton the order is H1, H2B, H4, H3, H2A.

2. What is the function of sodium bisulfite or dithiothreitol in H3 purifications? What would one see if such a reducing agent were left out?

REFERENCES

1. J. D. McGhee and G. Felsenfeld, *Ann. Rev. Biochem.*, **1980**, *49*, 1115.

2. I. Isenberg, *Ann. Rev. Biochem.*, **1979**, *48*, 159.

3. T. J. Richmond, J. T. Finch, and A. Klug, *Cold Spring Harbor Symp. Quant. Biol.*, **1982**, *47*, 493.

4. J. T. Finch, M. Noll and R. D. Kornberg, *Proc. Nat. Acad. Sci. (USA)*, **1975**, *72*, 3320.

5. S. Zamenhof, *Methods Enzymology*, **1957**, *12*, 700.

6. N. A. Nicoli, A. Fulmer, A. M. Schwartz and G. D. Fasman, *Biochemistry*, **1978**, *17*, 1779; R. H. Simon and G. Felsenfeld, *Nucleic Acids Res.*, **1979**, *6*, 689.

7. K. Murray, G. Vidali and J. M. Neelin, *Biochem. J.*, **1968**, *107*, 207.

8. D. R. Westhuyzen, E. L. Bohm and C. Von Holt, *Biochem. Biophys. Acta*, **1974**, *359*, 34.

9 A. Panyim and R. Chalkley, *Arch. Biochem. Biophys.*, **1969**, *130*, 337.

10. N. V. Beaudette, A. W. Fulmer, H. Okabayashi and G. D. Fasman, *Biochemistry*, **1981**, *20*, 6526.

11. A. Zweidler in *Methods in Cell Biology*, ed. G. Stein, New York: Academic Press, 1978, p. 223.

12. R. Hardison and R. Chalkley, *Methods in Cell Biology*, ed. G. Stein, New York: Academic Press, 1978, p. 235.

13. M. A. Keene and S. C. R. Elgin, *Cell*, **1981**, *27*, 57; I. L. Cartwright and S. C. R. Elgin, *Nucleic Acids Res.*, **1982**, *10*, 5835.

14. R. W. Allington, *Anal. Biochem.*, **1976**, *73*, 78.

15. P. N. Lewis, E. M. Bradbury, C. Crane-Robinson, *Biochemistry*, **1975**, *14*, 3391.

16. W. M. Bonner, M. H. P. West and J. Stedtman, *Eur. J. Biochem.*, **1980**, *109*, 17.

17. J. H. Waterborg and H. R. Matthews, *Biochemistry*, **1983**, *22*, 1489.

18. J. H. Waterborg, I. Winicov and R. E. Harrington, *Archiv. Biochem. Biophys.*, **1987**, *256*, 167.

Chapter 25

DNA Thermal Denaturation

The temperature and ionic strength dependence of the dsDNA-to-ssDNA transition is determined by monitoring the absorbance at 260 nm of a dilute DNA solution as the temperature is slowly stepped up from 30°C to 95°C. The melting temperature T_m is taken as the temperature at which 50% of the maximum absorbance increase is observed. The absorbance vs. temperature data are used to prepare a first derivative plot of dA/dT vs. T. As an optional class project the melting temperature is determined as a function of $NaClO_4$ concentration over the range 0.01 to 4 M. The reassociation of rapidly cooled ssDNA can also be followed at 260 nm.

KEY TERMS

Differential thermal-denaturation profiles
DNA "melting" temperature T_m   *Hypochromism*
Reassociation kinetics   *Zipper mechanism*

BACKGROUND

The strong UV absorbance of all nucleic acids arises almost exclusively from the purine and pyrimidine bases, which have a λ_{max} of around 260 nm. However, this UV absorbance is strongly dependent on how the bases interact through space. Base stacking interactions markedly reduce this absorption band. This phenomenon is known as *hypochromism*. Determining temperature and salt effects on hypochromism exhibited by a DNA solution is one of the oldest known physical chemical approaches to studying nucleic acid structures in solution. Although the quantum mechanical explanation is clearly beyond the scope of this course, in simple terms it is due to the through space interaction between an electronic excited state of a given chromophore in one base with different electronic states of chromophores on neighboring bases.[1] For example helical double-stranded DNA with stacked base pairs absorbs light at 260 nm but the absorbance per mole of base in such a DNA molecule is only 60% that of observed with the same concentration of free bases.

In double-stranded DNA (dsDNA) the bases are stacked one above the other; this interaction very effectively suppresses the absorption of UV light. In single-stranded DNA (ssDNA) the bases are oriented more randomly and the interaction is less important (see Figure 25-1). By monitoring the absorbance as a function of temperature, one can see the relative abundance of dsDNA and ssDNA. At a temperature sufficient to overcome the hydrogen-bonding and hydrophobic interactions that stabilize helical dsDNA, this structure breaks down and ssDNA dominates. For complete conversion to ssDNA the increase in absorbance is about 37% and careful monitoring of the absorbance of a DNA solution at 260 nm as a function of temperature can be used to study the dsDNA to DNA transition.[2]

Correctly speaking this is not a shift in equilibrium because, over the time span of this experiment, the dsDNA-to-ssDNA transition is not necessarily reversible; consequently, reproducible data can be obtained only if the temperature is stepped up gradually and the temperature is held constant for about 10 minutes at each temperature.

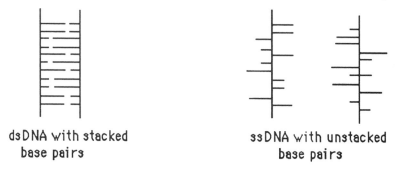

dsDNA with stacked base pairs   ssDNA with unstacked base pairs

Figure 25-1 Base stacking interactions in dsDNA and ssDNA

Figure 25-2 illustrates the absorbance vs. temperature plots for two succesive melting experiments with a DNA sample. Notice that the two plots are not identical because some ssDNA regions do not reassociate quickly. The Absorbance vs. Temperature data can be analyzed in several ways. The traditional plot of A vs. T gives an S-shaped curve, as shown in Figure 25-2. The *DNA "melting" temperature T_m* value is taken as the temperature corresponding to one-half the increase in relative absorbance upon going from the low-temperature dsDNA to the high-temperature ssDNA . If careful measurements are made, T_m is a reproducible parameter characteristic of the DNA composition. Because G-C pairs are more tightly H-bonded than A-T pairs, higher temperatures are required to dissociate the chains of GC-rich DNA and T_m is linearly related to the average DNA base composition.[2]

A higher G-C content confers a higher thermal stability. Table 25-1 gives the melting point for DNA from various sources and the mole % G-C in such DNA. The melting temperature is related to the latter by empirical equations[2] such as the following (for 0.2 M Na$^+$):

$$T_m = 69.3 + 0.41(\%G\text{-}C) \qquad\qquad (25\text{-}1)$$

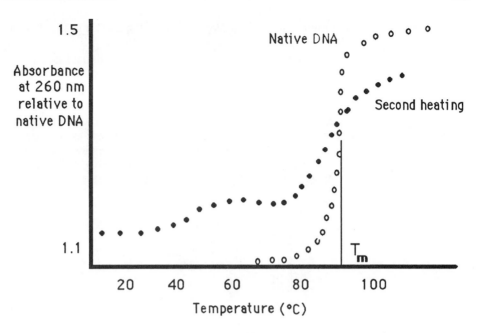

Figure 25-2 Absorbance vs. Temperature plots for two successive melting experiments

Table   25-1
T_m AND MOLE % G-C FOR DNA FROM VARIOUS SOURCES[a]

Source	T_m	mole % G-C
E. Coli (K12)	90.5	50.1
Chicken	87.5	41.7
Calf	87	41.9
Human	86.5	41.4

[a] Source: Ref. 2

The dependence of melting temperature T_m on base composition predicted by equation (25-1) is a good approximation but a more detailed analysis of the melting of heterogeneous DNA shows that the dsDNA→ssDNA transition is more complex than that predicted based solely on overall base composition. Short pieces of DNA melt from the ends first but longer pieces also begin to melt at internal sequences. AT-rich regions melt at lower temperatures than GC-rich regions, to give rise to localized single-strand regions.[3,4] Once AT-rich ssDNA "joints" form, they serve as the starting point for further melting at higher temperatures (i.e., the single strand regions enlarge as additional base-pairs are opened with the twist taken up in the remaining helical regions). The melting process depends on the base sequence, especially the relative frequency of AT-rich strings of base pairs. *Differential DNA thermal-denaturation profiles,* in which the differential change in absorbance, dA/dT, is plotted versus temperature, permiting a more detailed analysis of the melting process. Plots of dA/dT vs T can be obtained with sufficient resolution to resolve the melting process into a number of steps, each corresponding to the dissociation of individual dsDNA domains that melt at distinct temperatures.

Other factors, including salts also contribute to the fine structure of DNA thermal denaturation profiles. Added salts influence the dsDNA-to-ssDNA transition. At low salt concentrations (0.01 to 1 M) added cations shield the phosphate backbones and minimize the interchain repulsions that destabilize the dsDNA molecule. This results in an increase in T_m with increasing ionic strength.[5] At very high salt concentrations, the added anions break the inter-base hydrogen bonds so that dsDNA is destabilized and T_m decreases with increasing electrolyte concentration over the range 1 to 4 M. The decrease depends on the the anion and is particularly pronounced with perchlorate salts. The salt dependence of T_m is illustrated in Figure 25-3.

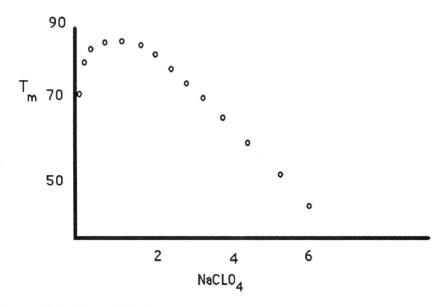

Figure 25-3 Effect of $NaClO_4$ on the melting temperature of DNA

The relative abundance of dsDNA and ssDNA depends on the rates for both the chain dissociation and chain association processes. Factors that favor dissociation and factors that disfavor association will both lower the T_m value. The reassociation of ssDNA can be followed by monitoring the decrease in absorbance as hypochromism sets in. In order for ssDNA molecules to reassociate to form dsDNA, the ssDNA molecules must diffuse together in an orientation that permits hydrogen bonding between complimentary base pairs; that is, the rate should be a second-order process depending on the concentrations of complementary ssDNA molecules. Helix formation follows a cooperative *zipper mechanism* that requires a three-base-pair nucleus.[6,7] The first step in the association is the formation of a single base pair which is rather unstable but the addition of two more stacked next-neighbor base pairs serves as a nucleus from which further base pairs in the helix are formed just like a zipper closing.

In the zipper model, the formation of the first base pair is rate determining since this process depends on the probability that complementary regions meet each other as the result of random collisions between ssDNA molecules. This probability is proportional to the product of the molar concentrations of the two complementary chains. If concentrations are expressed on a weight per unit volume basis, then chain reassociation is more probable with shorter chains since there will be more copies of a shorter sequence per unit weight.   Short viral ssDNA molecules reassociate more rapidly than equal concentrations (on a mg/mL basis) of longer bacterial DNA molecules, which, in turn, reassociate faster than very long eukaryotic ssDNA pieces. Length also has another effect. The formation of fully base-paired dsDNA from ssDNA with a number of closely related but nonidentical sequences requires a more complicated process. As the ssDNA chains begin to reassociate, those regions with maximum base-pairing opportunities form stable dsDNA regions most rapidly. The probability of a correct match is less if the chains are quite long since weaker partial or incorrect matches become more probable as the chain length increases. Reassociation of ssDNA to dsDNA is most rapid for sections containing repeated sequences, as more copies implies a higher probability of complementary sections finding each other; that is, sections of eukaryotic ssDNA with multiple copies of a particular sequence are more likely to re-form dsDNA with complementary hydrogen-bond-pairing bases and reassociate faster than single-copy sections. Consequently, a long eukaryotic DNA piece will not reassociate in one smooth process; some regions with multiple copies of sequences will become double-stranded quickly while other single copy regions will anneal much more slowly. This accounts for the observation pictured in Figure 25-2; after one melting experiment some complementary ssDNA sequences may not find each other for a long time, so that the absorbance never drops back down to the original dsDNA level within a convenient waiting period. The specific recombination of single DNA strands to form the native helical structure will not be complete even after hours; the optical density will return to only about 75% of the way from the denatured to native density. Moreover when DNA is thermally denatured, allowed to cool and partly restore the native helical structure, then reheated, the second thermal transition does not resemble the first. Imperfect helical regions melt out at much lower temperatures (see Figure 25-2).

EXPERIMENTAL SECTION

Students can work in groups of three or four, depending on the availability of UV spectrometers. Every group can determine a melting curve for each of several salt concentrations; alternatively, each group can select one unique salt concentration so that the whole salt dependences emerges from a class effort. The cuvette must also be tightly stoppered to minimize evaporation. The stopper must be removed from a hot cuvette before it is allowed to cool. Every group should monitor the decrease in absorbance as a function of time for a rapidly cooled cuvette of DNA at least once. For eukaryotic DNA such as chicken erythrocyte (see Chapter 24) or calf thymus DNA (Sigma), the cooling A vs. T curve does not overlay the melting curve.

The simple method of Marmur and Doty[2] is recommended. Bacterial DNA (chapter 10), chicken erythrocyte DNA (Chapter 24) or commercially available calf thymus DNA (Sigma) can be used, although the latter should be deproteinated by a phenol extraction or a chloroform-isoamyl alcohol extraction (see Chapters 19 and 23). The DNA is dissolved in dilute $NaClO_4$ containing 0.015 M trisodium citrate (pH 7.0) to give a final concentration of 20 μg/mL. The sample in a 1-cm quartz cuvette is allowed to equilibrate to the temperature of the spectrometer cell compartment, which is controlled by a constant temperature circulating water bath. *Caution*: All plumbing connections will be hot. The absorbance should be measured at 25°C then the temperature raised quickly to about 5° below the estimated onset of the melting region. The temperature is then raised about one degree at a time, with 10 minutes allowed between the temperature adjustment and the absorbance measurement; this time is required for thermal equilibration between cuvette and compartment. The temperature of the DNA buffer solution within the cuvette must be measured with a thermalcouple or a small but accurate thermometer since the temperature of the water bath will be higher than the cuvette temperature. The temperature drop between the bath and cuvette can be minimized by heavy insulation of the water hoses and cell compartment. The maximum absorbance observed at high temperatures must be measured, with care taken to ensure that prolonged heating near 100 °C does not damage the spectrometer optics and that the water circulation hoses remain stable and firmly attached when hot. This experiment should not be attempted on spectrometers without shielded cell compartments. The spectrometer must never be left unattended during this experiment and all hose connections should be secured with hose clamps.

CAUTION: WEAR GOGGLES WHEN WORKING WITH SEALED CUVETTES AT HIGH TEMPERATURES!

The absorbance reading at 260 nm can be corrected for thermal expansion of the solution (see Table 25-2).[8]

Table 25-2
DENSITY OF WATER AS A FUNCTION OF TEMPERATURE[a]

Temperature	Density
25°	0.9970
50°	0.9880
60°	0.9832
70°	0.9778
80°	0.9718
90°	0.9653
95°	0.9619
100°	0.9584

[a] Source: Ref. 8

DISCUSSION QUESTIONS

1. Explain why dsDNA is unstable in distilled water and denatures rapidly to ssDNA.

2. If a compound such as spermine binds to dsDNA with greater affinity than to ssDNA, small amounts of this compound not only increase the T_m of a heterogeneous DNA sample but also increases the difference in melting temperatures of various sequence domains.[9] Explain.

REFERENCES

1. C. A. Bush in *Basic Principles in Nucleic Acid Chemistry*, ed. P. O. P. Ts'o, New York: Academic Press, 1974, p.92; C. R. Cantor and P. R. Schimmel, *Biophysical Chemistry*, Vol. II, San Francisco: W. H. Freeman, 1980, pp. 399-405.

2. J. Marmur and P. Doty, *J. Molec. Biol.*, **1962**, *5*, 109.

3. H. C. Spatz and D. M. Crothers, *J. Molec. Biol.*, **1969**, *42*, 191.

4 W. Saenger, *Principles of Nucleic Acid Structure*, New York: Springer-Verlag, 1985, p. 145.

5. R. J. Britten and D. E. Kohne, *Science*, **1962**, *161*, 529.

6. D. Porschke, *Mol. Biol. Biochem. Biophys.*, **1977**, *24*, 191.

7. D. Porschke, *Biopolymers*, **1971**, *10*, 1989.

8. *CRC Handbook of Chemistry and Physics*, 57th ed., Boca Raton, Fla: CRC Press, 1976, p. F-2.

9. J. E. Morgan, J. W. Blankenship and H. R. Matthews, *Arch. Biochem. Biophys.*, **1986**, *246*, 225.

Chapter 26

The Alu I Family of Repetitive DNA Sequences

Human placental DNA is freed from associated proteins by phenol and chloroform/*iso*-amyl alcohol (24:1) extractions and ethanol precipitation, then sheared to approximately 2000 nucleotides in length by brief homogenization or repeated passage through a fine-gauge syringe needle. After dialysis in 0.01 M PIPES (pH 6.8) buffer, the DNA is denatured by boiling for 5 minutes, then renatured in 0.3 M NaCl at 60°C. After renaturation the ssDNA portions are cleaved away with S1 nucleases. The liberated blunt-end dsDNA is bound to hydoxylapatite in 0.12 M sodium phosphate. Elution with 0.3 M sodium phosphate frees the DNA, which can be concentrated with an Elutip-d (Schleicher & Schuell) system. Dialysis in 0.01 M TRIS HCl (pH 7.4) and electrophoresis on 1.4% neutral agarose with ethidium bromide staining gives a gel with a number of prominant bands standing out from a background smear of heterogeneous length DNA. The oligonucleotide length corresponding to the bands was determined with a Hae III digest of ϕX174 DNA in a control lane. The most prominent band corresponding to approximately 300 nucleotides is electroeluted from the gel, concentrated by the Elutip-d method and divided into two portions. The first portion is cleaved with Alu I in 0.01 M TRIS with 0.06 M $MgCl_2$ and 0.06 M 2-mercaptoethanol in a siliconized tube. After incubation overnight the restriction enzyme cleavage fraction is run on 2.5 % agarose in alternating lanes with the uncleaved 300 nucleotide fraction. The gel is vusualized with ethidium bromide. In an alternative procedure the 5'-ends of the 300 nucleotide fraction (after alkaline phosphatase treatment) are labeled with ^{32}P using T4 polynucleotide kinase. This can be used as a probe for Alu-sequence-containing human DNA. Human placental DNA is cleaved with a series of restriction enzymes and each restriction digest product mixture is run as a lane on an agarose electrophoresis gel. The gel pattern is then blotted to nitrocellulose or DBN paper and after hybrization with ^{32}P-labeled Alu sequences, the labeled DNA is observed by autoradiography. Radiolabeled Hae III digest fragments are used as molecular weight markers.

KEY TERMS

Autoradiography
DNA   strand   reassociation
Elutip-d   purifications
LINES
Repeated   base   sequences
Satellite   DNA
Southern   blotting

Cot   curves
Electroelution
Labeled   DNA   probe
Reassociation   kinetics
Repetitive   DNA
SINES

BACKGROUND

Chapter 24 focused on the presence of special nucleoprotein structures in eukaryotic chromatin, in contrast to prokaryotic chromosomes. Another major difference between eukaryotic and prokaryotic genetic material is the frequency and distribution of repetitive sequences in eukaryotic DNA. Only about half of the DNA in human cells consists of unique or nearly unique sequences coding for enzymes. Each enzyme sequence is encoded by only one or no more than a few gene copies per haploid genome. All the rest of the DNA is made up of various types of *repetitive DNA*. This includes multiple copies of genes coding for structural components of the cell, such as histone proteins. *Satellite DNA* is another type of repetitive DNA, which appears as clusters of millions of copies of short sequences organized as tandem arrays near the centromeres of some chromosomes, especially the Y chromosome. The function of satellite DNA is not known. There are also repetitive sequences which are not clustered together but scattered as individual copies throughout the genome. These dispersed repeats fall into two classes: short interspersed nuclear elements (*SINES*) and long interspersed nuclear elements (*LINES*).[1,2] The SINES are represented by the *Alu* family of sequences which are found only in human cells, where they are repeated 300,000 times in the human haploid genome. Not infrequently Alu sequences are present close to one another but with opposite orientations so that such DNA can renature almost instantly after brief heating above the helix-to-coil transition temperature. The procedure described below can select these interspersed sequences from among the many sequences in the human genome because the multiple "foldback" copies of this sequence leads to an unusually rapid ssDNA→dsDNA reassociation rate and that dsDNA, formed immediately upon cooling a thermally denatured DNA sample, can be bound selectively to hydroxyapatite.[3,4] Alu sequences about 300 bp long can be isolated and characterized by their single site for the restriction enzyme Alu I about 130 bp from one end--hence the name for the sequences.

Because the Alu sequence is specific to humans it can be used as a hybridization probe for DNA of human origin. The purpose of this experiment is to isolate from human placental DNA Alu sequences and then label them with ^{32}P so that they can then be used as a probe for human genomic material in a blotting experiment.

Also important is the demonstration of a number of concepts that have played an important role in the development of molecular biology. These include the importance of sequence frequency on the reversibility of DNA thermal denaturation, Cot curves, S1 preferential cleavage of ssDNA, hydroxyapatite retention of dsDNA, electroelution, Elutip-d purifications, hybridization, autoradiography, blotting methods, and the use of labeled DNA probes in Southern blotting of DNA.

DNA Reassociation Kinetics

The rate of reassociation of thermally denatured human DNA can be followed by observing the loss of absorption at 260 nm due to the hypochromicity that accompanies the conversion of ssDNA to dsDNA, as described in Chapter 25. Alternatively, the fact that dsDNA binds more tightly to hydroxyapatite than ssDNA can be employed to separate dsDNA from ssDNA and to determine the amount of renaturation (ssDNA→dsDNA) occurring before a briefly heated DNA sample is applied to the column. Either experimental approach demonstrates the same principle. The kinetics of reassociation reveal that a portion of human DNA renatures much faster than the rest because this portion contains many repeated sequences.

The reassociation kinetics are second-order, depending on the concentration of the complementary single-stranded molecules. The rate law is described by the differential equation:

$$\frac{dC}{dt} = -k \, C^2 \qquad (26\text{-}1)$$

where C is the ssDNA concentration at time t and k is the reassociation rate constant. Integrating this equation between the limits C_0, at initial time ($t = 0$), and C, at some later time $t > 0$, gives an expression for the concentration of ssDNA remaining at time t:

$$\frac{C}{C_o} = \frac{1}{1 + k \cdot C_o t} \qquad (26\text{-}2)$$

Equation (26-2) predicts a characteristic "reverse S-shaped" curve describing the decreasing percentage of the total DNA remaining as ssDNA with time. Reassociation of DNA follows the form of a "*Cot curve*," introduced by Britten and Kohne.[5] A comparison of the Cot curves for *E. coli* DNA and calf thymus DNA [5] is shown in Figure 26-1, where C/C_0 (as a percentage) is plotted as a function of the product of the DNA concentration and the incubation time ($C_0 t$), the parameter most often used to describe the reassociation. The $C_0 t$ value is defined as the initial concentration (C_0) in moles nucleotide per liter multiplied by the time (t) in seconds. One-half the thermally denatured DNA has reformed dsDNA at one reassociation half-life $t_{1/2}$. If all portions of the DNA reassociate at the same rate, all are characterized by the same half-life and the Cot curve has a sharp inflection at $C_0 t_{1/2}$. On the other hand, if the DNA contains regions that reassociate faster than other regions, the Cot curve will not have a single inflection point. **Portions of eukaryotic DNA containing repetitive sequences will reassociate faster than equivalent concentrations of shorter nonrepetitive prokaryotic DNA.** In Figure 26-1 prokaryotic *E. coli* DNA reassociates along a single "Cot curve" defined by a single k value ($k = 1/C_0 t_{1/2}$) whereas the renaturation of eukaryotic calf thymus DNA exhibits a more complex behavior. There is a fast component making up about 20% of the DNA, which renatures very rapidly, $C_0 t \leq 10^{-2}$ moles nucleotide x sec/liter, an intermediate component that renatures over the range $10^2 < C_0 t < 10$ and the remaining 60% of the DNA is a slow component, $C_0 t$ ranging from 10^2 to 10^4 mol nucleotide x sec/liter.

The slow component has a $C_0t_{1/2}$ 4000-times greater than that for the bacterium, indicating the effect of DNA length. It is the concentration of each reassociating sequence that determines its rate of reassociation.

The rapidly and moderately rapidly associating components contain SINE sequences, including the Alu family. Most single-copy sequences are separated by repetitive elements that are typically 300 bp long. There are a number of families of repetitive sequences characterized by their unique cleavage patterns with restriction enzymes.

Alu Sequences

The Alu family of 300-nucleotide sequences makes up about 5% of the genome and occurs with a repetitive frequency of at least 300,000 appearances in a haploid genome. There is one member of this family for every 6 kilobases of DNA. These 300-nucleotide repeated sequences occur interspersed between, but not within, longer (approximately 1000 nucleotides) single-copy genes. Hence the sequences of the Alu family are said to be *interspersed moderately repetitive sequences*. The individual members of the family are not identical in sequence, but they have an average homology of 87%.

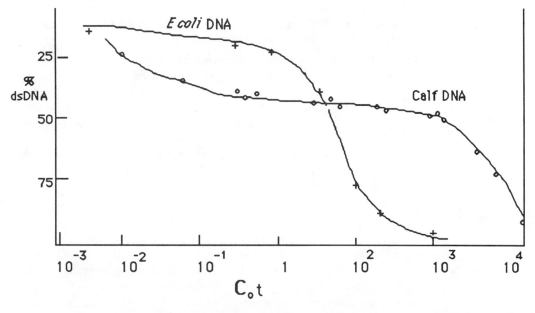

Figure 26-1 Reassociation kinetics: "Cot curves" for prokaryotic and eukaryotic DNA

Each Alu sequence is a dimeric structure composed of two related, tandemly arranged monomeric units, one about 130 bp long and the second 170 bp. Each copy is flanked by short direct repeats, and thus they resemble transposons, but the repeat lengths differ for individual members of the family. Related sequence families are present in the mouse (B1 family), Chinese hamster (Alu-equivalent family) and other mammalian genomes. Although they comprise a substantial portion of the total genome, the function of these interspersed moderately repetitive sequences of DNA is unknown. Their short length and high degree of repetition tempts one to assign to them a structural or regulatory role and it has been suggested that they may serve as binding sites for specific regulatory proteins or some structural function.[4] On the other hand, they may not have a function; "useless" DNA, resulting from some ancient transposition event, will survive as long as it does no harm to the host.[6-8] The Alu sequence is transcribed and it is believed to have become dispersed throughout the genome by a process involving reverse transcription and reintegration. Alu sequences have been termed retroposons.[9] They have been transcribed *in vitro* with RNA polymerase III.

Under conditions of stringent hybridization, Alu sequences can be used to establish the presence of human DNA.[10-13] Almost every human gene has an Alu sequence nearby, so that this sequence is a natural marker for human genomic material. Hence the Alu sequences provide a natural tag permitting screening of plasmid or λ-phage gene libraries with ^{32}P-labeled Alu probes. The probe will hybridize only with human gene fragments that contain Alu sequences. The Alu sequence has been incorporated into a plasmid.

Specialized Laboratory Methods

The following is a brief background account of the specialized techniques employed in this project. The specific protocols are described at length in the experimental section but the following background is important for a complete understanding of this project and its possible extensions. The reader is advised to review also a number of preceding projects, especially Chapter 19, for descriptions of agarose electrophoresis, DNA denaturation and S1 and restriction enzyme digests.

Purification of Commercial Human Placental DNA
It is necessary to remove the proteins associated with commercially available DNA (Sigma D 7011), which is a fibrous preparation containing less than 3% protein. Proteins are removed by extraction with freshly water saturated phenol then with phenol/chloroform/*iso*-amyl alcohol (24:1, v/v) and finally with chloroform/*iso*-amyl alcohol (24:1, v/v) alone. The phenol must be equilibrated with aqueous buffer containing 0.1% hydroquinone and 0.2% β-mercaptoethanol. The extractions can be carried out on a small scale in a polypropylene Eppendorf microfuge tube and centrifugation can speed up the phase separations. The DNA can be recovered by ethanol precipitation as described in Chapters 21 or 10.

Recovering DNA from Agarose Gels
A frequent problem encountered in molecular biology is the isolation and subsequent concentration of a DNA fragment that can be identified only after electrophoresis. Electroelution followed by reversible binding to a small ion-exchange column is one of the most common ways of getting DNA back from an agarose gel. The electrophoresis is run on 1.4 % agarose, as described in Chapter 10, with the exception that the desired band is removed by the trick described in detail in the experimental section. The quality of the agarose is reported to be critical for the recovery of usable electroeluted DNA.[14] It is also important to extract the desired DNA fragment immediately after electroelution. Siliconized pipettes and tubes should be used.

Blotting Methods
DNA restriction fragments, which have been fractionated according to molecular weight in agarose gels, can be transferred to an immobilizing membrane or chemically activated paper in such a way as to retain the original pattern. Buffer is allowed to perculate up through the gel and through the blotting material and any DNA within the gel is carried vertically through the gel to become attached to the blotting surface. The

immobilizing matrix can be nitrocellulose, charge-modified nylon membranes (eg., Bio-Rad Zeta-Probe), or DBM, DPT or DEAE sheets. This blotting method was developed by E. M. Southern[15] who used this method to transfer DNA from agarose gels to thin nitrocellulose membranes. The resulting pattern is called a "Southern blot." (An analogous procedure for transfering RNA is humorously referred to as "Northern blots.") Transfer to the surface of a membrane makes it possible to perform analyses that are difficult, if not impossible, within an agarose gel. This includes hybridization reactions, where the DNA must be denatured by brief heating or pH changes, followed by reaction with a labeled ssDNA or RNA probe. Hybridization to ^{32}P-labeled DNA or RNA followed by autoradiography is used to locate any bands complementary to the probe. In an extension of the Alu-sequence isolation project, the Southern blotting method can be used to demonstrate the use of nitrocellulose (or DBM paper) and hybridization with radiolabeled DNA. Nitrocellulose is the traditional matrix used in blotting, but high blotting efficiency requires high levels of salt and rather lengthy probes. As an alternative to nitrocellulose, chemically modified ABM paper containing covalently attached aminobenzyloxymethyl groups which can be activated with nitrous acid to form diazobenzyloxymethyl paper (DBM paper) can be used. The nucleic acid bases become covalently bonded to the paper through diazo linkages to the benzyloxymethyl groups. The diazo linkage is a covalent attachment that is not readily reversible. The reader is referred to Chapter 19 or Ref. 16 for a discussion of the chemistry involved with activating DBM paper. Because the DNA fragments are covalently bound to the DBM paper, small fragments are retained efficiently. This is not the case with nitrocellulose which is best used only with large fragments or glyoxylated DNA. A more recent innovation, Zeta-Probe, is a positively charged nylon membrane that is reported to be an ideal blotting surface for the analysis of nucleic acids, especially in low salt and for short oligonucleotides (<500 bp), conditions where nitrocellulose is not efficient.[17]

Autoradiography is very sensitive and gives a high-resolution nondestructive way of locating radiolabeled nucleic acids in polyacrylamide gels or on paper or membrane supports. The strongly β-emitting isotope ^{32}P gives very good autoradiographs on X-ray film from the procedure described in the experimental procedure section.

DNA Modifying Enzymes

A number of DNA-modifying enzymes are introduced in this experiment:

T4 polynucleotide kinase can be used for rapid 5'-end labeling of DNA. The enzyme catalyzes the transfer of ^{32}P from γ-labeled ATP to the 5' hydroxy terminus of polynucleotides.[18] Maximum activity is obtained at 10 mM Mg^{2+} and 5 mM dithiothreitol with 0.1 mM spermidine which stimulates incorporation of ^{32}P while inhibiting nuclease impurities. The kinase is supplied (New England Bio-Labs) as100 units in 100 μL of 50 mM KCl, 10 mM TRIS HCl (pH 7.4), 0.1 mM EDTA, 0.1 μM ATP with 1.0 mM dithiothreitol, 200 μg/mL bovine serum albumin and 50% glycerol. It should be stored at -20°C and cared for with the same sense of respect as restriction enzymes (see Chapter 23). Ammonium ions are strong inhibitors of polynucleotide kinase.

S1 nuclease is used to cleave denatured single-stranded regions that are attached to the renatured double-stranded DNA. It is necessary to trim back all ssDNA tails because all molecules containing duplex regions will be retained by the hydroxyapatite column even though only part of their length is dsDNA. The use of S1 nuclease is described in Chapter 23. S1 nuclease leaves behind 5'-phosphoryl groups which must be removed with *E. coli* **alkaline phosphatase** if the double stranded DNA is to be labeled with ^{32}P for autoradiography. Ideally the amount of alkaline phosphatase required should be titrated by the method of Maxam and Gilbert[19] but a shortened version employing phosphatase followed by kinase will be used in this experiment.

Alu I is a restriction enzyme isolated from *Arthrobacter luteus* which gives blunt-ended cuts in the sequence:[20]

$$5'\text{........A G } \wedge \text{ C T........}3'$$
$$3'\text{........T C } \wedge \text{ G A........}5'$$

Alu I stock solution (Bio-Labs) contains 3 units/μL and is stabilized with the same ingredients (KCl, TRIS, BSA, etc.) as T4 polynucleotide kinase (see above).

EXPERIMENTAL SECTION

This is a lengthy project with several optional elements. Second-semester laboratory practitioners who have already completed the projects on DNA agarose gel electrophoresis, restriction enzymes, and DNA denaturation may find that it nicely caps off a year of laboratory with an emphasis on molecular biological techniques.

First Laboratory Period

DNA Preparation

Commercial human placental DNA (Sigma D 7011) is cleaned up by the phenol and phenol/chloroform/*iso*-amyl alcohol extractions followed by cold ethanol precipitation as described in Chapters 10 or 22. Only a few milligrams (5 mg) of this costly DNA are required for a group of ten students. The cleanup procedure may be done ahead of time. The DNA should be precipitated with ethanol (two-fold volume excess) and stored frozen. One cannot use chicken erythrocyte or lambda phage DNA for this experiment; they do not contain Alu I repeats. Calf thymus DNA does give bands after partial renaturation and S1 cleavage suggesting some repeat elements.[6] The DNA precipitate is placed in a 5-mL dialysis bag and dialyzed into 0.01 M PIPES (pH 6.8) buffer; the dialysis bag is swirled occasionally to ensure that all the DNA goes into solution. At this stage the DNA solution is divided into five 1 mL aliquots and distributed to pairs of students.

The concentration of DNA is determined by measuring the absorbance at 260 nm of a 25-μL aliquot diluted to 2.5 mL. An absorbance of 1.0 corresponds to 50 μg/mL. Then the DNA is sheared by 5 to 10 minutes of vigorous stirring with a sterile polyethylene Pellet Pestle (VWR) in a 1.5 mL microfuge tube or by repeated passage through a fine-gauge needle. Failure to shear the DNA completely is the most common problem that students encounter with this project.

Renaturation and Recovery of dsDNA

The DNA solution is brought to a rapid boil for a few minutes (2 to 5 minutes), then adjusted to 0.3 M with NaCl and allowed to renature at 60°C until $C_0t = 68$ mole x sec/liter. The absorbtion spectrum of a diluted aliquot of the DNA sample should be measured before and after heating to boiling and the absorbance before heating should be used to calculate the DNA concentration (see Chapter 19). A marked increase in absorbance must be observed. After renaturation an equal volume of 0.05 M sodium acetate buffer pH 4.5 is added and the solution is brought to 0.2 mM $ZnCl_2$, and 0.025 M in 2-mercaptoethanol (see chapter 23) at 37°C and cleaved with the requisite number of units of S1 nuclease. The reaction buffer contains 30 mM sodium acetate pH 4.6, 50 mM NaCl, 1 mM $ZnCl_2$, 5 % glycerol and enzyme. One unit of enzyme produces 1 μg of acid soluble material per minute at this temperature. The reaction is incubated at 37°C. The reaction is halted by adjusting the solution to 0.12 M phosphate buffer.

The separation of native and denatured DNA on hydroxylapatite can take place at any temperature below the melting range of the DNA; fractionation at 60° is convenient and often used. Bio-Gel HTP is suspended in 0.01 M sodium phosphate buffer (pH 6.7) and a 1.0-mL bed volume is packed into a Pasteur pipette plugged with glass wool. The column is kept warm by circulating hot water through an improvised water jacket (see Figure 26-2). BioGel HTP has a capacity for 100 to 200 μg of native DNA per milliliter of bed volume, and a larger volume sample than indicated by this limit should not be applied to the column. The sample in 0.12 M phosphate buffer is loaded onto the column at room temperature and then eluted with 8 to 10 mL of 0.12 M phosphate buffer at 60°C, which washes out denatured DNA, followed by four 1-mL aliquots of 0.3 M phosphate buffer to elute renatured dsDNA. The DNA content of each fraction is determined from the UV absorbance at 260 nm and the DNA from the richest fraction(s) is precipitated by adding 0.1 volume of 3.0 M sodium acetate and 2.5-fold volumes of very cold ethanol.

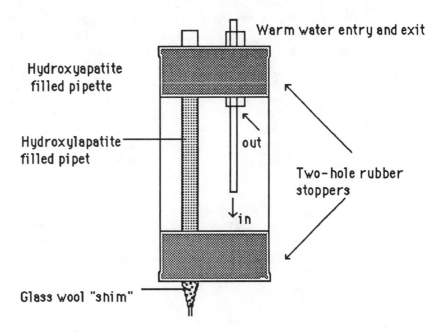

Figure 26-2   Thermostated hydroxylapatite column

Second Laboratory Period

Electrophoresis and Electroelution

The DNA recovered from the column and precipitated with ethanol is redissolved in TE buffer to a concentration of 1 µg/µL and run on a neutral 2.5% agarose slab with 0.5 µg/mL ethidium bromide staining. The 300-bp band is recovered by electroelution and purified and concentrated by the Elutip-d method as described below.

The gel run is terminated and the band of interest is identified with a short-wavelength UV lamp.

UV PROTECTING GOGGLES MUST BE WORN!

A trough is cut with a sterile razor blade just to the positive side and about 2 mm wider than the desired band. The trough is filled with electrolysis buffer containing 50% (v/v) glycerol and voltage is applied (20 V/cm) again so that the band is *electroeluted* into the trough.[21,22] Three changes of trough buffer after three minutes each are suggested. **The voltage must be disconnected each time** the trough is drained with an automatic pipette. A useful modification of this procedure involves inserting a small piece of DEAE paper (Whatman DE-52) behind the band to be electroeluted to prevent other bands from entering the trough and contaminating the sample. The gel is photographed before and after electroelution under UV illumination as a permanent record, alternatively only half the desired lanes are electroeluted. Because the gels are visible and can be photographed only in the dark, the electrophoresis portion of this experiment should be done in a closet or separate room so that the whole class is not interrupted by lights going off and on.

The Schleicher & Schuell Elutip-d can be used directly to purify and concentrate very small quantities of dsDNA from 50 bp to 50,000 bp with 95% recovery over the range 10 ng to 100 µg.[23] The Elutip-d is preferred for recovering DNA that has been electroeluted from an agarose gel. The Elutip-d is a small disposable anion-exchange column that can be attached to a Luer lock syringe. The column is designed for only one direction of flow: out from the syringe. The syringe must not be filled with the column in place! The column is first prepared by passing a syringe full of high salt (1.0 M NaCl, 0.02 M TRIS, pH 7.4, 0.001 M EDTA) solution followed by a low salt (0.2 M NaCl, 0.02 M TRIS, pH 7.4, 0.001 M EDTA) solution. A fresh syringe is loaded with about 5 mL of the DNA solution of interest and fitted with a disposable filter (Schleicher & Schuell FP 27/2) and the Elutip-d. The DNA solution is passed through the filter and into the tip. The 0.45 µm cellulose acetate filter is necessary to remove gel fragments, and so on. DNA absorbs to the Elutip matrix very nearly quantitatively if the flow is slow. The syringe may be filled once again with 2 to 3 mL of low-salt buffer and this passed through the column to insure against loss of DNA in the syringe and filter. This also serves to remove sulfated polysaccharides which are extracted from the gel with the DNA; these compounds are potent inhibitors of restriction enzymes. The DNA can be removed from the Elutip with 0.4 mL of high-salt buffer. The DNA could be further concentrated by precipitation with a twofold volume excess of very cold ethanol.

Third Laboratory Period

The DNA recovered by electroelution is redissolved to about 2 μg/μL in 0.01 M TRIS and the concentration of DNA determined spectroscopically (or by comparing the fluorescence of a spot made by combining 2 μL of DNA solution with 2 μL of 5 μg/μL ethidium bromide on a piece of Parafilm, with the fluorescence of an equivalent spot of known DNA of 2 μg/μL concentration).[24] The recovered DNA isolution is divided into two portions. One portion is digested with Alu restriction enzyme and then characterized by electrophoresis on a 2.4% gel, with alternating lanes for Alu digested and non-digested DNA. Restriction with Alu I is performed by combining the recovered DNA solution with 0.1 volume of 0.06 M $MgCl_2$ and 0.06 M 2-mercaptoethanol and 2 units of enzyme per microgram of DNA . The digestion can be run for 3 hours for partial digestion or overnight at 37° for more complete digestion if a sterile stoppered microfuge tube is used. After ethidium staining the Alu-digested sequences show up further down the gel than the untreated sequences.

PROJECT EXTENSIONS

1. Alternatively the 300-bp Alu fraction can be removed from a low-melting agarose gel by an alternative to electroelution. If the agarose gel electrophoresis of Alu sequences is done on low-melting agarose (Sea-Plaque FMC corp.), the appropriate band can be cut out with a fresh razor blade, and removed from the gel either by melting the gel and extracting the DNA[25] or by the following procedure.[26] The gel is transferred to an Eppendorf tube and combined with two volumes of 6.0 M NaI 0.12 M Na_2SO_3, and warmed to 37°C for about an hour (in the dark) until the agar is completely melted. The melt is shaken with 100 μL of a slurry of glass fines (IBI) in water and stored on ice for 90 minutes followed by centrifugation. The DNA is reversibly bonded to the recovered glass fines. Decant away the supernatant but do not dispose of it until determining the DNA content of the glass extract. The glass is washed once with cold NaI solution and once with cold buffered 50% ethanol (10 mM TRIS pH 7.4) then the DNA is extracted from the glass by incubation with 25 μL of NaCl, Na_3 citrate buffer at 37°C for an hour. (If ^{32}P labeled DNA is used the ^{32}P count of the extract is used to determine the recovery of DNA.)

2. The 300-bp Alu sequences recovered by electroelution or the extraction method above can be ^{32}P end-labeled with T4 polynucleotide kinase and then used as a probe for Alu sequences in heterogeneous human DNA. The following procedures for the preparation of a labeled probe, Southern blotting of human DNA, and blot hybridization to the probe are recommended for this project extension.

Optional Protocols

Probe 5'-end Labeling with [32]P

Alkaline phosphatase is used to remove DNA 5'-terminal phosphates by reaction with calf intestine alkaline phosphatase followed by labeling with T4 polynucleotide kinase in the presence of inorganic phosphate, which inhibits alkaline phosphatase.[27] In a microfuge tube are combined:

| DNA (50 pmoles 5'ends) | 10 μL |
| Calf intestine alkaline phosphatase | 0.1 unit |

The concentration of 5' ends is estimated from the total nucleotide concentration calculated from the absorbance at 260 nm (see Chapter 10) or the spot test described above and an average oligonucleotide length of 300 bp. The mixture above is incubated at 45°C for 45 minutes, then enough sodium phosphate should be added to give 1 mM phosphate. Then the following is added:

| 10x kinase buffer | 5 μL |

(500 mM glycine-NaOH at pH 9.2, 100 mM dithiothreitol, and 50 mM $MgCl_2$)

[γ-^{32}P]ATP (at 1000 Ci/mmol)	to give 20 μM ATP
T4 polynucleotide kinase	4 units
Water	to 50 μL total volume

CAUTION: WEAR GLOVES AND USE A RADIATION SHIELD WHEN WORKING WITH ^{32}P.

The above is incubated at 37°C for 15 minutes, then stopped with EDTA stop buffer. The DNA is deproteinated by phenol extraction and ethanol precipitation. Unlabeled carrier tRNA will facilitate quantitative precipitation; 1 μL of 1 mg/mL tRNA is used as a carrier to facilitate precipitation at -70°C for 5 minutes followed by centrifugation in the Eppendorf at 4° (in cold room) for 5 minutes. The activity of labeled probe can be determined by liquid scintillation counting.

Southern Blotting of Heterogeneous Human DNA

An agarose gel of partially sheared heterogeneous human DNA is blotted onto nitrocellulose sheet, by the following procedure, and permanently bonded by baking in a vacuum oven for 2 hours at 80°C. The gel is first soaked twice in 1.5 M NaCl/0.5 M NaOH for 15 minutes each time. This denatures the DNA, facilitating its transfer. The gel is then neutralized by soaking in blotting buffer: 1.0 M NaOAc (pH 4), twice for 30 minutes each time. The gel is layered carefully onto several thicknesses of blotting paper (Schleicher & Schuell No. 470) or Whatman 3MM filter paper in a glass dish. The paper serves as a wick, transferring buffer from the dish to the gel. Trapped bubbles must be removed. The gel is then covered with nitrocellulose cut to the right size and smoothed with a glass rod. Additional saturated blotting paper and at least a 1-inch stack of clean white paper towels are layered onto nitrocellulose membrane. The whole affair is capped off with a light weight, usually a sheet of glass. The gel is surrounded with paraffin strips to keep the lower blotting paper from contacting the nitrocellulose and the pan is partially filled with blotting buffer, not enough to touch the gel directly. Saran wrap is used to cover the whole stack and the pan.

The transfer is complete within 3 hours but can be left overnight. The next day the towels and wick are removed without disturbing the gel and nitrocellulose sandwich. The location of the gel lanes is indicated by pinholes in the nitrocellulose. The gel is removed and the left-to-right orientation of the nitrocellulose filter is marked by cutting off a corner. The nitrocellulose is placed between two pieces of filter paper and baked for two hours at 80°C

Blot Hybrization Using Labeled Probe

The nitrocellulose filter is freed of the filter paper and pretreated with hybridization buffer at 42°C for 3 hours; the hybridization buffer contains 0.3 M NaCl and 0.02 M sodium citrate at pH 7.0. The bag is drained and the filter removed; nonspecific background hybridization can be reduced by adding 1 mg of salmon sperm DNA in 0.5 mL of water to the filter followed by 50 μL of 10 M NaOH. The filter is then neutralized with 1 M TRIS HCl solution. A fresh bag is filled with hybridization buffer containing the radioactive probe (ca. 10^7 cpm) and the filter is hybridized by incubation with the labeled Alu sequences at 42°C in the resealed plastic bag overnight. The DNA-DNA hybridization can be visualized by autoradiography of the washed and dried nitrocellulose sheet.

Kodax XOMAT-AR film is placed next to the DNA sample bound to nitrocellulose (or DBM paper), and this pair is sandwiched between two calcium tungstate phosphor screens (shiny sides toward the film). The whole thing is wrapped in lead foil and taped tightly and stored at -70°C. The events recorded on film are the long-wavelength fluorescence emissions that occur when β-particles strike the screen. The film sensitivity is increased 10-fold by the use of two fluorescent screens. The exposure is carried out at -70°C, if possible, in order to prolong the period of fluorescence. The film is developed with Kodak GBX developer (5 minutes), a 3% acetic acid stop bath (30 seconds) and Kodak GBX fixer (2 to 4 minutes).

DBM paper hybridization can be used rather than nitrocellulose. ABM paper is converted to DBM-paper by immersing a 10 x 12 inch sheet in 120 mL of ice cold 1.2 N HCl and adding 3.4 mL of freshly prepared sodium nitrite, $NaNO_2$ 10 mg/mL. The paper should be immersed for at least 30 minutes at 0°C. Just prior to transfer, the paper is washed twice in ice-cold water, then twice in 20 mM sodium acetate buffer, pH 4.0. The DBM paper is soaked in 0.5 M NaOH for 30 minutes at room temperature to inactivate any remaining diazonium groups. The paper is then neutralized in blotting buffer and subjected to autoradiography.

DISCUSSION QUESTIONS

1. How would you determine the nucleic acid-to-protein ratio of placental DNA before and after cleanup with chloroform/*iso*-amyl alcohol?

2. Human placental DNA is much more expensive than Calf-Thymus DNA. Why must one use this more expensive type of DNA for this project?

3. Why is it necessary to run the strand reassociation at 60°C and in 0.3 M salt?

4. Why does one use siliconized glass- or plasticware when working with small amounts of DNA.

5. One unit of DNA has an absorption of 1.0 at 256 nm for a 1-cm path length. How many units of DNA are in a 200 mL sample with an absorption of 0.5 (for a 1-cm path length)?

6. A 1-mL sample contains 5 μg of DNA. If this sample is reacted with T4 polynucleotide kinase with ATP labeled with a specific activity of 6×10^9 dpm/mole, to yield a labeled DNA solution affording 10^3 dpm, what is the molecular weight of the DNA sample?

7 What is the purpose of the high-salt wash step with the Elutip-d purification?

8. Explain the purpose of the blue filter below the gel dish and the red filter on the camera when one is taking photograph of an ethidium bromide-stained gel.

9. What problem will be encountered if the alkaline phosphatase activity is too low?

10. What is the purpose of baking a nitrocellulose Southern blot before attempting a hybridization with a labeled DNA probe?

REFERENCES

1. R. L. P. Adams, J. T. Knowler and D. Leader, *Biochemistry of Nucleic Acids*, 10th ed., New York: Chapman and Hall, 1986, p 42.

2. M. F. Singer and J. Skowronski, *Trends in Biochem. Sci.*, **1985**, *10*, 119; M. F. Singer, *Cell*, **1982**, *28*, 433.

3. W. R. Jelinek et al, *Proc. Natl. Acad. Sci. (USA)*, **1980**, *77*, 1398.

4. W. R. Jelinek and C. W. Schmid, *Ann. Rev. Biochem.*, **1982**, *51*, 813.

5. R. J. Britten and D. E. Kohne, *Science*, **1968**, *161*, 529.

6. L. E. Orgel and F. H. C. Crick, *Nature*, **1980**, *284*, 604; J. H. Rogers, *Int. Rev. Cytol.*, **1985**, *93*, 188.

7. A. M. Weiner, P. L. Deininger and A. Efstatiadis, *Ann. Rev. Biochem.*, **1986**, *55*, 631.

8. R. J. Britten, W. F. Baron, D. B. Stout and E. H. Davidson, *Proc. Natl. Acad. Sci. (USA)*, **1988**, *85*, 4770.

9. N. Hastie, *Trends Genetic*, **1985**, *1*, 37.

10. R. L. Neve and D. M. Kurnit, *Gene*, **1983**, *23*, 355.

11. J. S. Rubin, A. L. Joyner, A. Berstein and G. F. Whitmore, *Nature (London)*, **1983**, *306*, 206.

12. J. F. Gusella et al, *Proc. Natl. Acad. Sci. (USA)*, **1980**, *77*, 2829.

13. S. Wood, *Bio-Techniques*, **1984**, *4*, 314.

14. K. Struhl, *BioTechniques*, **1985**, *3*, 452.

15. E. M. Southern, *J. Mol. Biol.*, **1975**, *8*, 503.

16. M. Bittner, P. Kumpferer, and C. E. Morris, *Anal. Biochem.*, **1980**, *102*, 459.

17. Bio-Rad Laboratories Bulletin 1110, *Zeta Probe Blotting Membranes*.

18. A. M. Maxam and W. Gilbert, *Methods Enzymology*, **1980**, *65*, 499.

19. A. M. Maxam and W. Gilbert, *Methods Enzymology*, **1980**, *65*, 520.

20. R. J. Roberts, P. A. Meyers, A. Morrison, and K. Murray, *J. Mol. Biol.*, **1976**, *102*, 157.

21. R. C.-A. Yung, J. T. Lis, and R. Wu, *Anal. Biochem.*, **1979**, *98*, 305; *Methods Enzymology*, **1979**, *68*, 176.

22. R. F. Schleif and P. C. Wensink, *Practical Methods in Molecular Biology*, New York: Springer-Verlag, 1981, p. 123.

23. Schleicher & Schuell Technical Bulletin 204, Mar 1983.

24. R. Ogden and Debra Adams, *Methods Enzymology*, **1987**, *152*, 73.

25. L. Weislander, *Anal. Biochem.*, **1979**, *98*, 305.

26. L. G. Davis, M. D. Dibner, and J. F. Battey, *Basic Methods in Molecular Biology*, New York: Elsevier, 1986, p. 123.

27. F. Cobianchi and S. W. Wilson, *Methods Enzymology*, **1987**, *152*, 99.

Chapter 27

Carbohydrate Chemistry

Classical colorimetric reactions in carbohydrate chemistry are used to characterize an unknown carbohydrate. The identity of the unknown is confirmed from the melting point of its osazone. Silica gel thin layer chromatography is used to characterize the malto-oligosaccharides in corn syrup. In the third part of the experiment, the liver of a painlessly sacrificed rat is digested by boiling with 30% potassium hydroxide. The dissolved glycogen is precipitated from the KOH solution with ethanol and separated by centrifugation. The glycogen is quantitatively determined using an anthrone-thiourea-sulfuric acid reagent.

KEY TERMS

Aldoses   *Colorimetric   assays*
Disaccharides   *Furanoses*
Glycosidic   linkages   *Ketoses*
Monosaccharides   *Polysaccharides*
Pyranoses   *TLC spray reagents*

BACKGROUND

Carbohydrates are polyhydroxy aldehydes or ketones or compounds that can be hydrolyzed to them. *Monosaccharides* cannot be hydrolyzed into simpler compounds whereas *disaccharides* liberate two monosaccharide moieties upon hydrolysis; *polysaccharides* liberate many. If the monosaccharide contains an aldehyde group it is an *aldose*, and if it has an internal carbonyl group it is a *ketose*. In most aldoses the aldehyde group is involved in intramolecular hemiacetal formation. Glucose like most aldoses exists primarily as a cyclic hemiacetal:

Similarly fructose like most ketoses exists as cyclic hemi-ketals. The equilibrium strongly favors the cyclic compound in each case but much of the solution chemistry, including the colorimetric tests described here, is most easily rationalized in terms of mechanisms involving the free carbonyl group. The cyclic form of carbohydrates with a six membered ring including one oxygen is known as the *pyranose* form, and carbohydrates existing in such a form are referred to as pyranoses. Similarly the five-membered single oxygen form is the *furanose* form. A carbohydrate may in fact exist in both pyranose and furanose forms; β-fructose exists both as β-fructopyranose (57±6%) and as β-fructofuranose (31 ± 3%).[1] The aldehyde or α–ketone group liberated from the cyclic forms by mild acid or base hydrolysis of the cyclic carbohydrate can easily be oxidized and such carbohydrates are said to be reducing sugars. Often one or more of the monosaccharide hemiacetal or ketal OH groups are not present in a di- or polysaccharide.

α-D-(+)-Glucopyranose
(mp 146°, [α]$_D$ =+112°)

β-D-(+)-Glucopyranose
(mp 146°, [α]$_D$ =+18.7°)

β-Frutofuranose

The linkage between the two monosaccharide units (*glycosidic linkage*) involves the hydroxyl group of one with the carbonyl group of the other. A disaccharide in which both monosaccharide aldehyde or ketone groups are involved in the glycosidic bond cannot easily be oxidized; it is a non-reducing carbohydrates. Sucrose is such a compound: In principle such a compound will not give a positive test for the presence of an easily reduced carbonyl group, however partial hydrolysis, especially in strong acid, will release some more easily reduced monosaccharide such as glucose which will give rise to a positive reducing sugar test.

The qualitative and in some cases quantitative determination of carbohydrates often employs a colorimetric analytical method that can discriminate between broad categories of carbohydrates (e.g., reducing versus nonreducing sugars, pentoses versus hexoses, etc).

Glucopyranose moiety

Fructofuranose moiety

The objective in qualitative tests is to accentuate the differences in reactivity of sugars to produce readily distinguishable colored products. This often involves some artful variation in reaction conditions to pull out the maximum information. Most of the colorimetric methods date from the days before mechanistic organic chemistry and were discovered empirically; little is known, even today in some cases, of the mechanism of the reaction or the detailed nature of the colored species.

In the following experiment a number of such *colorimetric tests* will be used to characterize an unknown carbohydrate. One or two known carbohydrates will also be tried in each case so that definite positive and negative reactions can be seen for comparison purposes. This comparison approach is necessary because the color changes are often hard to describe in words and much significance can sometimes be attached to minor variations in hue or color intensity. A number of tests are very general (e.g., Molisch test) whereas others are quite specific (e.g., Elson-Morgan test for hexosamines). By comparison of the observed test results with those given for the known compounds listed in Table 27-1 one can deduce the chemical identity of the unknown carbohydrate. The unknown will be one of the compounds listed in the table. Confirmation of this assignment requires determination of the melting point of the osazone derivative.

Table 27-1

TEST RESULTS FOR CARBOHYDRATES

Carbohydrate	Test Results[a]						
	Molisch	Bial	Resorcinol	Benedict	Elson-Morgan	Iodine	Barfoed
Ribose	+	bg	-	+	-	-	+
Fructose	+	y	r	+	-	-	+
Galactose	+	y	-	+	-	-	+
Glucose	+	y	-	+	-	-	+
Glucosamine+	y	–	+	+	-	+	
Sucrose	+	-	(+)	-	(+)	-	-
Lactose	+	-	-	-	-	-	-
Maltose	+	-	-	+	-	-	-
Amylose	+	-	-	-	-	bl	-
Glycogen	+	-	-	-	-	br	-

[a]+ indicates positive test result, bg=bluegreen, y=yellow, r=red, bl=blue, br=brown;(+) indicates positive due to hydrolysis products.

Colorimetric Tests Run in Acid

Most of the colorimetric reactions of carbohydrates depend on the formation of acid or base hydrolysis products and their subsequent reaction by dehydration, condensation and decarboxylation. Sulfuric acid is a strongly dehydrating material that can reduce solid sugar to a mass of charcoal (a well-known elementary science demonstration). Even in solution sugars are profoundly dehydrated by sulfuric acid. In concentrated mineral acid, monosaccharides can often be dehydrated down to furfural derivatives which then form colored adducts with phenolic reagents. The colored species is not known for certain for most of these reactions and there is presumably a mixture of condensation products in most cases. The furfural derivative may or may not be unique to the original carbohydrate, hence a large number of carbohydrates may give comparable mixtures of colored products, differing only slightly in hue or stability. This accounts for the usefulness of such reactions. They give similar results for similar compounds.

The results are most easily observed as color changes if care is taken to use small clean glass test tubes that can be held up against a white background In some cases the colored reaction product is soon air oxidized to a colorless product; reproducible results require prompt attention to the color formed immediately upon reaction, since this may fade or change.

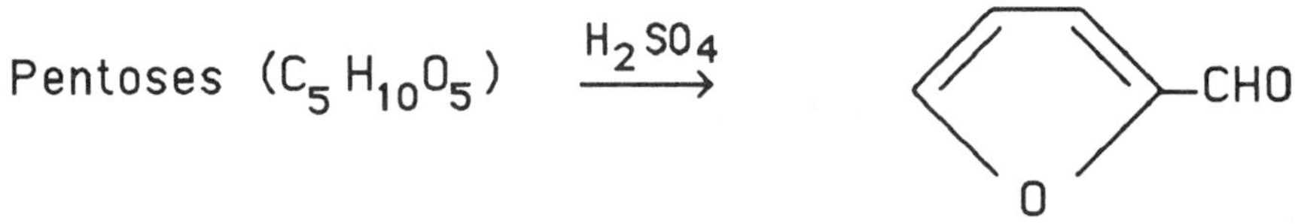

Pentoses $(C_5H_{10}O_5)$ $\xrightarrow{H_2SO_4}$

Furfural

Hexoses $(C_6H_{12}O_6)$ $\xrightarrow{H_2SO_4}$ $HOCH_2$—

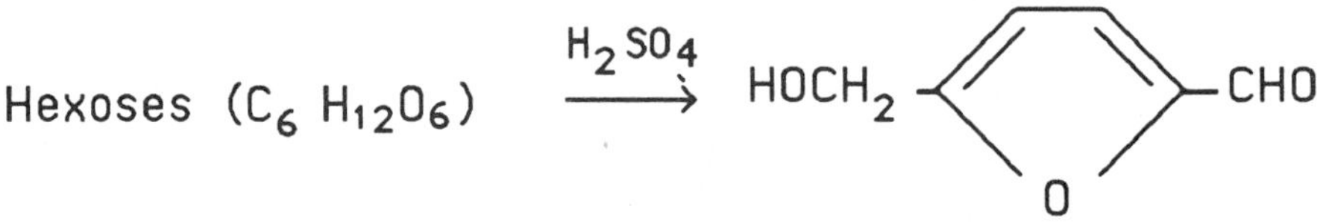

5-Hydroxymethylfurfural

The furfural derivative is then condensed with some phenolic reagent such as α-naphthol, resorcinol and orcinol. Not all sugars give colored products and the identity of the color and even the exact hue are important observations.

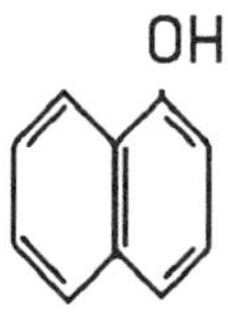

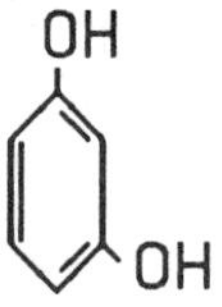

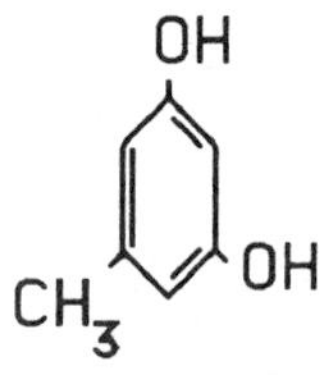

α-Naphthol   Resorcinol   Orcinol

The three important tests described below can be used to identify a general carbohydrate, distinguish a hexose from a pentose, and to discriminate between a ketose and an aldose.[3]

The *Molisch test* is routinely used as a qualitative test for carbohydrate material. A dilute aqueous solution of carbohydrates is combined with α-naphthol (in ethanol) and overlayered onto concentrated sulfuric acid. Red or violet coloration at the interface is taken as a positive reaction.

Bial's test is used extensively for the determination of hexoses (yellow-brown products) and pentoses (blue-green products). A dilute solution of the carbohydrate is combined with orcinol in concentrated HCl containing ferric chloride and heated briefly. The solution is allowed to cool and is diluted with an equivalent volume of water and the colored product extracted by shaking with a small portion of amyl alcohol.

The *resorcinol test* can be used to distinguish between ketoses and aldoses. Ketoses react more readily than aldoses with this reagent. A deep red color after brief heating is a positive test for ketohexoses. Prolonged heating is required for aldohexoses to produce a red color. A green color indicates a pentose. The dilute carbohydrate solution is combined with resorcinol in 4 M HCl and the mixture boiled for one minute. The colored product developed in the original reaction can be extracted into an overlayer of amyl alcohol as in Bial's test. Since some disaccharides like sucrose hydrolyze in aqueous acid they will give partial tests because of partial hydrolysis. For example, sucrose gives a positive resorcinol test because of liberated fructose.

Tests Run in Alkali

Treatment of carbohydrates with alkali produces a complex mixture of reaction products due primarily to carbanion-carbonyl condensation reactions, usually followed by base-catalyzed dehydrations to form colored unsaturated compounds. Also amino sugars and their acylated derivatives are converted to pyrrole derivatives by condensation with β-dicarbonyl compounds. The pyrrole derivative is combined with p-dimethylaminobenzaldehyde (Ehrlich's reagent) to give a pink color. In the *Elson-Morgan reaction*, hexosamines are converted to 3-acetyl-2-methylpyrrole by reaction with 2,4-pentanedione in base followed by treatment with an acidic p-dimethylaminobenzaldehyde reagent. This method is the most specific of the available methods for the determination of the 2-amino-2-deoxy sugars.

Tests of Reducing Ability

Reducing sugars (both aldoses and ketoses) participate in complex redox reactions in which cupric ion is reduced to cuprous ion which precipitates as red cuprous oxide.

$$Cu^{2+} \longrightarrow Cu_2O$$

Two variations of this method are the **Benedict's test** and **Barfoed's test**. Benedicts' test gives a colored precipitate (red, yellow, or green) of Cu_2O from alkaline $CuSO_4$ with reducing sugars after 5 minutes. The color of the precipitate depends on the size of the Cu_2O particle, with the smaller colloidal particles appearing green and the larger red. The size of the particle depends in part on the rate of the precipitation reaction. Barfoed's test involves precipitation from acidic $Cu(OAc)_2$ and permits discrimination between monosaccharides and disaccharide reducing sugars. Only monosaccharides give plentiful precipitate of red Cu_2O after 1 minute.

Specific Tests for Glucose

Quantitative determinations of carbohydrates often employ colorimetric determinations where the color intensity or rate of color formation can be related directly to the concentration of a specific carbohydrate.

For example there are two common methods of determining glucose. The first method involves glucose oxidase which catalyzes the formation of the glucose delta-lactone and hydrogen peroxide; this reaction is very specific for glucose. The resulting concentration of hydrogen peroxide is directly proportional to the original glucose concentration. Although H_2O_2 is colorless, it can be converted to a colored product (Trindar's color reaction) by the action of a second enzyme peroxidase.[4,5] Phenol and 4-aminoantipyrane are condensed together at the expense of hydrogen peroxide.

β−D−Glucopyranose   D−Gluco−δ−lactone

Colored product

A second more classical method involves the reaction of anthrone, thiourea and glucose in concentrated sulfuric acid. This reaction is not specific for glucose since it will give a positive color test for any di- or polysaccharide that hydrolyzes in concentrated sulfuric acid to form hexoses, pentoses, deoxy sugars, uronic acids, and even heptuloses. It is a very versatile reagent but quantitative determination is possible only when the identity of the sugar is known since not all sugars give the same color intensity. Control experiments are necessary to rule out interference from other substances. Tryptophan and salts at high concentration are known to interfere in the reaction.[6] In our experiment glycogen extracted from KOH-digested rat livers will be quantitated by the anthrone method.

Anthrone   Thiourea

Three other reagents to determine reducing sugars are widely used. The *neocuproine assay*[7] can be used to detect both carbohydrate and noncarbohydrate agents capable of reducing cupric ion. The tetrazole blue reagent gives a blue violet color with reducing sugars and some nonreducing ones (e.g. sucrose)[8]. An alkaline solution (1 M NaOH) of 0.07% tetrazole blue in water/methanol (2:1) can be used as a spray reagent for silica-gel thin-layer chromatography (TLC) plates. Alkaline dinitrosalicylic acid can also be used as a reagent for reducing sugars; this reagent is described in Chapter 28.[9]

Thin-Layer Chromatography of Carbohydrates

Carbohydrates can be separated by thin -ayer chromatography on silica-gel or cellulose with butanol/ethanol/water (5:3:2) as the eluting solvent. The separation is primarily on the basis of molecular weight, so that the mono-, di-, and higher-oligonucleotides released upon partial hydrolysis of polysaccharides such as starch can be separated on a single plate. In the experiment described here the malto-oligosaccharides in corn syrup produced by the hydrolysis of starch are visualized by a nonselective charring technique. The chromatogram is sprayed with 10% sulfuric acid in ethanol, then placed silica gel side up on a hot hotplate for 1 minute. The identity of the oligonucleotides is confirmed by direct comparison with authentic samples run on the same TLC plate as shown in Figure 27-1.

SPRAY AND DEVELOPE TLC PLATES IN THE HOOD!

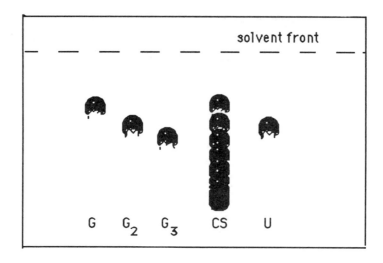

Figure 27-1 TLC of glucose G, maltose G2, maltotriose G3 corn syrup CS and an unknown on silica gel, visualized by charring method

Commercial separations of oligonucleotides are now mostly done by HPLC with a 4% cross-linked cation-exchange column (e.g., Bio-Rad Aminex HPX-42A); the elution profile observed with refractive index detection is shown in Figure 27-2.

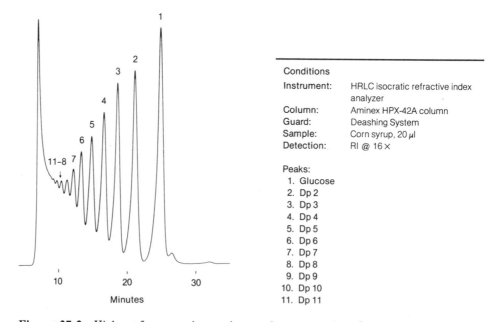

Conditions

Instrument:	HRLC isocratic refractive index analyzer
Column:	Aminex HPX-42A column
Guard:	Deashing System
Sample:	Corn syrup, 20 μl
Detection:	RI @ 16 $\times$

Peaks:
1. Glucose
2. Dp 2
3. Dp 3
4. Dp 4
5. Dp 5
6. Dp 6
7. Dp 7
8. Dp 8
9. Dp 9
10. Dp 10
11. Dp 11

Figure 27-2 High performance ion exchange chromatography of corn syrup containing malto-oligonucleotides. A Bio-Rad Aminex HPX-42A column and refractive index detection was used. (Courtesy of Bio-Rad Laboratories).

The interaction between the oligonucleotide and the resin is not really ion exchange since the oligonucleotide is not charged. Although trace carboxylic acid impurities present in sugar mixtures (due to air oxidation are very tightly bonded to the resin), the neutral oligonucleotide is much more loosely bound and can be eluted off the column with larger molecules coming off first. Cation-exchange resins containing silver or calcium counterions have been shown to be quite effective in separating the homologous series of malto-oligonucleotides, containing up to 10 glucose residues. The silver form of the resin increases compound-matrix interaction and column efficiency. Such methods are often used to analyze mixtures prepared by the partial hydrolysis of starch; commercial corn syrup is such a mixture.

Solid Derivatives of Carbohydrates

The traditional structure proof of an organic compound usually involved melting-point determinations. This approach remains valid for carbohydrates, at least with monosaccharide derivatives, but some problems must be kept in mind. Carbohydrates, particularly when impure, can be crystallized only with difficulty. Imagine the difficulty of obtaining sucrose crystals from syrup. Fortunately many carbohydrates react with phenylhydrazine to form bright yellow crystalline derivatives known as *osazones*. These derivatives can be used to identify carbohydrates, most of which form unique osazones with characteristic crystalline forms and decomposition (melting) temperatures (see Table 27-2). The osazone is prepared by combining phenylhydrazine in sodium acetate buffer with an aqueous solution of the carbohydrate and warming in a boiling-water bath for 15 minutes. Upon cooling crystals of osazone derivative will form for reducing sugars. The crystals may be recrystallized from aqueous alcohol and, when dry, the temperature of decomposition is characteristic of the original carbohydrate. The osazone formation destroys the configuration about C-2 of an aldose, so that aldoses differing only in the C-2 configuration will give the same derivative (e.g., glucose and mannose).

Glucose   Common osazone   Mannose

Table 27-2
MELTING POINT OF OSAZONE DERIVATIVE

Carbohydrate	Osazone mp (°C)
Ribose	166
Fructose	205
Galactose	201
Glucose	205
Sucrose	(205)
Lactose	200
Maltose	206

Polysaccharides

Polysaccharides are compounds made up of many monosaccharide units per molecule. The two most important glucose-based polysaccharides occuring in nature are *cellulose* and *starch*. Starch occurs in two fundamentally different forms: a completely linear glucose polymer linked by α-1,4-glycosidic linkages and branched forms containing 1,6- as well as 1,4-linkages. Plant starch has both types of polymer known as *amylose* and *amylopectin,* respectively. Amylose is water-soluble and comprises about 20% of plant starch, and amylopectin is water-insoluble, making up about 80% of the polysaccharide in plant starch granules. Animal starch is composed of α–glucose units linked by 1,4- and 1,6-glycosidic bonds to form a spherical branched polymer with a molecular weight of several million. For any one glycogen molecule there is only one glucose with a free reducing end A; the remaining glucose subunits are either part of the 1,4-linear portions B, branch points with 1,4- and 1,6-bonds C or nonreducing termini D. The hypothetical structure of glycogen is summarized by the formula:

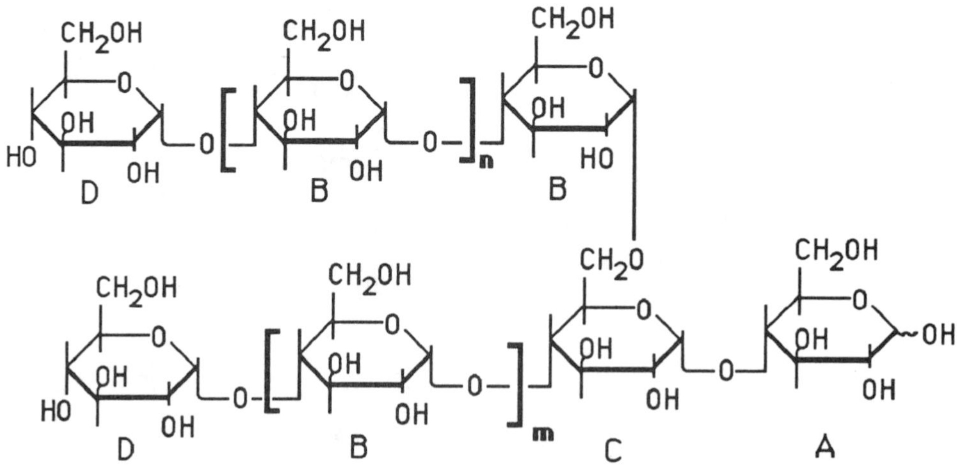

It is no easy chore to determine glycogen quantitatively because it lacks a chromophoric group and its branching structure precludes the use of the blue iodine stain used with amylose (see below). Its reducing-end is buried within the polymer, so most of the classical carbohydrate tests are not decisive. Also, glycogen has an indefinite structure; the degree of branching is variable depending on a variety of physiological parameters which are poorly understood at present. The differently branched glycogens differ in their sedimentation properties and glycogen can be characterized by centrifugation, but that is not convenient for routine quantitative analysis. The rather unspecific anthrone colorimetric method is commonly used. Practical limitations of teaching laboratory time preclude a detailed study of all the parameters governing the use of anthrone reagent to determine liver glycogen but it must be pointed out that the procedure described in this experiment is incomplete. For research purposes the experimental procedure for the colorimetric determination of glycogen from liver would have to be accompanied by control experiments. Whenever an unspecific colorimetric method is applied to quantitatively determine a constituent in a complex biological system, several factors must be investigated to ensure the reliability of the method.

It must be shown that the isolation from the biological matrix is quantitative and there must be no specific interference by factors in the matrix. For example, liver from starved rats supplemented with weighed amounts of glycogen can be used to check the reproducibility of the KOH extraction. These additional control experiments may be required by the instructor to insure that the glycogen is precipitated quantitatively and that it can be determined without interference from liver proteins that are also precipitated. The effect of variations in boiling time, ethanol volume, and color development time should also be systematically studied.

Most samples of plant starch can be distinguished by the simple iodine test, in which a 1% solution of the carbohydrate is combined with a solution of iodine and potassium iodide to give a blue color for amylose. Glycogen gives a red-brown color but it cannot be used for quantitative analysis because the color intensity is very dependent on the presence of short starch fragments and on the degree of branching. The blue color observed with amylose is very vivid and has been used as an indicator endpoint for redox reactions converting iodide to iodine. The blue color is presumably due to a polymeric iodine-iodide complex (I_5^-, I_7^-) which is stabilized within the helical structure of amylose. Amylopectin and glycogen have 1,6-branches and cannot form the helical structure necessary to stabilize the blue species.

EXPERIMENTAL SECTION

In the first part of this experiment each student is given 10 mg of an unknown to characterize by colorimetric tests and by osazone formation and melting point. The tests should be run on known compounds and the results compared to that observed with the unknown. Some of the stock reagents can be prepared ahead of time or each student can prepare one reagent to be shared by the whole class. In the second part of the experiment, each student must identify an unknown malto-oligonucleotide by silica-gel TLC using glucose, maltose, maltotriose, and corn syrup as reference standards. In an optional third part, students work in groups of two or three to determine liver glycogen levels. The instructor may also require a number of control experiments in support of the liver glycogen analysis.

Parts 1 and 2 can be completed in one 3-hour laboratory session while the liver glycogen project requires an additional lab period.

WEAR SAFETY GLASSES WHEN USING STRONG ACID.

Colorimetric Characterization of Unknown Sugar

Molisch Test

The Molisch reagent contains 1.25 g of reagent grade α-naphthol in 25 mL of 95% ethanol. A small amount of this reagent (20 μL) is added to 1 mL of 0.1% carbohydrate (1 mg/mL) solution in a small test tube. The tube is then tilted and 0.5 mL of concentrated sulfuric acid is poured down the side of the tube. Do not use an automatic pipette to measure concentrated acids. A glass rod can be used to prevent spillage. A red-violet layer at the interface between the acid (bottom) and aqueous (upper) layers is a positive test for a carbohydrate. *Remember*: Always add acid to water.

The reagent may be made up in advance by the stockroom assistant and stored in a foil-wrapped bottle.

Bial's Test

The reagent contains 1.0 g of reagent-grade orcinol in 300 mL of concentrated HCL also containing 0.2% ferric chloride. The reagent must be stored in a foil-wrapped glass-stoppered bottle. A sample of 0.1 % carbohydrate solution in water is combined with a fivefold excess of Bial's reagent and heated to just boiling. When the solution has cooled, an equal volume of water and an overlayer of n-amyl alcohol is added and the mixture transferred to a small test tube. A blue-green color in the upper layer indicates pentoses; a yellow color indicates hexoses.

Resorcinol Test

The reagent contains 0.1 g of reagent-grade resorcinol in 200 mL of 4 M HCl. The reagent must be stored in a foil-wrapped bottle. The test is run in the same way as Bial's test except that the tube is heated near boiling for <u>only 1 minute</u>. A deep red color after 1 minute indicates ketohexoses, and a green color indicates a positive test for pentoses. Sucrose gives a positive ketohexose test because of partial hydrolysis to fructose.

Benedict's Test

The reagent contains 17.3 g of trisodium citrate (dihydrate) and 10.0 g of anhydrous sodium carbonate in 60 mL of warm water. The solution is clarified by filtration and diluted to 85 mL with water. To an additional 15 mL of water is added 1.73 g of copper sulfate (pentahydrate) and, when dissolved, the blue solution is added slowly, with stirring, to the citrate-carbonate solution. A 1% carbohydrate is combined with a five-fold excess of Benedict's reagent and heated in a water bath for 5 minutes. Precipitation indicates a positive test.

Barfoed's Test

The reagent, containing 6.6 g of cupric acetate (monohydrate) and 1.8 mL of glacial acetic acid in 100 mL of water, is combined with a one-half equivalent volume of 1% carbohydrate and heated in a boiling-water bath for exactly 1 minute, then cooled to room temperature. The carbohydrate solution should not stand in the water bath for more than 1 minute before the reagent is added. An ample amount of brick-red precipitate indicates a reducing monosaccharide. Some hydrolysis of disaccharide may lead to trace precipitates which are false-positive tests.

The Iodine Test

A few drops of 0.01 M iodine in 0.12 M KI is added to a 1% solution of the carbohydrate in question. The immediate formation of a vivid blue color indicates amylose.

Osazone formation[2]

A 2 mg sample of carbohydrate is dissolved in 50 μL of water and combined with 4 mg of phenylhydrazine hydrochloride and 6 mg of sodium acetate. The solution is stirred thoroughly and heated for up to 20 minutes in a hot-water bath then cooled to room temperature. Do not heat the solution to dryness; add additional water if necessary. The time heating is begun and the time of precipitation should be noted. The time for osazone formation is a valuable aid in distinguishing between sugars: fructose, 2 min; glucose, 5 min; galactose, 15 to 19, min; lactose and maltose osazones are soluble in hot water. Sucrose gives an osazone after 30 minutes due to the formation of glucosazone.[2] The precipitate should form after twenty minutes of cooling. The precipitate can be recrystallized from ethanol-water and air dried.

Thin-Layer Chromatography of Oligosaccharides

A solvent containing butanol/ethanol/water (5:3:2) is used to develop a chromatogram such as that pictured in Figure 27-1. Five microliters of corn syrup is added directly to the plate with a microcapillary TLC applicator; see Chapter 29 for details. The very viscous solution must be slowly added to avoid spreading. The samples of glucose, maltose and maltotriose and an unknown are dissolved in a minimum of the developing solvent (4 to 5 mg/mL) and applied to the column with capillaries as 5 μL samples as well. About an hour and a half is required to get good separation on a 20-cm TLC plate. The plate is sprayed with 10% sulfuric acid in methanol, then placed silica gel side up on the hot surface of a hotplate. Only 90 seconds is required to "char in" spots. The spraying and visualization must be done in a hood.

WEAR SAFETY GLASSES.
CLEAN UP ANY SPILLS IMMEDIATELY.

Liver Glycogen[10]

A rat is quickly sacrificed by decapitation and the liver removed. The intact liver is digested for 1 hour in 3 mL of 30% KOH solution. The extract is clarified by centrifugation and the glycogen is precipitated from the clarified KOH digest by adding 25 mL of 95% ethanol followed by refrigeration for at least overnight.

STRONG ALKALI WILL DAMAGE A
CENTRIFUGE ROTOR.

DO NOT USE POLYCARBONATE CENTRIFUGE TUBES.

The supernatant can be removed by pipetting and the brown precipitate containing glycogen and protein is dissolved in anthrone-thiourea-sulfuric acid reagent and heated at 80°C for exactly 5 minutes. A dark blue color is formed. The total concentration of glycogen can be determined from the absorbance at 620 nm by comparison with a standard curve prepared from weighed samples of glycogen reacted with the reagent for exactly 5 minutes.

The reagent contains 0.05% anthrone, 1.0% thiourea and 72% (v/v) sulfuric acid. For each liter of reagent 720 mL of concentrated acid (sp. gr. 1.84) is cautiously added to 280 mL of distilled water. The anthrone (500 mg) and thiourea (10 g) are dissolved in the aqueous acid, which is warmed briefly to 80°C then cooled and stored in the refrigerator. This reagent can be made up ahead of time by the stockroom staff, but it cannot be stored for more than a week.

PROJECT EXTENSIONS

1. The control experiments where KOH concentration, boiling time, ethanol volume, and color development are systematically varied to improve the sensitivity and reliability of the anthrone test can be investigated. A comparison of the anthrone reagent value for the glycogen content of a sample of starved rat liver tissue spiked with added glycogen with that for the same-size sample of pure glycogen can be used to evaluate the interference due to the biological matrix. The extension of these methods to glycoproteins containing glucosamine, galactosamine, and sialic acid is quite useful. In the latter case the carbohydrate must be freed from the protein before analysis.

2. A crude preparation of salivary amylase can be had by chewing Parafilm and repeatedly expectorating into a cold test tube containing ice water. When the saliva has reached a concentration of about 5%, portions of this crude enzyme are added to an equal volume of 1% soluble starch (amylose) and the reaction mixture incubated at 37°C. The hydrolysis of amylose can be monitored by withdrawing samples and testing them with iodine solution. The hydrolysis of a more concentrated starch solution may be monitored by silica-gel TLC as above.

DISCUSSION QUESTION

What interaction is probably responsible for the binding of carbohydrates to a cation-exchange column? Remember that the column efficiency depends on the counterion and it is highest for tightly binding ions such as calcium or silver; also the elution order is inversely proportional to the number of residues (e.g., the largest oligonucleotide comes off first and the monomeric sugar last).

REFERENCES

1. A. Allerhand and D. Doddrell, *J. Amer. Chem. Soc.*, **1971**, *93*, 2777.

2. N. D. Cheronis and J. D. Entrikin, *Identification of Organic Compounds*, New York: Interscience, 1963.

3. D. Aminoff, et al, in *Carbohydrates, Chemistry and Biochemistry*, Vol. IIB, eds. W. Pigman and D. Horton, New York: Academic Press, 1970, pp. 739-807.

4. *J. Clin. Pathol.*, **1975**, *27*, 1015.

5. P. Trinder, *Ann. Clin. Biochem.*, **1969**, *6*, 24.

6. A. W. Devor, W. L. Baker and K. A. Devor, *Clin. Chem.*, **1964**, *10*, 597.

7 S. Dygert, L. H. Le, D. Florida and J. A. Thoma, *Anal. Biochem.*, **1965**, 28, 350.

8. M. F. Chapin and J. F. Kennedy, *Carbohydrate Analysis, a practical approach*, Washington, D.C.: IRL Press, 1986, p.12.

9. P. Bernfeld, *Methods Enzymology*, **1955**, *1*, 149.

10. N. V. Carrol, R. W. Longley and J. H. Roe, *J. Biol. Chem.*, **1956**, *220*, 583.

Chapter 28

Polygalacturonic Acid and Polygalacturonase

The viscosity of polygalacturonic acid solutions is determined as a function of Na^+ and Ca^{2+} ion concentrations at pH 4.5. Polygalacturonase is isolated from fresh red tomatoes and the polygalacturonate hydrolysis activity of this enzyme extract is monitored by viscometry and by a 3.5-dinitrosalicylate colorimetric determination of the reducing sugar released upon hydrolysis of the polysaccharide.

KEY TERMS

Chiten
Heteropolysaccharides
Intrinsic viscosity
Pectins
Polygalacturonic acids
Specific viscosity

Glucans
Homopolysaccharides
Pectinase
Polygalacturonase
Relative viscosity
Uronic acids

BACKGROUND

Polysaccharides are worthy of detailed study by chemists interested in the structure of such molecules in solution because they play an important role in determining the overall macroscopic features of intact cells and tissues. Some polysaccharides, including starch and glycogen, serve primarily as food storage compounds and play a structural role by virtue of taking up a large part of the intracellular volume. Other polysaccharides are exclusively structural elements and may feature special properties in solution that account for their structural role. In all cases the constituent monosaccharides are interconnected by glycosidic bonds to give either linear or branched polymeric products, sometimes bound to proteins to give a network structure. Some are *homopolysaccharides*, comprised of one type of monosaccharide, and others are

297

heteropolysaccharides, containing several constituent sugars. Homopolysaccharides are discribed by the suffix "an"; thus a polyglucose is termed a *glucan*. The mode of linkage more clearly defines the polymer; for example cellulose is a β[1→4]D-glucan, comprised of about 5000 glucose units. This linkage confers on cellulose, the most abundant biopolymer in nature, a linear structure and the ability to form bundles of parallel molecules. This accounts for its nearly crystalline properties and the insolubility, chemical inertness, and physical strength that determine the structure and mechanical properties of vascular plants. Wood is composed of cellulose fibers glued together by an amorphous lignin resin, and the mechanical rigidity of wood fibrils stems from the strong hydrogen binding between bundles of hydrated cellulose chains. Other polysaccharides play significant structural roles. The exoskeletons of arthropods are comprised of a polyglucosamine derivative *chiten*, which is one of the most mechanically rigid materials in nature.

Even soft tissue is held together and given form by carbohydrate polymers. The solidity of unripe fruit is due to polysaccharide pectic substances which form mechanically rigid gels in solution, and the breakdown of these gels upon ripening is part of the softening process. The gel breaks down as a consequence of enzymes that hydrolyze the *pectin* polymers. These pectic substances can be extracted from fruit with hot water and have the property of forming a gel when cold--this is how jam "sets." The following is a specific illustration of such a polysaccharide gel and how its features can be regulated by salts and by enzymes. The primary purposes of this project are (1) to determine the salt-induced gelation of dilute solutions of *polygalacturonic acid*, a homopolysaccharide related to the pectin of plants, and (2) to isolate and investigate the activity of a "pectinase" involved in fruit ripening, tomato polygalacturonase.

Uronic acids are usually prepared by oxidation of the primary alcoholic group of aldoses. Such compounds contain both carboxylic and aldehydo functions and the latter permits glycosidic bonds between uronic acid monomers to form polyuronic acids. Hexuronic acids are important constituents of many naturally occurring polysaccharides in plants, and at pH values above the average pK_a of the carboxylic acid groups, these polymers are polyanions. Unlike cellulose or chiten these polymers are very water soluble but they form aggregates in solution, especially in the presence of metal ions which interact with the negatively charged carboxylate groups. This aggregation leads to gelation which confers useful mechanical properties utilized by plants. Pectic acid or polygalacturonic acid is an α[1→4] D-galacturonan and the main constituent of pectin, a plant cell wall constituent. The stereochemical structure of this polymer is as follows.[1]

The individual pyranose rings orient to minimize the electrostatic repulsion between carboxylate groups of adjacent rings. The axial-axial configuration of the glycosidic linkages of such a polymer leads to a buckled ribbon with limited flexibility.

Examination of molecular models shows that large interstices exist between chains packed in this conformation.[2,3] Cooperative interactions between such buckled ribbons are strongest when these interstices are filled by cations. Added cations will lead to increased association between chains and an increase in the viscosity of the carbohydrate solution.

Viscosity of Polysaccharide Solutions

The flow properties of a solution often reveal significant details about the structure of the solute. The average shape, size, and degree of aggregation of a solute can be deduced by comparing the *viscosity* of a solution as the solute concentration is increased to the viscosity of the pure solvent. Aqueous polysaccharide solutions are frequently very viscous even when dilute because the large linear carbohydrate molecules move much more slowly in the direction perpendicular to their long axis, and at any one time, a sizable percentage of them will be oriented transverse to a given direction of flow. The aggregation of polysaccharides increases viscosity as well since the aggregated molecules must flow as a unit. The factors that influence aggregation can be monitored by their effect on solution viscosity.

Viscosity measurements can be used to characterize resistance to flow since viscosity is a function of the interactions between molecules in a flowing liquid. For instance, if a liquid flows down a narrow vertical tube, there is an internal resistance to flow, transmitted from the immobile walls of the tube to the flowing liquid. The narrower the tube, the more pronounced will be this resistance. Poiseuille showed that the the volume V of a liquid which flows through a capillary tube is proportional to the flow time t, the pressure p under which the liquid flows, and the fourth power of the capillary radius, r; moreover, the volume is inversely proportional to the length l of the capillary and to the viscosity η of the liquid.

$$V = \frac{\pi t r^4 p}{8 l \eta} \quad \text{or} \quad \eta = \frac{\pi t r^4 p}{8 l V} \tag{28-1}$$

Hence, at constant liquid pressure, the viscosity is proportional to the flow time. The pressure changes with the level of the liquid in the capillary and on its density, but if the starting level is kept constant and if only dilute solutions with nearly identical densities are considered, then the *relative viscosity* of a solution to that of pure solvent is merely the ratio of the flow times for the two liquids.

$$\eta_{rel} = \frac{t_1}{t_o} \tag{28-2}$$

A related parameter, the *specific viscosity*, is a measure of the increase in viscosity due to the dissolved substance and is calculated from expression. 28-3.

$$\eta_{sp} = \eta_{rel} - 1 = \frac{t_1 - t_o}{t_o} \tag{28-3}$$

The most popular device for measuring viscosity is the *Ostwald viscometer* (see Figure 28-1), which is a U-shaped glass tube; one upright portion of the U is a uniformly made capillary tube connected to the second upright portion, which is a liquid reservoir **D**. The solution of interest is drawn up the capillary by suction through rubber tube **C** until the liquid reaches a specified level indicated by a mark **A** on the tube. The solution is then released and the time of flow required for the liquid level to drop to a second mark **B** is measured. The apparatus is standardized by measuring the flow time of pure solvent: t_0.

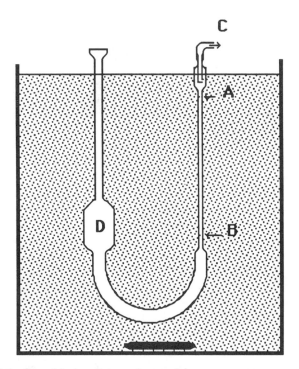

Figure 28-1 Ostwald viscometer submerged in a constant-temperature bath.

Specific viscosity is a complicated function of the concentration, molecular size and shape, and charge of the solute, and changes in η_{sp} for any solute, under a variety of conditions, reflect changes in these parameters. For example, the concentration dependence of a spherical colloidal particle is described by a linear function of concentration c, where [η], the *intrinsic viscosity*, is the intercept on the η-axis and K is the slope.

$$\frac{\eta_{sp}}{c} = [\eta] + Kc \qquad (28\text{-}4)$$

On the other hand, thread-like macromolecules exhibit a more complicated dependence.

$$\frac{\eta_{sp}}{c} = [\eta] + Kc[\eta]^2 \qquad (28\text{-}5)$$

The intrinsic viscosity is a function of the size and shape of the dissolved macromolecule; for random coil polymers it is proportional to the square root of the molecular weight.[4] For more rigid linear polymers the intrinsic viscosity is a steeper, albeit less precisely known function of molecular weight. Aggregation of roughly spherical or random coil polymers to form a more substantial network can have profound effects on viscosity and this aggregation can be monitored by viscosity measurements. In this experiment the aggregation of polygalacturonic acid will be studied as a function of Na^+ and Ca^{2+} ion concentrations since these cations, especially the latter, markedly enhance the association between extended ribbons of polysaccharide in solution. The student will be surprised to find that a 1% solution of this polysaccharide in water with 0.2 M $CaCl_2$ can gel so effectively that the solution will become too viscous to pour. It is understandable how pectin substances can account for the firmness of unripened fruit pulp.

Polygalacturonase and Fruit Ripening

One of the characteristic changes accompanying the ripening of fleshy fruits such as tomatoes is a decrease in firmness. Striking changes in the cell wall are noted during ripening, leading to the eventual disruption of the primary cell wall.[5] This is associated with an increase in the activities of enzymes that degrade cell walls, particularly polygalacturonase[6] and cellulase.[7] It is probable that the hydrolysis of pectic substances weakens the complex network of polysaccharides in the cell wall. Two isoenzymes of tomato polygalacturonase, PG I and PG II, can be extracted from tomatoes.[6] The two enzymes differ in molecular weight, thermal stability and in their chromatographic properties. SDS polyacrylamide gel electrophoresis of PG I results in two major bands corresponding to two subunits (47,500 D and 41,000 D) while PG II gives only one major band (47,000 D).[8] PG II is the more active isoenzyme and probably gives a different ratio of *endo*-to-*exo* cleavage of Galacturonan than the other one. The level of these two enzymes has been monitored in fruit as they ripen. Firm underripe fruit with an orange skin color contain both isoenzymes but fully ripe fruit contains only PG II.[8,9] It has been hypothesized that during the ripening process, when more polygalacturonase activity is required to soften the cell wall, a nondialyzable component of PG I is removed, converting PG I to PG II. A PG-converter factor that converts PG II to PG I has been isolated.[10]

Polygalacturonase Extraction

The enzymes are extracted from slices of tomato (100 g) in a 100-mL volume of cold water but preliminary steps must be taken to remove low-molecular-weight reducing sugars, which interfere with the enzyme assay before the enzymes are released from the cell membranes. The enzymes are not extracted by low salt solution at pH 3 whereas the sugars are.[11] The fruit (on ice) is blended at high speed 2 minutes followed by 1 minute (short bursts) with a polytron homogenizer. The pH of the homogenate is adjusted to 3 with 0.1 M HCl and the homogenized pulp is collected as a pellet by centrifugation at 11,500 x g for 10 to 20 minutes. The supernatant is assayed for carbohydrate content by the 3,5-dinitrosalicylate method (see below) and discarded. The pellet is resuspended in cold pH 3 water and washed once or twice again, until the supernatant gives a negative colorimetric test for reducing sugars. Once all the sugars have extracted from the pellet, it is resuspended in 150 mL of cold 1.2 M NaCl pH 6.5 which will release the enzymes from the fruit. The pulp is again homogenized for 1 minute (six 10-second bursts) with the polytron and stirred at 5°C for 20 minutes. The solubilized enzymes are freed from insoluble pulp material by centrifugation at 11,500 x g at 5°C for 20 minutes. Failure to release enzyme from cell membranes by adequate homogenization is the most common cause of low recovered activity.

Polygalacturonase Assays

Two methods can be employed to assay for enzyme activity. Hydrolysis of polygalacturonic acid leads to a marked decrease in the solution viscosity, especially if the enzymes give substantial *endo* cleavage, converting the long polymer molecule into two shorter ones with each cleavage reaction. The viscosity of a buffered sample of 1% polygalacturonic acid at pH 4.5 at 25°C is combined with an aliquot of enzyme solution and carefully mixed. The mixture is quickly transferred to a thermally equilibrated viscometer (at 25°C) in a viscometer water bath and the viscosity of the solution is measured repeatedly (as many times as possible) over 20 minutes. The reciprocal specific viscosity $1/\eta_{sp}$, expressed as $(t_\infty / t - t_\infty)$ where t is the flow time of the reaction mixture and t_∞ that of the reaction mixture after reaction is complete, is plotted against reaction time. The slopes of the resulting straight lines are proportional to activity. Care must be taken to avoid the formation of bubbles within the capillary section of the viscometer.

The hydrolysis of polygalacturonate also results in the formation of galacturaldehyde reducing groups which can be monitored by their ability to reduce alkaline 3,5-dinitrosalicylate to an orange-colored aminosalicylate product.[12] An aliquot of enzyme solution is combined with a fivefold volume of 0.5% polygalacturonate solution in sodium acetate buffer, pH 4.5 at 35°C. The enzyme activity is determined by removing aliquots of the reaction mixture at five minute intervals and quenching each one with a one-half volume of 2% 3,5-dinitrosalicylate in 0.5 M NaOH. The color that develops after heating the covered tube in a boiling water bath for 5 minutes is monitored at 450 nm. The slope of a plot of absorbance, relative to a blank prepared in the same manner without enzyme, is proportional to enzyme activity.

Because the PG I isoenzyme is more thermally stable than PG II, the ratio of these two enzymes in an extract can be determined. PG I is determined by measuring the residual activity at 35°C after heating a portion of the reaction mixture without substrate for 5 minutes at 65°C. PG II activity is calculated as the difference between total PG activity and PG I. A standard curve prepared by reacting known amounts of α–D-galacturonic acid (the monomer) with the reagent and plotting the observed absorbance change as a function of the monomer concentration; observed absorbance changes with polymer can be expressed in terms of μmole product produced per minute (units) or per 30 minutes (Pressey units).[12]

EXPERIMENTAL SECTION

The viscosity of 1.0% polygalacturonic acid (Sigma P-3889) at pH 4.5 with various concentrations of added salt ranging from 1 mM to 100 mM should be done at 25°C. The solutions should be prepared well in advance and all solutions well mixed. The polysaccharide must be dissolved in a small volume of warm 0.1 M NaOH; small portions of the carbohydrate are added at a time with considerable stirring. A slight orange coloration in alkali is not a problem. The polysaccharide requires at least 15 minutes to go into solution. The solution is then diluted to 1000 mL and the pH adjusted to 4.5 with more NaOH solution or with 0.1 M HCl. If acid is used, the initial NaCl concentration must be calculated.

The resulting polygalacturonate solution is then used to make up all the salt solutions. Again considerable stirring will be required to insure complete mixing. More concentrated $CaCl_2$ solutions are very viscous and may not be usable with some viscometers. The solutions are poured into the viscometer and at least ten minutes is allowed for thermal equilibration. All viscosity runs should be run in triplicate. The solution is then removed and the viscometer washed with water, air dried (not in an oven), and replaced upright in the bath for the next measurement.

Viscosity studies must be done at constant temperature since temperature affects solution density and viscosity. A stirred constant-temperature water bath large enough to submerge all but the open top of the viscometer must be used. The liquid should be slowly drawn up the capillary tube under reduced pressure by use of tube C (see Figure 28-1). Mouth suction can be used only if the solution contains no poisons and if the rubber tube connecting viscometer and mouth has a trap.

Two electronic timers are required if kinetics measurements are made by the viscosity method. One is a stop clock used to measure the flow time from A to B, and one is a continuously running clock to measure the reaction time for each repeat of the viscosity run. The reaction-time clock is started the moment 100 μL of enzyme solution is added to 10 to 15 mL of polygalacturonate solution in a thermally equilibrated volumetric flask, suspended in the bath. The flask is quickly inverted several times then the contents are poured into the viscometer. Care must be taken to avoid bubbles. Depending on the viscometer, flow times can take as long as several minute.

Since some time is also required to draw the liquid up past A for the next measurement, only four or five viscosity measurements can be made in any ten minute period. Data should be presented in both tabular and plotted form, including plots of specific viscosity as a function of [NaCl] and [$CaCl_2$].

PROJECT EXTENSIONS

1. The viscosity of polygalacturonic acid as a function of pH can also be measured. It appears that the viscosity increases markedly at low pH values. The effect of additional salts on the viscosity can be examined. The effect of the polycation spermine, in both the presence and absence of $CaCl_2$ is very interesting.

2. It may be that the two polygalacturonase isoenzymes have different specificities, with the PG II enzyme showing more *endo* cleavage. Since *endo* cleavage more profoundly affects viscosity than does *exo* cleavage, the relative PG I-to-PG II activity ratio measured by the viscosity method may not be equal to that measured by the colorimetric method which is independent of the position of cleavage. A silica-gel TLC study of the hydrolysis reaction products patterned after the corn syrup experiment described in Chapter 27 will also indicate the *endo*-to-*exo* activity of the enzyme.

3. The effect of salts on the relative activity of PG bed isoenzymes has not been examined. Salts that cause substrate aggregation may effect substrate binding to enzymes.

4. Activity staining of of polyacrylamide gels of the partially purified enzymes can be done by soaking the gels in 0.5% polygalacturonic acid in 0.05 M Na acetate

(pH 4.5) for 15 minutes and then overnight in 0.02% ruthenium red (Sigma R-2751) in water.[8]

DISCUSSION QUESTIONS

1. Some pectin substances are highly esterified, so that an esterase activity is also involved in the breakdown of pectin. These enzymes are called pectinesterases. How would pectinesterase affect the viscosity of pectin solutions?

2. Why is it necessary to dissolve the polygalacturonic acid first in dilute NaOH solution?

3. Some colloidal solutions are *thixotropic*, that is, the viscosity is a function of the shearing forces-- the rate of flow. How would you determine if these solutions are thixotropic?

4. Why is it necessary to mix all polygalacturonate solutions very well before beginning viscosity measurements?

REFERENCES

1. G. O. Aspinall, *The Polysaccharides*, Vol. 1, New York: Academic Press, 1982.

2. B. R. Morris, D. A. Rees, D. Thom and J. Boyd, *Carbohydrate Res.*, **1978**, *66*, 145.

3. I. T. Norton, D. M. Goodall, B. R. Morris and D. A. Rees, *J. Chem. Soc. Faraday Trans.*, **1983**, *79*, 2474.

4. P. J. Flory and T. G. Fox, *J. Polymer Sci.*, **1950**, *5*, 745; *J. Amer. Chem. Soc.*, **1951**, *73*, 1904 .

5. P. R. Crookes and D. Grierson, *Plant Physiol.*, **1983**, *72*, 1088.

6. G. A. Tucker, N. G. Robertson and D. Grierson, *Eur. J. Biochem.*, **1980**, *112*, 119.

7. E. Peisis, Y. Fuchs and G. Zauberman, *Plant Physiol.*, **1978**, *61*, 416.

8. M. Moshrefi and B. S. Luh, *J. Food Biochem.*, **1984**, *8*, 39.

9. G. A. Tucker, N. G. Robertson and D. Grierson, *Eur. J. Biochem.*, **1981**, *115*, 87.

10. R. Pressey, *Eur. J. Biochem.*, **1984**, *144*, 217.

11. R. Pressey, *Hort. Science*, **1986**, *21*, 490.

12. P. Bernfeld, *Methods Enzymology*, **1955**, *1*, 149

Chapter 29

Two-Dimensional PAGE of Erythrocyte Membrane Proteins

Hemoglobin-free human erythrocyte ghosts are prepared by single-stage hemolysis and washing in 10 mM (20 imOsm) phosphate buffer at pH 7.4. The membrane proteins are then separated in the first dimension by isoelectric focusing in Triton X-100/ urea polyacrylamide gels according to their isoelectric points. In the second dimension proteins are further separated according to their molecular weights using SDS-gel electrophoresis, with an acrylamide gradient gel to enhance resolution. The gels are stained with silver stain.

KEY TERMS

Ampholytes   *Erythrocyte ghosts*
Hypotonic solutions   *Isoelectric focusing*
O'Farrell two-dimensional gels   *Osmotic pressure*

BACKGROUND

One-dimensional electrophoretic methods often give insufficient resolution of complex protein mixtures. This is the case with membrane protein mixtures, which often contain 30 or more proteins. In such cases a two-dimensional (2D) electrophoretic separation is necessary. Each dimension involves a separation according to a different physical-chemical parameter. The very widely used method of O'Farrell[1] involves first isoelectric focusing, during which proteins are separated according to their isoelectric points. The second dimension is SDS-polyacrylamide gel electrophoresis (Laemmli), which separates according to molecular size.

The procedure here differs from the original O'Farrell protocol in a number of significant ways:

1. Membrane proteins are insoluble in normal buffers and are freed from the membrane and rendered soluble by the action of the nonionic detergent Triton X-100 and the chaotropic agent urea.

305

2. The first dimension gels are run in 1.5-mm-I.D. glass capillary gel tubes so that the first-dimension gel diameter is identical in thickness to the second-dimension slab gel. This eliminates the need for agarose embedding which was always a troublesome aspect of the original method.[2]

3. An acrylamide gradient separating gel is used to enhance resolution.[3,4]

The first laboratory period is spent preparing the membrane proteins and the first-dimension gel. The first-dimension gel can be run at low voltage overnight until early morning or at high voltage for 2 hours if the laboratory period can be extended. Otherwise, the first gel is run during the second laboratory period while the second-dimension slab gel is prepared. The second-dimension electrophoresis requires more than an hour to set up and 5 to 6 hours running time, followed by silver staining which may take an additional hour, so the final step requires almost a full day. Two well-planned afternoon sessions, a brief morning session to remove the first gel, and a Saturday for the final run is the usual schedule.

Isoelectric Focusing[5]

This method takes advantage of the fact that almost every protein has an unique isoelectric point pI, a pH where the sums of the cationic and anionic charges on the sidechain groups are equal, so the protein is electrically neutral. Proteins are separated according to their pI values by electrophoresis on a gel in which a stable pH gradient has been generated, with high pH at the cathode [(-)pole at top] and low pH at the anode [(+) pole at bottom of the gel]. The pH gradient is generated by a mixture of carrier ampholytes that distribute themselves along the tube in order of pI when subjected to an electric field. Each ampholyte maintains a local pH because of its strong buffering capacity. If the proteins are applied to the gel at a pH well above their isoelectric points, the anionic proteins will move toward the anode and into a zone of decreasing pH. When each protein passes down the gel, its anionic carboxylate (asp, glu, and C-terminus) groups become neutralized by protonation and its N-terminus and lys, arg, and his side chains are protonated to their ammonium forms. The protein becomes less and less negatively charged. When it reaches neutrality at its pI, it loses its electrophoretic mobility and becomes focused in a narrow zone. Diffusion is offset by the electric field, so the band does not broaden as in other methods, and proteins differing in pI by as little as 0.01 pH unit can be resolved.

Second Dimension: SDS Gel-Gradient Electrophoresis

This step requires a larger slab gel (13.5 cm x 16 cm x 1.5 mm) rather than the mini-gels described in Chapter 9. The gel apparatus is first filled with 1 cm of water-saturated butanol, then from the bottom to a height of 13.5 cm with a linear polyacrylamide gradient ranging from 20 to 8%. Polymerization is complete after 2 hours, so the butanol can be dumped out (gently!) and an upper stacking gel with its water-saturated-butanol overlayer can be applied.

Membrane Proteins

Hydrophobic membrane proteins are extracted from hemoglobin-free erythrocyte ghosts, which can be prepared from packed human erythrocytes. The erythrocytes are first freed from plasma and buffy coat as described in Chapter 11. The membrane prepared by washing in hypotonic buffer form structures called *erythrocyte ghosts*[6] because they assume roughly spherical shapes about the same size as intact erythrocytes, but they are empty and more or less transparent, ghostly vestiges of healthy red erythrocytes. The internal protein machinery of the cell is lost during the hemolysis and washing, but proteins intrinsic to the membrane remain in the ghostly membrane.

The *osmotic pressure* π of a dilute solution of total ion concentration c can be estimated from the expression

$$\pi = cRT \qquad\qquad (29\text{-}1)$$

A 0.155 M sodium phosphate (NaH_2PO_4) solution is 310 *ideal milliosmolar* (imOsm) and will have an osmotic pressure of 7.9 atm at 37°C (310°K). This is typical of the osmotic concentration of the cytoplasm inside animal cells. When such cells come in contact with more dilute (hypotonic) salt solutions or pure water, the osmotic pressure within the cells may cause destructive rupture of the cell membrane so that the cellular contents can be washed away. This happens in the hemolysis of red blood cells; blood serum ranges from 285 to 295 milliosmolar (mOsm) and erythrocytes lyse if exposed to salt significantly less than this salt level. In this experiment hypotonic 20 imOsm phosphate buffers are used to effect hemolysis.

The proteins within the ghost pellet must be solubilized using a detergent solution and this is a notorious difficulty of membrane protein biochemistry. The anionic surfactant SDS will not do at all and the nonionic surfactant Triton X-100 will not solubilize all proteins. Impurities in commercial-grade Triton X-100 also give troubles, including heavy background staining in the acidic portion of the 2-D gel and the peroxide oxidation of some sensitive proteins.[7,8] For critical research work the high-purity zwitterionic detergents CHAPS and CHAPSO are often used to solubilize intractable membrane components, but the expense of such detergents ($14 to 20 per gram) rules out their use in a teaching experiment.

CHAPS
CHAPSO (OH at arrow)

The ghost preparation should contain at least 100 µg of protein per microliter so a portion should be assayed by the Bradford method (see Chapter 4). A 2.5-µL sample of protein-rich erythrocyte ghosts in an Eppendorf centrifuge tube is made up about 50% (v/v) with glycerol to give 5 µL of a dense solution which is then combined with 3 mg of solid urea that is dissolved by gentle agitation. This solution is combined with 1 µL of the sample concentrate containing urea, Triton X-100, β-mercaptoethanol and the ampholytes Bio-Lyte 3.5/7 and Bio-Lye 3/10, which buffer over the pH ranges 3.5 to 7 and 3 to 10 respectively, and centrifuged to the bottom of the Eppendorf tube. The entire sample is applied to the gel surface of a gel tube via a very fine glass tube attached to a 10-µL automatic pipet tip with a fine piece of plastic tubing. Practice with just 50% glycerol solution may be necessary in order to optimize the pickup and delivery from this contrivance.

EXPERIMENTAL SECTION

First Laboratory Period

Students should work in pairs the first day with one student preparing the erythrocyte membrane ghosts and the protein sample while the other prepares the gels. The process will go faster if some of the solutions marked with * are prepared ahead of time by the stockroom assistants. The gel recipe given below is enough for several hundred thin gel tubes and can be made up by one student or by the stockroom staff.

Extraction of Membrane Proteins

Membranes are isolated according to the method of Dodge[6] from human erythrocytes. The packed erythrocytes are freed from plasma and the buffy coat by repeated washes in isotonic buffer (see Chapter 11), then repacked by centrifugation. The erythrocytes are first washed and centrifuged as in Chapter 11 then pipetted into a 30-fold volume excess of hypotonic buffer, mixed by gentle swirling and then centrifuged at 20,000 x g for 40 minutes. The hemolysis is performed by pipetting 2-mL aliquots of the washed red blood cell suspension into a plastic centrifuge tube containing 28 mL of 20 mOsm phosphate buffer and gently mixed by swirling then centrifuged at 20,000 x g for 40 minutes.The supernatant must be decanted very carefully and the ghost pellet resuspended and washed twice more with 20 imOsm buffer at pH 7.4. The supernatant after the last wash must be either colorless or very faintly pink.

The supernatant is decanted away carefully and replaced by an equivalent volume of fresh hypotonic buffer. The ghosts are to be washed at least three times or until the supernatant is nearly colorless. Careful decantation of the supernatant is the most critical feature of this preparation. The final pellet should be diluted with buffer until the protein concentration is about 100 µg/mL; a Bradford protein assay with prolonged shaking is necessary.

The membrane protein sample for isoelectric focusing is prepared by combining 2.5 µL of the foregoing membrane pellet suspension in an Eppendorf tube with enough glycerol to give a 50% (v/v) mixture, then combining the sample with 3 mg of solid urea. The urea is dissolved upon agitation. The 5 µL of sample is then combined with 1 µL of isoelectric focusing sample concentrate and all the liquid is brought to the bottom of the tube by centrifugation. Isoelectric focusing sample concentrate is prepared according to the formulation

10 % SDS*	0.1 mL
Bio-Lyte 3.5/10 (Bio-Rad)	0.02 mL
Bio-Lyte 5/7 (Bio-Rad)	0.18 mL
2-mercaptoethanol	0.1 mL
Triton X-100 (undiluted)	0.2 mL

It must be agitated in a vortex shaker prior to use.

First-Dimension Electrophoresis

In the procedure suggested here, 1.5 mm I.D. x 150 mm electrophoresis tubes are used. Each tube must be very clean (chromic acid overnight, followed by prolonged rinsing and oven drying) and is filled with gel solution from a syringe. To minimize bubble formation, the gel solution is delivered to the gel tubes by syringing through fine glass tubing which is inserted all the way to the bottom of the glass capillary electrophoresis tube, which has been sealed at the bottom with Parafilm. The fine tubing must be prepared ahead of time, by the laboratory staff; each tube will have to be drawn very thin and straight but not too thin or it will fill the tubes very slowly. One should be able to fill 10 tubes without fear that the gel will polymerize in the syringe. All air bubbles must be dislodged from the bottom. A 10-cm gel overlayered with 20 μL (1 cm) of overlay buffer is ideal. Air bubbles at the upper surface of the overlay are not a problem.

The isoelectric gel is prepared according to the following formulation:[2]

Isoelectric focusing gel solution:

Urea	48.6 g
Water	28.8 mL
Acrylamide/ BIS (30%) Stock	11.8 mL
10 % Triton X-100	20.3 mL
Bio-Lyte 5/7 (Bio-Rad)	4.5 mL
Bio-Lyte 3/10 (Bio-Rad)	0.5 mL

where the following solutions can be prepared in advance:

Acrylamide/BIS (30%) Stock*

Acrylamide	29.2 g
Bisacrylamide	0.8 g
Water	up to 100 mL

Triton X-100 (10%) Stock*

Triton X-100	10 mL
Water	90 mL

This is enough material for 250 to 280 gels. Long term storage of this solution is best[2] done in 5.5-mL aliquots frozen at -80°C (or at least in the freezer).

CAUTION ACRYLAMIDE IS A NEUROTOXIN. USE GLOVES AND A HOOD.

Enough gel solution to pour two dozen 10-cm gels is prepared by degassing 5.5 mL of the foregoing solution under reduced pressure for one minute then quickly adding 5.5 μL of TEMED and 7.25 μL of fresh 10% ammonium persulfate. The gel tubes must be very clean and dry and premarked at the 10 cm line. The tubes are stoppered at the bottom with parafilm. Care must be taken to avoid perforating the Parafilm with the capillary tubes; paraffin trapped in the bottom of the tube cannot be easily removed. The tubes are filled by injecting the gel solution through a very finely drawn-out hollow glass tube mounted on a syringe (see Figure 29-1).

It is always a good idea to pour more gels than required since many show bubbles or other imperfections upon hardening. Practice makes for perfection, but one should not dawdle as the gel will harden quickly in the fine tube. The gel solution must also be rinsed from the syringe before polymerization.

The gel in the tube should be overlayered with 20 µL (1 cm) of the following overlay buffer:

Urea	3.0 g
Triton X-100 10 %	2.0 mL
Bio-Lyte 5/7	0.45 mL
Bio-Lyte 3/10	0.05 mL
Water	up to 10 0 mL

The gel polymerization is complete within 1 hour. This is usually enough time to prepare the membrane protein sample. The water overlay is poured off and the protein sample is injected directly onto the gel surface with a fine glass tube attached to an Eppendorf automatic pipette. The sample is then overlayered with 10 µL of fresh overlay solution and filled to the top with 0.1 M NaOH. The gel is mounted in the gel apparatus with 0.06% phosphoric acid in the bottom reservoir, and all bubbles are drawn off from the bottom of the gel tube with a syringe. The upper reservoir contains 0.1 M NaOH.

Gels that are free of bubbles and where the bottom of the gel is nearly flush with the bottom of the tube (to avoid accumulating bubbles during the run) are placed in the gel apparatus. The aqueous overlayer is drawn off and replaced with sample in buffer applied directly on the gel surface with a fine drawn out glass tube attached to a 50 µL automatic pipette. The sample will not run down the side of the tube; it will bead up forming an air bubble. The tube is then layered to the top with 0.1 M NaOH, and then the upper electrode compartment is filled with then same solution. The lower reservoir is filled with 0.06% phosphoric acid. Great care must be taken to avoid leaking seals in the gel apparatus (see Chapter 9) as this will lead to an acid-base reaction and heat up the gels as well as shorting out the electrophoresis.

After electrofocusing for 2 hours at 800 volts constant voltage, the power is disconnected and the lower buffer reservoir is filled with ice to chill the gel tubes. The first-dimension gels are more easily removed from the gel tubes after they have been well chilled in ice. They are carefully blown out with air pressure after the gel has been loosened with a solution of 10% glycerol squirted around the gel within the tube with a very fine needle. The air pressure must be carefully controlled! Each gel is then equilibrated with 2-3 mL of SDS-reducing buffer for about 10 minutes, then either frozen at -20°C or applied to a second-dimension SDS slab gel.

Second Laboratory Period

Isoelectric focusing is carried out overnight (12 to15 hours) at 400 volts followed by two hours at 800 volts. The gels are chilled in ice and then carefully removed from the capillary tubes using a controlled flow of air as shown in Figure 29-1. The gel can be loosened by squirting 10% glycerol around the gel within the tube. A cotton plug should be used in the line if expired air is used to catch any proteinaceous material from the mouth. A T-tube permits fine regulation of the airflow. Connecting the T-tube to the capillary tube may require two concentric plastic tubes, one from the capillary fitting inside the one over the T-tube.

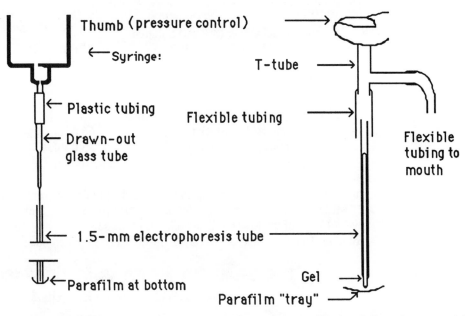

Figure 29-1 How the isoelectric gel capillary tubes are filled and the gels removed by gentle blowing

The gel is transferred to 2 to 3 mL of the following SDS-reducing buffer and equilibrated for 10 minutes:

0.5 M TRIS (pH 6.8)*	12.5 mL
2-Mercaptoethanol	5.5 mL
10 % SDS*	30.0 mL
0.1 % Bromophenol blue	1 mL

made up to 100 mL with water.

The equilibrated tube gel must be distinctly blue (because this serves as the tracking dye in the slab electrophoresis step). If it is not, it must be equilibrated once again with additional dye in the buffer.

If not used immediately the equilibrated gel should be frozen at -20°. If more than one gel rod is available, one can be used for pH-measurement. This gel is not equilibrated; instead, it is cut into 0.5-cm pieces and each piece soaked until the next lab period in CO_2-free water in a tightly stoppered vial. The pH can be measured with a combination electrode sometime during the next lab period.

Third Laboratory Period

The acrylamide gradient SDS-Slab gel is prepared according to the following formulations:[4]

Separating Gel Buffer*	
TRIS	18.2 g
10 % SDS	4.0 mL
Water	up to 100 mL

The pH is adjusted to 8.8 with concentrated HCl.

Separating Gel Upper (8% gel) Solution
Acrylamide/ BIS Stock (30%)   26.6 mL
Separating gel buffer (above)   25.0 mL
Water   up to 100 mL

Separating Gel Lower (20% gel) Solution
Acrylamide/ BIS Stock (30%)   66.6 mL
Separating gel buffer (above)   25.0 mL
Glycerol (87%, v/v)   up to 100 mL

Just before adding the foregoing solutions to the gradient mixer, one must add 10 μL of TEMED for every 16 mL of solution and 500 μL of 10% ammonium persulfate for every 17 mL of the lower gel, and 650 μL of the same for every 16 mL of the upper gel solution. The gel should be overlayered with water-saturated butanol until polymerized.

THE GRADIENT MIXER MUST BE WASHED IMMEDIATELY AFTER USE.

Stacking gel buffer*
TRIS   6.05 g
10 % SDS   4.0 mL
Water   up to 100 mL
The pH is adjusted to 6.8 mL with concentrated HCl

Stacking gel
Stacking gel buffer   8.25 mL
Acrylamide/BIS Stock (30%)   3.3 mL
Water   21.5 mL

The foregoing is combined with 35 μL of TEMED and 250 μL of fresh 10% ammonium persulfate and the gel is poured quickly. Polymerization should be complete within 1 hour.

Fourth Laboratory Period

The 2-D slab electrophoresis involves laying out a freshly equilibrated or thawed isoelectric focusing gel on a folded piece of Parafilm and drawing off all reducing buffer with a pipette. One should manipulate the gel only with a glass rod or the pipette tip. The gel is then placed carefully between the plates of the mounted slab gel. Start with one end and carefully slide the gel into place a section at a time. If the tube gel breaks in this process, it still can be used if the pieces are rather tightly linked together within the slab. Any trapped air bubbles must be removed with a gentle squirt of diluted reservoir buffer (below) and the gel must not be squeezed to deform its circular cross-section. After the tube gel is firmly in place on the surface of the slab gel, the slab plate is filled to the top with reservoir buffer.

Both the upper and lower reservoirs of the slab apparatus are filled with freshly diluted reservoir buffer stock solution.

Reservoir Buffer stock solution*

SDS	20 g
TRIS	60 g
Glycine	288 g
Water	up to 2000 mL

The solution is diluted 1:10 with water prior to use.

The electrophoresis is run at 25 mA/gel, constant current, until the bromophenol dye marker is at the bottom of the gel. This requires 5 to 6 hours. The gel is removed and stained by the silver stain method as described in Chapter 9

REFERENCES

1. P. H. O'Farrell, *J. Biol. Chem.* **1975**, *250*, 4007.

2. Bio-Rad Laboratories Bulletin **1983**, 1144

3. A. Boschetti, E. Sauton-Heiniger, J. C. Schaffner, and W. Eichenberger in *Chloroplast Development*, eds. G. Akoyunouglu et al, New York: Elsevier/ North-Holland, 1978, pp. 195-200.

4. A. Boschetti, E. Stauton-Heiniger, and K. J. Clemetson in *Membrane Proteins, a Laboratory Manual*, eds. A. Azzi, U. Brodbeck and P. Zahler, New York: Springer-Verlag, 1981, pp. 3-12.

5. J. P. Arbuthnot and J. A. Beeley, *Isoelectric Focusing*, London: Butterworths, 1975.

6. J. T. Dodge, C. Mitchell, and D. J. Hanahan, *Arch. Biochem. Biophys.*, **1963**, *100*, 119

7. H. W. Chang and E. Bock, *Anal. Biochem.*, **1980**, *104*, 112.

8. Y. Ashani and G. N. Catravas, *Anal. Biochem.*, **1980**, *109*, 55.

Chapter 30

Two-Dimensional TLC Lipid Characterization

One gram of soft tissue is homogenized with 18 mL of isopropanol/hexane (3:2, v/v) at room temperature. This solvent system dissolves virtually all the tissue lipids but very little pigment or protein. After filtration through sintered glass, the clear extract is concentrated, then spotted onto four 20 x 20 cm silica G plates. The plates are placed upright in a sealed chromatography tank and run in the ascending mode with chloroform/methanol/28% aqueous ammonia (65:35:5) until the solvent front reaches the upper edge. The plates are removed, dried in air, and turned clockwise 90° and set upright in a different tank containing chloroform/acetone/acetic acid/water (10:4:2:2:1). Again the solvent front is allowed to reach the upper edge before the plates are removed. The four dry plates are developed by the following methods: Plate 1 is sprayed either with 0.001% Rhodamine 6G in water, which reveals most lipids as pink spots on a yellow background when viewed with UV illumination, or with iodine vapor which reveals lipids as brown or grey spots on a pale yellow background. Plate 2 is treated with α-naphthol spray followed by 50% aqueous sulfuric acid. Brief oven drying reveals blue or purple spots for glycolipids, with most other lipids showing up as yellow spots. Plate 3 is treated with resorcinol spray test reagent to give purple spots for gangliosides, after brief heating, whereas other glycolipids give weak yellow-brown stains. Plate 4 is treated with modified Dittmer-Lester reagent, which gives blue spots for phospholipids.

KEY TERMS

Cerobrosides
Fatty acid esters
Glycolipids
Lipids
Saponification
Steroids

Developing solvent
Gangliosides
Glycosphingolipids
Phospholipids
Sphingolipids
Triacylglycerides

BACKGROUND

Together with proteins and carbohydrates, *lipids* form the bulk of the organic matter of living cells. But unlike proteins and carbohydrates, which are defined rather precisely on the basis of their structural units--amino acids and monosaccharides--the term "lipid" refers to a large heterogeneous collection of hydrophobic molecules more or less derived from aliphatic residues such as long-chain fatty acids. There is a wide variation in the structures and chemical properties of lipids. As a class they are defined by their solubility rather than structural motif; the class includes many chemically unrelated compounds. Lipids can be extracted from plant and animal tissue with a variety of organic solvents (e.g. benzene, chloroform, trichloroethylene), but almost all lipids are only sparingly soluble in water. However, lipids can form biochemically important aggregates such as micelles and bilayer membranes in water. Micellar aggregates are formed spontaneously when amphipathic lipids, e.g. lipids with hydrophilic polar "head groups" and hydrophobic "tails" are suspended in water. The polar head groups protrude into the solvent whereas the nonpolar tails associate by hydrophobic interactions and are excluded from the aqueous solvent. Such micelles play an important role in transporting nonpolar molecules from tissue to tissue. Biomembranes are comprised of both proteins and lipids, with the composition of the lipid portion rich in structural variety, including lipids that are strongly associated with proteins and with carbohydrates. Approximately 1500 different lipids have been identified in myelin of the central nervous system. About 30 lipids are present in substantial amounts.[1]

The class of lipids includes neutral fats (triacylglycerides), waxes, terpenes, carotenoids, steroids and complex lipids such as phospholipids and glycolipids. The components of lipids are linked in a variety of ways. *Triacylglycerides*, the common fats of adipose tissue and plant oils are esters of long-chain fatty acids with glycerol. This type of lipid is an important depot of stored chemical energy that is marshaled for ATP synthesis only when carbohydrate intake is reduced (Figure 30-1). Triacylglycerides are very nonpolar and water insoluble. Their hydrolysis or *saponification* in strong base solutions yields glycerol and three molecules of fatty acids, some of which contain one or more double bonds. Waxes are esters between fatty acids and hydrophobic alcohols; they seem even less polar and more water insoluble than triacylglycerides.

Terpenoids and carotenoids are highly unsaturated polymers derived biosynthetically from isopentylpyrophosphate, a five-carbon branched unit. Such compounds usually have empirical formulas with carbon atoms occurring in multiples of five, but beyond this the structural diversity of terpenoids and carotenoids is too involved for discussion here. *Steroids* are also synthesized from the same starting pyrophosphate and obey the five-carbon "isoprene rule." Steroids feature a 4-carbocyclic ring skeleton consisting of three fused six-membered rings and one five-membered ring. The most abundant steroid lipid is cholesterol with only a single OH group as a polar head group. Acylcholesterols are even less polar since the OH group of cholesterol is esterified to a fatty acid in such compounds.

Phospholipids are much more polar since they contain a negatively charged phosphate diester group usually linked to another polar residue. Phospholipids are always present in biomembranes and they make up 40 to 90% of the total lipid in most samples. Several types of phospholipids predominate. Some are derivitives of glycerol and are known as *glycerol lipids*. Phosphatidylcholine, phosphatidyl-ethanolamine, phosphatidylserine and phosphatidylinositol all are all phosphoric acid diesters of di-acylglycerol and a second polar residue. Cardiolipids or diphosphatidyl glycerols contain two diacylglycerol phosphates linked through a glycerol molecule. Phospholipids derived from an unsaturated amino alcohol, sphingosine, are known as *sphingolipids*.

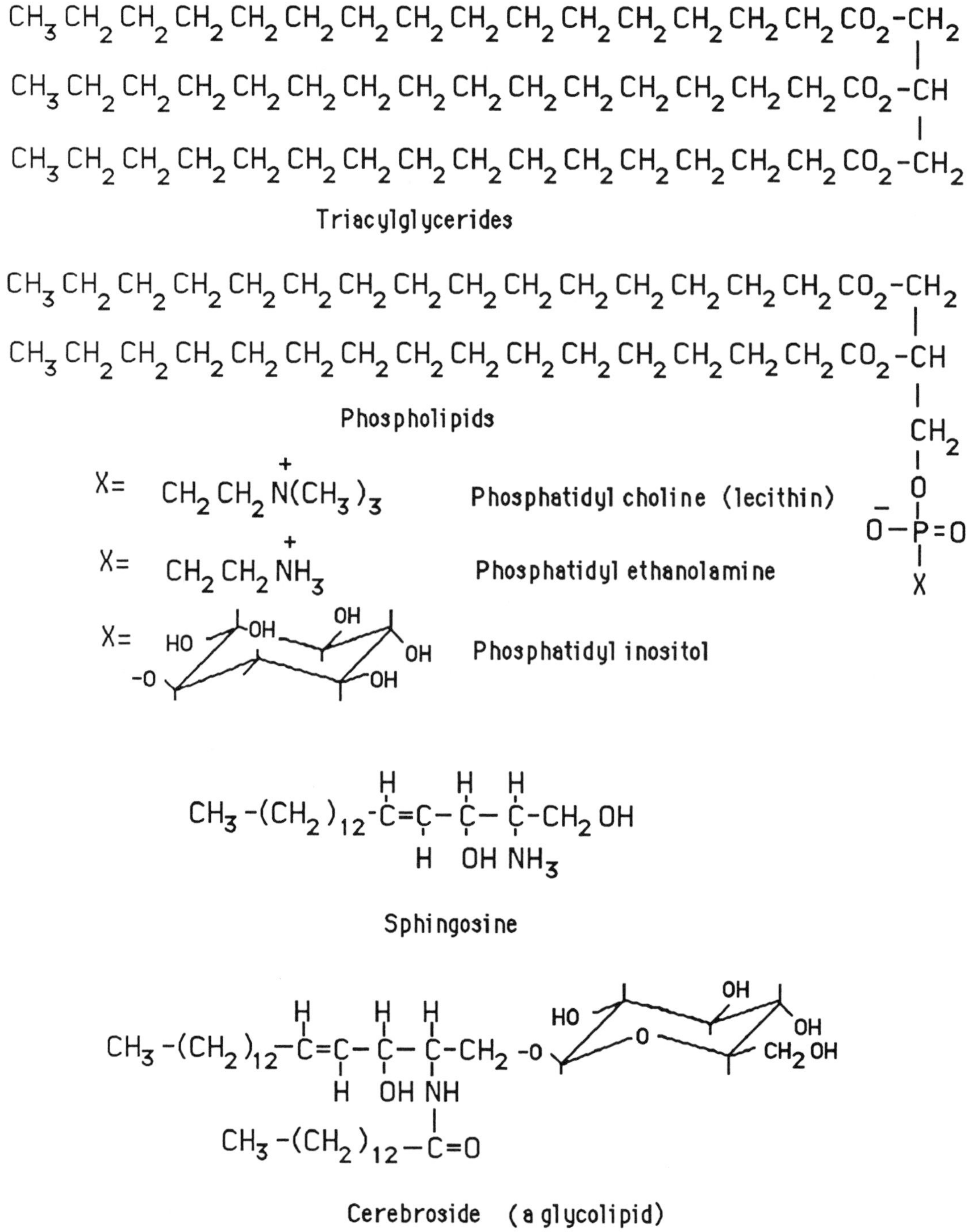

$$CH_3\ CH_2\ CH_2\ CH_2\ CH_2\ CH_2\ CH_2\ CH_2\ CH_2\ CH_2\ CH_2\ CH_2\ CH_2\ CH_2\ CH_2\ CO_2\text{-}CH_2$$
$$CH_3\ CH_2\ CH_2\ CH_2\ CH_2\ CH_2\ CH_2\ CH_2\ CH_2\ CH_2\ CH_2\ CH_2\ CH_2\ CH_2\ CH_2\ CO_2\text{-}CH$$
$$CH_3\ CH_2\ CH_2\ CH_2\ CH_2\ CH_2\ CH_2\ CH_2\ CH_2\ CH_2\ CH_2\ CH_2\ CH_2\ CH_2\ CH_2\ CO_2\text{-}CH_2$$

Triacylglycerides

$$CH_3\ CH_2\ CH_2\ CH_2\ CH_2\ CH_2\ CH_2\ CH_2\ CH_2\ CH_2\ CH_2\ CH_2\ CH_2\ CH_2\ CH_2\ CO_2\text{-}CH_2$$
$$CH_3\ CH_2\ CH_2\ CH_2\ CH_2\ CH_2\ CH_2\ CH_2\ CH_2\ CH_2\ CH_2\ CH_2\ CH_2\ CH_2\ CH_2\ CO_2\text{-}CH$$

Phospholipids

$X = CH_2\ CH_2\ \overset{+}{N}(CH_3)_3$ Phosphatidyl choline (lecithin)

$X = CH_2\ CH_2\ \overset{+}{N}H_3$ Phosphatidyl ethanolamine

$X = $ (inositol structure)   Phosphatidyl inositol

$$CH_3\text{-}(CH_2)_{12}\text{-}\overset{\underset{H}{|}}{C}=\overset{\underset{H}{|}}{C}\text{-}\overset{\underset{OH}{|}}{\overset{H}{\overset{|}{C}}}\text{-}\overset{\underset{NH_3}{|}}{\overset{H}{\overset{|}{C}}}\text{-}CH_2\ OH$$

Sphingosine

$$CH_3\text{-}(CH_2)_{12}\text{-}C=C\text{-}C\text{-}C\text{-}CH_2\text{-}O\ (sugar\ structure)$$
$$CH_3\text{-}(CH_2)_{12}\text{-}C=O$$

Cerebroside (a glycolipid)

Figure 30-1 Common types of lipids

A ceramide consists of a fatty acid linked by an amide bond to the amino group of sphingosine. In sphingomyelins the amino group of sphingomyelin forms an amide linkage to a fatty acid and the primary hydroxyl is esterified to phosphatidyl choline hence they are ceramide derivatives.

Glycolipids are compounds in which one or more monosaccharide residues are glycosidically linked to a lipid moiety, either a mono- or diacylglycerol, a shingosine or a ceramide derivative. The cerebrosides consist of sphingosine carrying a fatty amide or hydroxy fatty amide and a single monosaccharide unit glycosidically linked to its alcohol group. The sugar is usually glucose or galactose. *Cerebrosides* make up about 10% of the dry weight of the brain. Sulfatides are cerebrosides in which the sixth carbon of the carbohydrate is esterified to a sulfate residue.

More than one monosaccharide unit can be strung together off the primary alcohol group of sphingosine. A vast number of *glycosphingolipids* composed of some combination of gluucose, galactose, fucose, N-acetylglucosamine, and N-acetyl-galactosamine have been observed. If the chain contains sialic acid (N-acetyl-neuraminic acid), the lipid is a *ganglioside*. These compounds are especially abundant in the gray matter of the brain, and they are the characteristic components of some neuronal membranes of the central nervous system.

This project involves extracting the lipid compoments of a homogenized tissue into a mixture of hexane and isopropanol.[2] Because the extraction solvent consists of a low-polarity solvent and a moderately polar water-miscible solvent, it penetrates cell membranes and dissolves a wide range of lipids that differ considerably in their water solubility. Moreover it extracts almost no protein or other nonlipid tissue components and the solvent is not too volatile to be troublesome. The polar lipid components of this extract can be partitioned between silica gel and two organic solvent mobile phases in a two-dimensional thin-layer chromatography experiment (*2D TLC*). Because of the very large number of different lipids in a tissue even the best one-dimensional TLC cannot resolve all of them simultaneously. But if a two-dimensional technique with two widely different solvent systems is used, a large number of different lipids can be beautifully separated from each other. Silica gel ($SiO_2 \cdot xH_2O$) is a mildly polar solid adsorbant for TLC and will bind polar compounds more tightly than nonpolar ones. Polar molecules are bound by a combination of dipole-dipole interactions, hydrogen bonding, and coordination to the polar silicon oxide surface. The strength of these interactions varies for different compounds in subtle ways. The mobile phase or *developing solvent* is a mixture of organic solvents and different lipid solutes will have different solubilities in this mixture. As this solvent moves slowly up the plate, various lipids will be shuttled back and forth between the stationary and mobile phases in a process akin to extraction. Polar compounds that bind well to silica gel will reside in the stationary phase more of the time than will less polar lipids that move with the solvent mobile phase. Hence polar compounds will move up the plate more slowly than nonpolar ones but the actual mobility will depend on the solubility in the mobile organic solvent mixture. Because the mobility of lipids depends not only on their binding affinity to the absorbant but also on their solubility in the mobile phase, this mobility may be quite different for different solvent systems. In this experiment we use a basic first solvent (chloroform/methanol/28% aqueous ammonia; 65:35:5 by volume) and an acidic second solvent mixture (chloroform-acetone-methanol-acetic acid; 10:4:2:2:1 by volume).[3,4] The mixture of lipids is spotted in the right-hand bottom corner of the TLC plate and the plate is developed in one direction with the basic solvent. The plate is removed and dried when the solvent reaches the top edge, then turned 90° and developed in the second direction with the acidic solvent. The basic solvent must be used first rather than the acidic one since some lipids may decompose during drying of plates developed with acidic solvents.

After the second development cycle, the plate is removed, dried and then stained by some reagent that identifies specific spots with a characteristic color. These colors are the result of specific chemical reactions between the traces of lipids that make up each spot. Unfortunately no one has devised a reagent that serves to identify all types of lipids at one time on one plate. Most of the reactions are incompatible, and several identical chromatographs, prepared and developed under identical conditions, must be tested with a variety of reagents to identify the spots as to type of lipid. This technique requires considerable care and attention to detail inorder to get four nearly identical plates. Your instructor may elect to have each group of students prepare and test four plates with four different reagents, but if this is too expensive or time demanding, the entire class may be divided into four groups of students, with each group running only one plate and witnessing the effect of one reagent. The following four reagents are recommended because of the wide range of polar lipid types identified with this set:

1. A universal spray for lipids is 0.001% Rhodamine 6G in water which reveals most lipids as pink spots on a yellow background when viewed with UV illumination. Another method is to expose plates to iodine vapor, which stains all absorbed compounds either brown or gray.

2. Glycolipids are revealed by reactions analogous to the chemical tests for carbohydrates described in Chapter 27. A spray of α-naphthol in 95% ethanol, followed by 50% aqueous sulfuric acid and brief oven drying, reveals blue or purple spots for glycolipids with most other lipids showing up as yellow spots.

3. Resorcinol spray test reagent[5] gives purple spots for gangliosides, after brief heating, whereas other glycolipids give weak yellow-brown stains.

4. Modified Dittmer-Lester reagent,[6] a molybdenum-based reagent, gives blue spots for phospholipids. Hence only plates stained by the methods above will give a pattern of lipid spots for all lipids, glycolipids, sialic acid containing gangliosides, and phospholipids. Hence four plates give a pattern of lipid spots, partially identified by lipid type. In careful hands, the 2D chromatographs are highly reproducible and reveal interesting distributions of lipid types from tissue to tissue.

EXPERIMENTAL SECTION

Students work in groups of two or three with a 1-g sample of lipid containing tissue provided by the instructor (see below). The extraction solvent is prepared from dry, reagent grade or freshly distilled isopropanol and hexane in the ratio 3:2 by volume.[2] One gram of soft tissue is suspended in 18 mL of solvent and homogenized in a tissue grinder, either a Brinkman Polytron or a Dounce or Potter-Elvehjem homogenizer, for at least 1 minute. Some preliminary chopping may be required before homogenization. The suspension is filtered through a sintered-glass funnel.

The insoluble residue is resuspended in 3 mL of additional extraction solvent and left standing for an additional minute, then filtered as before. This re-extraction can be repeated a second time to afford a total of about 24 mL of clear extract containing virtually all the tissue lipids, but very little pigment or protein. The extracts of lipid-rich tissues (e.g., brain) can be used as is, but extracts of most tissues have less total lipid and should be concentrated by evaporation in a stream of dry air.

2D Thin Layer Chromatography

Chromatographs should be run on 20 x 20 cm squares of <u>fluorescent</u> <u>indicator-free</u> silica gel on glass (J.T. Baker, BC 7000-4) or a glass microfiber sheet (Gelman ITLC, 61886). The latter medium is developed very rapidly. Each plate is handled as follows. The lipid is applied with a glass capillary to the lower right corner about 1 to 2 cm from either edge; several applications to the same point, after the previous spot has dried, may be necessary for low-lipid extracts. The application spot should be kept less than 2 mm in diameter. The plate should then be placed in an equilibrated chromatography tank containing a 1-cm layer of the first eluting solvent, chloroform/methanol/28% aqueous ammonia (65:35:5), and eluted until the solvent front reaches the upper edge. The chromatography tank must not be left standing open while all four plates are spotted and loaded. Spot all the plates first then open the tank and quickly place the plates upright with the absorbant sides not touching anything.
Each plate is removed, dried in air (5-10 minutes in a hood), then turned clockwise 90° and placed in a fresh equilibrated tank containing the second eluting solvent, chloroform/acetone/methanol/acetic acid/water (10:4:2:2:1). The chromatogram is developed in the second direction, then removed and air dried as before.

Spraying Techniques

Many common spray reagents are poisonous when inhaled and can cause serious burning of the skin and eyes.

ALL SPRAYING MUST BE CONFINED TO A
DEDICATED, WELL VENTILATED   HOOD.
STUDENTS MUST WEAR SAFETY GLASSES
AND GLOVES.

All liquid left after the spraying must be cleaned up promptly. The spray bottle or container must be clean before filling with a test reagent. There are commercially available spraying units employing aerosol propellents but such devices should not be used with solvents that interact with the spraying unit, especially sulfuric acid. A rapid flow of air from an air compressor can be used to generate a fine mist by capillary flow. An airstream is forced past the capillary opening on the spray bottle as illustrated in Figure 30-2; the liquid is drawn up the tube and exits as a fine spray. The spray is applied evenly with constant motion of the aerosol can, if one is used. Do not allow spray to puddle or run on a chromatogram.

Specific Detection Tests

Rhodamine Spray
Plates sprayed with 0.001% Rhodamine 6G in water are viewed under UV illumination while still damp. Lipid spots show up on sprayed silica gel as pink, orange or bright yellow spots on an orange background. Because the spots are not visible to the unaided eye, they must be circled with a soft pencil right after staining.

Iodine Vapor Spray
Plates are placed in a sealed glass tank or large jar containing a few iodine crystals. The iodine vapor is brought to the pale purple stage by briefly warming the bottom of the container and the plate is stained within a minute. The stain is not permanent so the

spots must be circled with a soft pencil right after staining. Rhodamine staining followed by iodine treatment is a very sensitive method, detection levels in the 0.1 μg range are possible.[7]

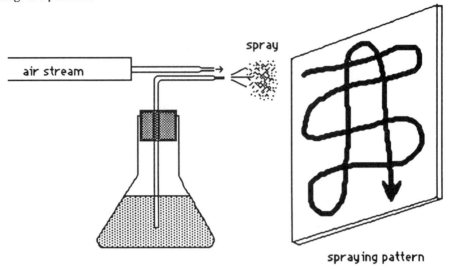

Figure 30-2   Reagent Sprayer and suggested spraying pattern

α-Naphthol Test for Glycolipids

A 0.3 % solution of a-naphthol in 95% ethanol is sprayed until the plate is damp. The plate is dried in warm air and sprayed lightly with a fine mist of 50% sulfuric acid and then heated at 100° for 5 minutes.

**CAUTION: SULFURIC ACID SPRAYS
MUST BE DONE IN A HOOD.
SAFETY GLASSES AND GLOVES MUST BE WORN.**

Glycolipids and gangliosides give purple bands and phospholipids give yellow bands. Spots due to splattered reagent or sulfuric acid show up as yellow or brown stains with very definite borders; they are usually scattered in an obvious pattern on the plate if the reagent was improperly applied.

Resorcinol Test for Gangliosides

The reagent takes some time to prepare and should be made by stockroom personnel at least 1 day before the chromatography lab; it can be stored at 4° but it must be used at room temperature. Ten milliliters of 2% aqueous resorcinol solution and 0.25 mL of 0.1 M copper sulfate (2.5 g pentahydrate per 100 mL) are added to concentrated HCl and all is brought up to 100 mL with water. The reagent is sprayed as a mist and the wet plate is covered with a glass plate, held in place with a wood clothes pin.The plate sandwich is heated in an oven at 100°C for 20 minutes. Sialic acid gives deep purple spots whereas other glycolipids give a less intense yellow brown color.

Dittmer-Lester Reagent for Phospholipids

This reagent must be made several days before the experiment is begun. A. Eight grams of molybdic anhydride (MoO_3) is suspended in 200 mL of 70% sulfuric acid and stirred and boiled gently until it is homogeneous, then it is cooled to room temperature. B. To

100 mL of the solution above is added 0.4 g powdered molybdenum metal and the mixture is boiled 1 hour, then cooled to room temperature. Equal volumes, 100 mL, of solutions A and B are combined and filtered through sintered glass, <u>not filter paper</u>. The resulting solution is diluted with 200 mL of water and 100 mL of glacial acetic acid. Plates are sprayed evenly until lightly damp. Phospholipids give blue bands almost immediately or upon brief heating (1 to 2 seconds).

PROJECT EXTENSIONS

The source of lipids is left to the instructor. The author has had success with brain tissue, which is rich in glycolipids, and with trout muscle, which has little lipid. Erythrocyte membranes, used in Chapter 29, are a source of lipids, but more extraction solvent, 30 mL per gram of sample, must be used because the membrane fraction is usually quite wet. There is a voluminous literature on the distribution of lipids within tissues and, if reproducible 2D TLCs can be obtained, meaningful comparisons, between tissues from different species (e.g., beef, pork, and fish muscle) or from developmentally different tissues of the same species, are possible.

DISCUSSION QUESTIONS

1. The TLC solvent system described in this Project will not separate nonpolar triglycerides. Where would such compounds appear in a 2D TL chromatogram run by this protocol?

2. How would you change the procedure to analyze triglycerides by 2D TLC?

3. How do adsorbed lipids change the fluorescence of Rhodamine dyes, can you give a mechanism?

4. A number of the glycoprotein spray reagents are derived from the carbohydrate specific color tests decribed in 27. Given the information in chapter 27, how does the α−naphthol spray reagent work; i.e., what does the hot sulfuric acid do to carbohydrates?

REFERENCES

1. P. Mueller, D. O. Rudin, H. T. Tien and W. C. Wescott, *Nature (London)* , **1962**, *94, 979*.

2. N. S. Radin, *Methods Enzymology*, **1981**, *72*, 5.

3. O. Renkonen and A. Luukonen in *Lipid Chromatographic Analysis*, Vol. I, ed. G. V. Marinetti, New York: Marcel Dekker, 1976.

4. G. Rouser et al, *Methods Enzymology*, **1969**, *16*, 310; G. Rouser, S. Fleischer and A. Yamamoto, *Lipids*, **1970**, *5*, 494.

5. A. N. Siakotos and G. Rouser, *J. Am. Oil Chem. Soc.*, **1965**, *42*, 913.

6. J. C. Dittmer and R. L. Lester, *J. Lipids Res.*, **1964**, *5*, 126.

Chapter 31

Rat Liver Subcellular Fractionation

An adult male rat is painlessly sacrificed and the liver removed, homogenated in buffered 0.25 M sucrose (10 mM HEPES, pH 7.5) and the homogenate is filtered, then fractionated using several sucrose step gradient centrifugations. This procedure permits the purification of plasma membrane, nuclei, mitochondria, Golgi apparatus and microsomes. An optional rapid mitochondrial isolation procedure is also introduced so that electron transport and oxidative phosphorylation can be studied with an oxygen-sensitive electrode in the presence and absence of inhibitors and uncouplers.

KEY TERMS

Buoyant densities	*Clark oxygen electrode*
Homogenization	*Marker enzymes*
Mitochondria	*Oxidative phosphorylation*
P/O ratios	*Step gradient centrifugation*
Uncouplers and inhibitors	

BACKGROUND

As the first step in fractionating subcellular particles, cells must be disrupted by a gentle *homogenization* procedure in a medium that ensures that the subcellular components retain their native structure or characteristics without osmatic disruption. The disruption is usually accomplished in Dounce or Potter-Elvehjem homogenizers. Both devices employ rotating pistons within glass holding vessels; the first is a hand-driven glass piston device, the second is motor-driven with a Teflon piston. The clearance between piston and vessel wall is small in either apparatus (0.05 to 0.6 mm) and the shearing forces will cleave most cells without major damage to the subcellular organelles and membranes. Isotonic sucrose is very frequently used as a homogenization medium and has the advantage that it can also be used as gradient medium, although at high concentrations sucrose is hypertonic [>0.9% (w/v)] and very viscous. Biological particles differ in their buoyant densities and will spin down at different rates depending on the density of the centrifugation medium. The sedimentation rate s is a function of viscosity η, the particle radius r_p and the difference in density between the particle ρ_p and the medium ρ_m.

$$s = \frac{2r_p^2(\rho_p - \rho_m)g}{9\eta} \qquad (31\text{-}1)$$

In *step gradient centrifugation* the particle will sediment down only if its density is greater than that of the medium; often, organelles can be separated because the less dense particles will settle at the interface between low-and high-density layers whenever the bottom layer has a density greater than the less dense particle but not that of the second. Table 31-1 gives the *buoyant densities* of biological particles in sucrose solutions.

Table 31-1

BUOYANT DENSITIES OF SUBCELLULAR PARTICLES IN SUCROSE[a]

Particle	Buoyant Density (g/cm^3)
Liver plasma membranes	1.13-1.18
Mitochondria	1.19
Lysosomes	1.21-1.22
Microbodies (peroxisomes)	1.23
Nuclei	>1.30

[a]Source: Ref. 1.

The fractionation procedure recommended below is that of Fleischer and Kervina[2] and gives generally satisfactory results if the particles are used for microscopic or protein chemistry studies. Centrifugation of rat liver homogenate (0.25 M sucrose, 0.01 M HEPES, pH 7.5) at 1000 x g for 10 minutes gives a pellet that can be used for isolation of plasma membrane, nuclei, and mitochondria; the supernatant can be used for the isolation of Golgi complex, lighter mitochondria, peroxisomes and lysozomes, rough and smooth microsomes, and cell supernatant.

Purification of Plasma Membrane and Mitochondria

The initial pellet is resuspended in buffered 0.25 M sucrose/1 mM $MgCl_2$ with a hand-driven Dounce homogenizer and centrifuged at 70,900 x g for 70 minutes through a two-layered step gradient (0.25 M on 1.6 M sucrose) to give a band at the interface rich in plasma membrane and mitochondria and a pellet that can be used for nuclei. The interfacial band can be further cleaned up by two brief centrifugations (for 10 minutes each) at 1200 x g through 0.25 M sucrose then a two-layer step gradient (0.25 M on 1.45 M sucrose) at 68,400 x g for 60 minutes. The plasma membrane appears as a band at the interface and the mitochondria as a pellet. The membrane fraction at the interface is collected with a Pasteur pipet and resuspended in 0.25 M sucrose, 10 mM HEPES, and 1 mM EDTA (pH 7.5) and centrifuged first through 0.25 M sucrose and then through a 0.25 M to 1.35 M sucrose step gradient. About 5 mg of whitish pellet of purified plasma membrane is obtained. The mitochondrial pellet can be further purified after resuspension and centrifugation at 25,300 x g for 10 minutes through buffered 0.25 M sucrose-EDTA solution. This procedure gives mitochondria good enough for microscopic and protein isolation work, but it is lengthy and leaves little time in a normal laboratory period for careful measurements of the coupling of respiration to phosphorylation. Mitochodria uncouple spontaneously if stored for any length of time and they cannot be frozen.

Rapid Method for Mitochondria Isolation

The foregoing procedure for mitochondrial isolation is unsuitable for P/O ratio measurements but can be shortened appreciably by the method of Guerra.[3] The rat livers are ground in a mortar with specially treated sand (see Experimental Section) in order to disrupt the cells. The isolation medium contains 0.33 M sucrose buffered with 15 mM TRIS (pH 7.4) and 0.025 mM Na_2EDTA. The nuclei, whole cells, cell debris, and red blood cells are removed by filtration through glass wool followed by 800 x g centrifugation for 10 minutes. The supernatant containing mitochondria is then "spun down" by centrifugation at 8200 x g for an additional 10 minutes, resuspended in 0.33 M sucrose by stirring with a glass rod and immediately recentrifuged at 8200 x g for 10 minutes more. A yield of 15 mg of mitochondria per gram of fresh liver protein can be expected.

Purification of Nuclei

The nuclear pellet from the first sucrose gradient (above) is gently resuspended in about 25 mL of 2.2 M sucrose, 5 mM HEPES, and 3 mM $MgCl_2$ with the aid of a hand-rotated loose-fitting glass-Teflon Potter-Elvehjem homogenizer. The resuspension is diluted to 60 mL and centrifuged at 70,900 x g for 1 hour. Nonnuclear material floating on the buffer is discarded and the nuclear pellet is resuspended in about 26 mL of buffered 2.2 M sucrose (as above) and centrifuged for an additional hour to yield a white pellet.

Fractionation of the First Supernatant

The supernatant (approximately 70 mL) is filtered through double layers of cheese cloth and centrifuged at 25,000 x g for 10 minutes. The pellet can be used for separation of mitochondria, lysosomes, and peroxisomes by procedures given in Ref. 2. The second supernatant can be used in the purification of Golgi complex, microsomes, and cell supernatant. Centrifugation at 34,000 x g for 30 minutes sediments a "heavy" microsomal fraction enriched in Golgi complex. The pink sediment can be further fractionated by multi-layer step gradient centrifugation in sucrose.[2] Samples for electron microscopic studies of structure must be fixed with buffered glutaraldhyde to prevent structural damage by osmotic shock and mechanical handling.

Characterization of Liver Fractions

Cell fractions can be assayed for protein, by total dissociation and by the Bradford protein assay (see Chapter 4), and for phospholipids by the standard colorimetric assay for total phosphorus.[4] Microscopic and electron microscopic characterization can also be used. Fresh mitochondria can be used to measure phosphorylation efficiency (*P/O ratio*) with an oxygen electrode and this option is spelled out in more detail below and in the experimental section. There are also a number of assays for *marker enzymes* that can be used to estimate contamination. To assess the distribution of the various organelles, the relative amounts of these marker enzymes can be determined. For example, mitochondria contain succinate dehydrogenase and plasma membrane 5'-nucleotidase. Detailed protocols for these assays are given in Refs. 1 and 6 and are not repeated in the experimental section.

Characterizing the Efficiency of Oxidative Phosphorylation

The student should review lecture material on electron transport and oxidative phosphorylation. The stoichiometry of oxidative phosphorylation in terms of molecules ATP produced per atom of oxygen reduced is known as the P/O ratio. When NADH is oxidized and the released free energy coupled to ATP synthesis at full efficiency, three molecules of ATP are synthesized (P/O = 3); when $FADH_2$ is oxidized under optimum conditions, only two ATP molecules are made (P/O = 2). The consumption of dissolved oxygen as a function of time following the addition of a known amount of ADP can be measured with the *Clark oxygen electrode*. The pattern of oxygen consumption can be used to characterize the efficiency of substrate utilization in the presence of various inhibitors of electron transport.[5]

An oxygen electrode is a widely used device for directly measuring oxygen consumption as a function of time and with the convenience approaching that of a simple pH measurement. The probe utilizes a polarographic sensor covered with a Clark type membrane. The thin permeable membrane stretched over the sensor isolates it from most easily reduced components in the environment but permits diffusion of oxygen and other gases. When a polarizing voltage sufficient to cause oxygen reduction (usually about 0.6 volt depending on the reference electrode) is applied across the sensor, oxygen that has passed through the membrane reacts at the cathode causing a current flow. In principle the membrane passes oxygen at a rate proportional to the oxygen pressure difference across it. However, since oxygen is rapidly consumed at the cathode surface, the oxygen pressure within the membrane compartment is vanishingly small. This means that the diffusion across the membrane and the resulting current is directly proportional to absolute pressure of oxygen outside the membrane. The electrode technique is analogous to that of pH measurement except that the variable measured is current at constant potential whereas pH measurements are electrical potential measurements under conditions of minimum current.

EXPERIMENTAL SECTION

It is impossible to complete the entire fractionation scheme and also make meaningful P/O measurements or enzyme assays within one 6- to 8-hour day. Those most interested in isolating organelles and characterizing them by microscopic or marker enzyme assays are advised to run through the scheme as indicated (Figure 31-1) with some of the chemical characterizations postponed until the second laboratory period. Either the separation of plasma membranes and mitochondria or the isolation of nuclei can be done in a single 4-hour laboratory. Those most interested in characterizing mitochondrial respiratory efficiency should use the rapid isolation procedure and set to work with the dissolved oxygen electrode as soon as possible. The special preparation of sand for this method must be done ahead of time--possibly by the stockroom staff. The following work is all done at 0 to 4°C with ice baths, refrigerated centrifuges, and overnight storage (if necessary) in the cold room.

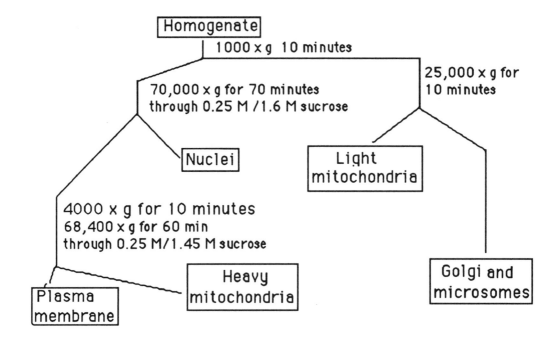

Figure 31-1 Flow diagram for fractionation of rat liver

Preparation of Rat Liver Homogenate[2]

A well-fed adult rat of 200 to 275 g is sacrificed by decapitation as quickly and as painlessly as possible and the liver is quickly removed, immersed in ice-cold 0.25 M sucrose, and cut into pieces. The pieces are placed on blotting paper, then weighing paper and weighed. The tissue is further minced with a pair of clean sharp scissors and combined with 5-volumes of 0.25 M sucrose with 10 mM HEPES pH 7.5. Six passes of about 5 seconds each with a 50-mL Teflon pestle Potter-Elvehjem homogenizer gives a good homogenate. The homogenate must be filtered through cheese cloth and chased with about 10 mL of cold homogenization buffer.

Fractionation   Details

The resulting homogenate is fractionated as described in the background sections and in refs. 1 and 2. The sucrose solutions are prepared ahead of time and, for very reproducible work, the concentrations of sucrose should be adjusted to ± 01 % using a refractometer to estimate the sucrose concentration.[1,2] The g-values given above are calculated for R_{av} = 70 mm. Pellets are resuspended by gentle agitation with a glass rod or a Dounce homogenizer pestle.

Rapid Mitochondria Isolation[3]

Quartz sand, 500-mesh, is soaked in concentrated HCl at 60°C for 2 hours then washed clean with copious quantities of tap water. The sand is then heated at 80°C with 20 mM Na EDTA for 1 hour to remove bound metal ions; the liquid is changed once after 30 minutes. The sand is then washed six times by shaking with distilled deionized water and stored at -20°C. The cold sand is used "right from the freezer" in the homogenization procedure. The liver is minced and transferred to a chilled mortar (no buffer) and ground for about 5 minutes with a 12-fold volume of cold sand. The resulting pulp is placed over glass wool in a column and 10-volumes of cold (<4°C) 0.33 M sucrose/15 mM TRIS, 0.025 mM Na_2-EDTA is perculated through the mass.

The subsequent fractionations resulting from centrifugation at 800 x g and 8200 x g (twice) are as described above and in Ref. 3.

Operating Instruction for the Oxygen Electrode

The Clark oxygen electrode consists of a platinum cathode and a silver anode separated by a thin oxygen permeable teflon membrane. Both electrodes are in saturated KCl. Oxygen is reduced at the platinum cathode in a complex process that consumes four electrons.

$$O_2 + 4e^- + 2H_2O \rightarrow 4OH^-$$

And at the silver anode the required four electrons are released in a oxidation when silver ion is released by the reaction:

$$4Ag + 4Cl \rightarrow 4AgCl + 4e^-$$

The current flowing through the electrode system is directly proportional to the concentration of oxygen diffusing through the membrane, this in turn is related to the pressure of oxygen above the sample when last exposed to the atmosphere, since the solubility of oxygen is proportional to its partial pressure (Henry's law).

The dissolved oxygen meter must be calibrated. The concentration of oxygen at any oxygen partial pressure pO_2 is a function of the solution composition and its temperature. Tables 31-2 and 31-3 give the solubility of oxygen in water as a function of temperature and atmospheric pressure, respectively. We shall assume that the solubility of oxygen is unaffected by the other solutes--a questionable assumption if very accurate data are required. Some instruments have automatic temperature corrections and salinity adjustments that take into account differences in temperature and salinity between the calibration solution and measurement samples (i.e., the YSI Model 57 Dissolved Oxygen Meter). The O_2 electrode is calibrated against well aerated water (stirred in air for 15 minutes). The calibration procedure varies with the instrument used and will be explained in detail in lab but the following is generally applicable. The instrument is first switched off and the meter or recorder is adjusted to mechanical zero. (The YSI model is then switched to the red line and the red line knob is adjusted until the meter needle aligns with the red mark at the 31°C position. The instrument is then switched to zero and adjusted to zero with the zero control knob.) The probe is attached and the sample chamber is filled with water and the calibration potentiometer set to establish a voltage span proportional to the current produced by oxygen reduction at the electrode. For example, at 25°C the solubility of oxygen is

8.24 mg/L or $0.00824/32 = 258$ μmolar. The temperature and atmospheric pressure are measured if the meter readings are to be corrected for these effects. The salinity knob should be zero unless buffers containing high levels of salt are used.

Table 31-2

SOLUBILITY OF OXYGEN IN DISTILLED WATER

Temperature (°C)	Dissolved Oxygen (mg/L)	Temperature (°C)	Dissolved Oxygen (mg/L)
0	14.60	23	8.56
2	13.81	24	8.40
5	12.75	25	8.24
10	11.27	26	8.09
15	10.07	27	7.95
20	9.07	28	7.81
21	8.90	29	7.67
22	8.72	30	7.54

Table 31-3

PRESSURE CORRECTION FACTOR ON OXYGEN SOLUBILITY[a]

Atmospheric Pressure mm Hg	Correction Factor
775	1.02
760	1.00
745	0.98
730	0.96
714	0.94
699	0.92
684	0.90

[a]The correction factor can be used to correct for the effects of atmospheric pressure on oxygen solubility.

Membranes will last indefinitely depending on care and usage but new membranes should be installed by the stockroom or teaching staff just before this experiment is introduced in the laboratory. Should electrolyte evaporate so that excessive bubbles form under the membrane, the reservoir should be flushed with KCl and the membrane replaced by the teaching staff. Erratic or unstable readings also indicate that the membrane should be replaced.   The gold cathode should always be bright and untarnished.

Oxygen-uptake assays are performed at room temperature as follows: The reaction is run in a small (5 to 10 mL) sealed flask equipped with a airtight opening for the electrode, a septum for injecting substrate and other material, and a magnetic stirrer bar. The electrode should be emersed in the reaction mixture even when stirring. A typical run of experiments requires 1.5 mL of reaction buffer, 1.3 mL of water, and 0.1 mL of mitochondrial suspension in buffered sucrose with EDTA (as above). The reaction buffer contains 0.5 M sucrose, 100 mM potassium phosphate (pH 7.2), 2 mM $MgCl_2$, 0.2 mM NAD, 0.2 mM thiamine pyrophosphate (TPP), and 0.02 mM CoA. The electrode is inserted and the sample is sealed and the stirrer restarted. The recorder is started and the background oxygen loss rate (state 1) is measured. The increase in oxygen uptake (drop in dissolved oxygen) with added 1 M succinate (30 μL via a syringe) is observed (state 2). The effect of added ADP (state 3) and its depletion (state 4) can be observed by syringing in 5 μL aliquots of 0.1 M ADP. The effect of another substrate, 30 μL of 1.0 M malate, and various inhibitors and uncouplers (amytal, cyanide or 2,4-dinitrophenol) can be observed as well. A scheme of experiments is indicated in Table 31-4 and Figure 31-2.

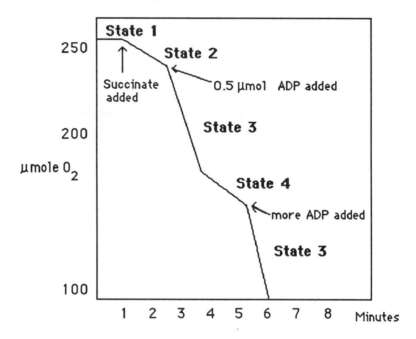

Figure 31-2 Typical oxygen electrode trace of succinate oxidation by mitochondria

Table   31-4

POSSIBLE EXPERIMENTS ON OXIDATIVE PHOSPHORYLATION[a]

	Expt. 1	Expt. 2	Expt. 3	Expt. 4
Reaction Mix				
Reaction buffer (mL)	1.5	1.5	1.5	1.5
Water (mL)	1.4	1.4	1.4	1.4
Mitochondria (μL)	100	100	100	100
Substrate				
1.0 M succinate (μL)	30	--	30	--
1.0 M malate (μL)	- -	30	--	30
0.1 M ADP (μL) added later	5	5	5	5
Inhibitors etc.				
50 mM Amytal	--	--	100	100
300 mM KCN	added after constant state 3 has been obtained in a second series of experiments **(Expts. 5-8).**			

[a]Adapted from the protocol of R. Criddle, University of California, Davis.

PROJECT EXTENSIONS

1. The rapid isolation of spinage chloroplasts and the characterization of their metabolic processes including photophosphorylation[7] is a very challenging extension of the general methods established in this project.

2. The use of 2-D electrophoresis to characterize the plasma membrane and nuclear proteins is also a reasonable extension of this project.

REFERENCES

1. D. Rickwood, *Centrifugation, a Practical Approach*, Washington, D. C.: IRL Press, 1984

2. S. Fleischer and M. Kervina, *Methods Enzymology,* **1974**, *31*, 6

3. F. C. Guerra, *Methods Enzymology*, **1974**, *31*, 299.

4. P. S. Chen, T. Y. Toribara and H. Warner, *Anal. Chem.*, **1956**, *28*, 1756; G. Rouser and S. Fleischer, *Methods Enzymology*, **1967**, *10*, 385.

5. A. L. Lehninger, *Biochemistry*, 2nd ed. 1975, New York: Worth Publishers, Chap. 19.

6. R. Douce, E. L. Christensen and W. O. Bonner, *Biochem. Biophys. Acta*, **1972**, *275*, 148.

7. P. S. Nobel, *Methods Enzymology*, **1974**, *31*, 600; D. C. Lin and P. S. Nobel, *Arch. Biochem. Biophys.*, **1971**, *145*, 622.

Index